GLOBAL WARMING

INTERNATIONAL ENCYCLOPAEDIA OF SUSTAINABLE DEVELOPMENT SERIES

GLOBAL WARMING

Edited by

P. C. SINHA

ANMOL PUBLICATIONS PVT. LTD.

NEW DELHI-110 002 (INDIA)

ANMOL PUBLICATIONS PVT. LTD.
4374/4B, Ansari Road, Daryaganj
New Delhi-110 002

Global Warming

First Edition 1998
ISBN 81-7488-954-X

PRINTED IN INDIA

Published by J. L. Kumar for Anmol Publications Pvt. Ltd., New Delhi-110 002 and Printed at Mehra Offset Press, Delhi.

Contents

Preface

THIS twenty volume series on sustainable development is a very timely effort undertaken by Anmol Publications Pvt. Ltd. The contours of this concept were discussed elaborately in Earth Summit, 1992, in light of the facts that:

- world population was increasing swiftly;
- stocks of many resources were declining sharply;
- deforestation was epidemic;
- species extinction was accelerating;
- deserts were expanding;
- soils were deploting swiftly;
- fist stocks were diminishing;
- wildlife habitat was disappearing;
- toxic poisons circulated from air to water to food to soil and bioaccumulated up the food chain;
- the protective ozone shield was thinning rapidly; and
- the buildup of greenhouse gases threatened to alter climatic systems—perhaps sending them into chaos.

Without intending to, just by doing better and better that which people had always done, humans were changing the way the planet's life systems worked.

Better laws and better technology alone were unlikely to reverse those accelerating trends. It became clear to many in the 1970s and 1980s that continued environmental damage was rooted in the fundamental beliefs and values of modern society and that the society itself must be transformed. But how should society be transformed? This planet wide summit of national leaders was soon dubbed the "Earth Summit" and firmly established environmental concerns and the sustainability of society and its ecosystems at the top of the world's agenda.

No doubt, we face a huge relearning task that is being forced on us by the present absence of long-term sustainability. The continued good functioning of life systems depends on how thoughtful and foresighted our learning will be.

One of the crucial insights underlying the sustainable development concept is the realisation that there are severe environmental costs associated with the absence of economic development. Desperate, hungry people will till any soil regardless of the long-term consequences; they will be tempted to cut any tree for fuel if they have no other option but cold; and many of them will poach any wild animals for which anyone will pay. This insight is at the core of the concept of sustainable development; the concept implicity asserts that both development and environmental protection are essential. In the view of sustainable development advocates these are not contradictory objectives. Sustainable development, therefore, is also a recognition that no one can ignore any longer the concerns advanced by environmentalists.

To cover all the said points the twenty volumes are as followed:

1. Sustainable Development
2. Green Movements
3. Green Philosophy
4. Environmental Policies
5. Global Warming
6. Biodiversity Depletion
7. Sea-Level-Rise
8. Atmospheric Pollution and Climate Change
9. Marine Pollution
10. Desertification
11. Ozone Depletion
12. Small-Island Development
13. Protected Area Networks
14. Wildlife and Forest Conservation
15. Wilderness
16. Toxic and Hazardous Wastes
17. Nuclear Wastes
18. Energy Crisis
19. Population Explosion
20. Habitat

This Encyclopaedia of Sustainable Development aims at enriching knowledge about the key issues related to sustainable development of our socio-economic and environmental systems. Our goal is to provide a realistic view point touching deeply the core issues and suggest solutions in different perspectives. This encyclopaedia would certainly help create knowledge and inculcate wisdom at all levels of society.

P.C. SINHA

1. Greenhouse Effect and Global Warming

THE EARTH'S climate is driven by solar radiation. In the long term the energy absorbed from the sun must be balanced by outgoing radiation from the earth and the atmosphere. Part of this outgoing energy is absorbed and re-emitted by radio-active atmospheric gases ('greenhouse gases'), thereby reducing net emission of energy to space. To maintain the global energy balances, both the atmosphere and the surface will warm until the outgoing energy equals the incoming energy. This is the greenhouse effect.

The main natural greenhouse gases are water vapour (the largest contributor to the greenhouse effect), carbon dioxide, methane, nitrous oxide, and ozone. There are also purely man-made greenhouse gases, including many ozone-depleting substances such as CEF, which are controlled under the Montreal Protocol. The main greenhouse gases, differ in the intensity of their heat trapping (or 'radio-active forcing') and atmospheric lifetimes and thus in their ability to affect the radioactive balance of the earth. CFE and nitrous oxide are many times more potent than the same quantity of carbon dioxide or methane.

The additional carbon dioxide that human activities put into the atmosphere between 1980 and 1989 came principally from fossil fuels. Additions from changes in land use, such as deforestation, are estimated to have been one-half as large. All these net additions from human activity are dwarfed by the natural exchanges of carbon between the earth and the atmosphere. The largest sources of methane in the atmosphere are natural wetlands, rice paddies, and livestock. Natural gas production (drilling, venting, and transmission), biomass burning, termites, landifills, and coal mining also release methane. Nitrous oxide is released by the oceans and soil, but human activities such as biomass burning and the use of fertilisers play a role that is not yet fully understood or quantified. Much uncertainty sounds the role size of the sources of the both methane and nitrous oxide.

One of the most important stages in the thousand million year long history of planet Earth was the appearance of organisms capable of carrying out-photo-synthesis. Nowadays we generally call these organisms-plants and they range in size from tiny one called algae to giant trees, dozens of storeys high. Plants are capable of taking up simple substances like carbon dioxide and converting them into much more complex materials using the sun's light as an energy source. Therefore plants accumulate carbon and other elements in their tissues during their growing period.

Three hundred million years ago plants covered much of the Earth's land surface. When these plants died they fell in swamps and were later covered by layers of mud. Ages later the layers of trees and other plants were converted into the hard substance we nowadays call coal. In other places different organisms were also covered with layers of mud to give rise to petroleum oil and natural gas. All these organisms took out of the atmosphere vast quantities of carbon dioxide gas and converted it into solid compound of carbon.

Later on in the history of the Earth, about a million years ago an organism called Man appeared and he slowly learned to dig up and use the vast energy stores laid down as fossil fuels. Subsequently, some two hundred years ago at the start of the industrial revolution Man started to use fossil fuels much more heavily. This started the process of liberating into the atmosphere, in a short span of time much of the carbon dioxide that had taken millions of years to be removed.

If the Earth had no atmosphere any heat received from the sun would simply be reflected back into space. However certain naturally occurring substances like carbon dioxide, water vapour and methane, present in the atmosphere in small quantities trap some of the sun's energy because they can absorb radiation in the infrared region. This effect causes an average increase in surface temperatures of about 33°C so that the average temperature at the Earth's surface is some 15°C. The activities of Man have also added other greenhouse gases like nitrogen oxides (NO_2) and chloro-fluoro-carbons (CFCs), but the most significant contributor to global warming appears to be carbon dioxide.

There is strong evidence suggesting the implication carbon dioxide in global warming, because it was found that during the last ice age the concertration of this gas in the atmosphere was 200 ppm[2] (parts per million). The concentration of this gas has increased from 315 to 355 ppm from 1958 to 1990. Consequently the 1980's has been the warmest decade since records have been kept.

In a study sponsored jointly by the United Nations Environment Programme and the World Meteorological Organisation and conducted under the auspices of the Intergovernmental Panel on Climate Change (IPCC), it was established that extensive climate change is probable. The report concludes that if current emission trends continue, global temperatures will increase at the rate of 0.3°C per decade. This will result in an increase of 1°C by the year 2025 and 3°C by the end of next century. The temperature increases predicted are based on mathematical models currently available but other models predict temperature rises of upto 0.8°C per decade. The latter models take into consideration such factors as melting snow cover which will lessen the reflectance of the suns's rays back into space and thus increase the amount of energy absorbed by the planet's surface. Such figures for temperature change may appear insignificant at first glance but it must be remembered that the temperature difference between an ice age and an inter-glacial period is between 3-6°C.

An increase in the greenhouse effect will not give rise to major problems for most of the people living today. However nations are now starting to tackle the issue. This shows that the effects of our activities on the life of future generations is widely recognised. It is an acknowledged fact that all the efforts of this generation regarding global warming can only be felt by the next generation. This is because atmospheric concentrations of long lived greenhouse gases adjust in a very slow way to change in emissions. The IPCC report says that if current emission trends continue we would be burdened with high concentrations of greenhouse gases for centuries ahead. The report suggests immediate reductions of carbon dioxide and CFC's by up to 60 per cent if the concentrations of these gases are to stabilise at today's levels.

Higher surface temperatures will increase the saturation pressure of water vapour so that the air will be able to carry more moisture. This will lead to a faster evaporation rate, assuming that there is water to evaporate. This rapid evaporation will in turn lead to drier soils. Summer droughts will last longer in the interior of continents. Such an effect will seriously jeopardise the food production of many nations. Forest fires will become more frequent and even more carbon dioxide will be liberated from the trees which act as a reservoir of carbon. It is also thought that the increased surface atmosphere moisture exchanges will lead to greater atmospheric turbulence, making weather more unstable. Tropical perturbations like hurricanes will possibly shift to higher latitudes leading to disastrous flooding and devastation of coastal areas and small islands. Such effects will be worsened because sea level has been constantly rising over the past century by as much as 2mm per annum. Inundation and shoreline recession are also predicted for small Islands like Malta.

As the ocean level rises, salt water will start invading water courses which even now are over exploited. This will spell disaster for whole nations which depend on a single source of water for their survival. Water tables may become increasingly contaminated with salt water so that the reservoir of potable water stored underground will become progressively useless.

A change in climate which is too fast will destroy whole forests and other ecosystems because organisms will not be able to adapt or migrate fast enough. However some cold and barren regions might well become suitable for farming and for colonisation by Man and other organisms.

The greenhouse effect attributable to anthropogenic sources is caused by four gases and each is responsible to different degrees for global warming.

Carbon dioxide is the main culprit, followed by CFC's and methane. As most CFC's are already being phased out of production and methane is a by product of certain forms of agriculture, carbon dioxide has been pin-pointed as the gas whose emissions must be reduced. Also, carbon dioxide production is most often associated with the production of other harmful substances like sulphur dioxide and nitrogen oxides.

A look at table 3 will show that the counters with the most elevated per capita emissions of carbon dioxide are affluent countries, especially those which have large petroleum deposits. This can be seen when looking at countries such as United Arab Emirates which has a GNP per capita of over $18,000 and which in 1989 had a relative per capita emission of 14.5. On the other hand a nation such as Austria with a GNP per capita similar to that to score a relatively diminutive 3.5 on the index.

Fossil fuels have shaped civilisation as we know it and their careless use might also destroy it. This fact is slowly starting to be recognised by a number of nations and of individuals who are starting to take action-which action will be felt by future generations. However many environmentally risky practices are still being justified by the 'energy dependence' of economies. This apology is not used to justify the expense of installing solar energy systems.

Fuel efficiency is seen by many as a way to decrease emissions but this is only a short term solution, because eventually all fossil fuels must be phased out. Meanwhile all experts agree that fuel efficiency in all aspects of our lives will repay initial costs. Cars, for example can be made to run more efficiently, but as always, in such issues there are many vested interests. The USA for example, a nation notorious for its gasoline guzzling cars and with 5 per cent of the world's population uses 25% of the world's energy.

Corporation which in one way or another are producing carbon dioxide should make it a point to continuously plant trees. Some years ago a Maltese environmental group launched a tree planting campaign where individuals, for a small fee would have a tree planted in their name or in any other person's name. Each contributor would receive a certificate showing 'ownership' of the tree. This campaign was seen as an alternative idea to the giving of expensive and sometimes meaningless gifts. By planting trees, some of the carbon dioxide that we have released is put back into a reservoir. The trees only draw carbon dioxide from the air while they are growing and will release it again if they die and decompose or if they are burnt. Meanwhile an acre of forest land is being destroyed every second.

Nuclear power, which is still being mentioned as an alternative poses problems which are in some aspects similar to those involved in the burning of fossil fuels. With fossil fuels we are leaving for some time behind us the burden of carbon dioxide, while with nuclear fuels we leave for a very long time behind us long lived radio-active waste. Carbon taxes have been suggested as 'incentives' for burning less carbon-rich fuels. The funds generated by these taxes can be targeted at reducing emission in third world countries. The distribution of such a tax load must be carefully considered so as not to cripple economic growth. This implies an uneven distribution of energy costs, so that households receiving non-carbon generated energy will pay less.

The long-term solution is to cutout carbon dioxide production by using renewable energy sources, which ultimately, all depend on the sun. Wind and solar energy are two energy sources which have been shown to be potent alternatives to fossil fuels. Another alternative biomass energy is produced by growing plants and burning them as fuel. In this way carbon dioxide is passed through a cycle so there is never any net emission of the gas. The land used for biomass production may however, be better used for cultivating protein rich food for a growing world population

Unlike other forms of pollution the effects of greenhouse gas emission are not immediately obvious. The fact that predictions on the consequences of our emissions are based on mathematical models which sometimes disagree as to the exact outcomes is noted in the UN Convention on Climate change. Countries where fossil fuels and the technologies utilising them are cheap, are reluctant to consider change. Because of these factors and of the inertia inherent in most human actions, progress, up to now has been very slow and consequently the Convention stresses the need for ". . . developed countries to take immediate action . . . ."

Human activities are now thought to have the potential to alter significantly the Earth's climate on a global scale. This potential derives from increasing atmospheric concentrations of greenhouse gases (GHGs) which are predicted to warm the earth's lower atmosphere and surface by reducing the efficiency with which it cools to space. The most important GHGs are water vapour, carbon dioxide, methane, nitrous oxide, halocarbons (e.g., the chlorofluorocarbons), and upper-tropospheric and lower-stratospheric ozone.

The two primary sources of the observed increase in atmospheric carbon dioxide are combustion of fossil fuels and land-use changes (primarily deforestation) accounting for approximately three-quarters and one-quarter respectively. The main sources of methane emissions are natural wetlands, rice paddies, enteric fermentation from cattle, landfills, coal mining, oil, natural gas. There are still many uncertainties in accurately quantifying the magnitude of anthropogenic plus natural methane emissions from individual sources. But, human activities are thought to currently account for about 60-70 per cent of the total emissions. Recent methane studies suggest that 20 per cent of the total methane source is of fossil origin, largely from the coal, oil, and natural gas industries.

A key issue is to relate emissions of GHGs and their precursors to *future* concentrations of GHGs in order to assess their impact on the radiative balance and thereby the Earth's climate. A number of different types of carbon cycle and tropospheric chemistry models have been developed for this purpose. However, it should be noted that *all* carbon cycle models are subject to considerable uncertainty. Similarly, tropospheric chemistry models exhibit substantial differences in their predictions of changes in ozone, and in other chemically important active gases.

Gases can exert a radiative forcing both directly and indirectly—direct forcing occurs when the gas itself is a GHG, and indirect forcing, when chemical transformation of the original gas produces a gas or gases with greenhouse impacts. The concept of the Global Warming Potential (GWP) has been developed for policy-makers as a measure of the possible warming effect on the surface-troposphere system arising from the emission of each gas relative to carbon dioxide. For the concept to be most useful, both the direct and indirect components of the GWP need to be quantified. Relative to carbon dioxide, the direct GWPs of the other GHGs are: methane, 11; nitrous oxide, 270; CFC-11, 3400; CFC-12, 7100; HCFC-22, 1600 and HCFC-134a, 1200.

Combining the 1990 emissions of GHGs with their global warming potentials suggests that the most important GHG directly influenced by human activities is carbon dioxide, with methane being the second most important. More than half of the enhanced greenhouse affect can be attributed to carbon dioxide, with methane being responsible for up to another quarter; hence, it is important that any intervention such as GEF places an emphasis on these two gases.

Future emissions of GHGs will depend upon a wide range of economic, demographic and policy conditions and are inherently controversial to predict because they reflect different views of the future. Considerable uncertainties arise from the evolution of the types and levels of human activities, technological advances, and human responses to possible environmental, economic and institutional constraints. It is, clear, however, that if present trends persist, it is very likely that the resulting impacts on the global atmosphere will lead eventually to changes of the global climate that would seriously perturb human societies and perhaps even endanger human life. Further, the response time of the climate system is such that, by the time significant changes are detected, it may be too late to rectify the situation.

This is why many countries are committed to putting precautionary policies in place, of which the GEF is a part. The idea is that, as further evidence on global warming and its consequences is gathered, the investments of the GEF will leave the international community better placed to reduce GHG accumulations to safe levels, over the long term, should the need arise. The approach is equivalent to an insurance policy. The various measures (technologies, practices) and policies that the GEF should support are those that would need to be turned to on a large scale if it becomes certain that GHG accumulations have to be restricted appreciably.

A distinction must be made between the objective of *reduction of net GHG emissions* and the objective of *stabilisation of GHG accumulations*. By focussing only on the *reduction* of, for example, CO_2 emissions, the global warming problem would only have been delayed by a decade or two, but would have been substantially unaddressed. Delaying global warming is not the same as preventing it. The reduction of net emissions is a necessary step, but not sufficient; in addition, the magnitude of reduction should be sufficient to achieve the stabilisation, if not reduction, of GHG accumulations in the atmosphere at some safe level over the long term. Thus, the effectiveness of interventions cannot be divorced from objectives and benefits cannot be separated from targets. However, virtually all these interventions involve investments and therefore there has to be strategy for reducing net GHG emissions into the atmosphere.

The Greenhouse Effect has caught the imagination of the general populace in the last decade. What's more, the respected, generally conservative scientific establishment has become associated with relatively dire predictions of future climate changes the Greenhouse Effect may cause. But how much do we actually know about the Greenhouse Effect? Can we really establish how much the climate will change, and when? Perhaps by separating the "hard" science—that which can be verified and is considered well-understood—from scientific theory or estimates, we *can* investigate the likelihood of near-term climate changes that have been projected. The series of questions which follow will help us explore what we currently know, or think we know, about the Greenhouse Effect.

The "Greenhouse Effect" is the name for the physical process whereby energy from the sun passes through the atmosphere relatively freely, while heat radiating from the earth is partially blocked or absorbed by particular gases in the atmosphere. Because the sun is warmer than the earth, its energy is radiated at a higher frequency which is not absorbed well by gases such as carbon dioxide (CO_2) or water vapor. In contrast, these triatomic gases (gases with three atoms per molecule) are effective absorbers of the lower-frequency energy radiated by the earth. Since the gases responsible for this selective absorption make up only about one per cent of the atmosphere, they are known as "trace" gases. In general, we can calculate very accurately the energy absorbed by different gases, although there are some uncertainties, and when the concentration of a gas changes, we know how much more energy is being absorbed. This additional absorption by itself warms the planet: for example, doubling the concentration of CO_2 in the atmosphere would eventually lead to a global air temperature increase of 1.2 Centigrade (C)—about 2.2 Fahrenheit (F)—if there were no other changes in the climate system.

However, what we do not know is exactly how the rest of the system will react. The current numerical computer models of the arth's climate predict that the warming due to the increase in CO_2 will lead to more evaporatioin of water vapor from the ocean. Water vapor itself is a "greenhouse" gas, so as its concentration increases in the atmosphere, the planet will warm even further. With rising temperatures there will be less snow and ice to reflect energy from the sun back to space (snow and ice are very good reflectors). This promotes further warming because more of the sun's heat is retained in the earth. These are examples of "positive feedbacks" in which the system responds to a warming climate with changes which amplify the warming even further. Both of these system responses are very likely to occur, although we cannot be sure of the magnitude of the changes. The models also predict cloud cover changes that will provide even more warming, but clouds are not modeled in a very sophisticated way because they are not well understood. Thus, the likely impact of cloud cover changes is quite uncertain.

The net result of these different processes in the various models is the tripling of the warming caused by the doubled CO_2 levels alone, producing a total warming of about 4°C (or 7°F) for the global annual average. Yet it is only the initial Greenhouse Effect due to increased CO_2 or increases in other trace gases, which we know with great confidence. The atmosphere of nearby planets validate the general concept of the greenhouse theory, especially in a qualitative sense, but they cannot tell us what the magnitude of the changes on earth will be. Venus, with its massive atmosphere composed essentially of CO_2 has a surface air temperature close to 500° warmer than would be expected without a Greenhouse Effect. Mars, with a very thin atmosphere and thus little atmospheric capacity to absorb radiation, has an observed temperature close to the expected.

The earth, with intermediate amounts of greenhouse gases in its atmosphere, is about 30°C (54°F) warmer than it would be otherwise, The differences among the planets are very large, and cannot really be used to estimate sensitivity to relatively small changes in greenhouse gas levels. Furthermore, as noted above, the big uncertainty lies inthe magnitude of the climate system response (or feedbacks). The mast important feedbacks involve the reaction or processes related to water, and the other planets have no free-standing water.

Since the establishment of an atmospheric monitoring system in 1958, we have observed the concentration of CO_2 growing systematically. During the past 28 years, CO_2 values in our atmosphere have increased from 315 parts per million (ppm) to 350 ppm. These values are especially significant since air bubbles trapped within the ice in Greenland and Anatartica have been use to measure what CO_2 concentrations were like over the past several hundred thousand years. During that time, up to just prior to the industrial revolution, CO_2 levels had not exceeded 280 ppm (and thus were well below current values). The rising CO_2 concentrations are believed to be associated with the widespread use of fossil fuels such as gas and oil and by the denuding of the world's tropical and other forests, a process which lowers the earth's ability to use trees as a CO_2 absorbent.

Chlorofluorocarbons (CFCs), better known for their impact on atmospheric ozone levels, are artificially generated gases that also have the capacity to contribute to the Greenhouse Effect, and which are known to be increasing. They have no natural sources and probably did not exist in the atmosphere prior to the last few decades. Recent measurements indicate that other contributing greenhouse gases, such as methane and nitrous oxide, are also increasing; however, since we are not sure of the reason for their increase, we have less confidence in their long-term trends.

It is estimated that the average surface air temperature has increased globally by about 0.6°C (or 1°F) in the past century, but there is some uncertainty as to how accurately the change can be estimated because there were far fewer temperature recording stations 100 years ago. Large portions of the globe were poorly sampled, especially in the Southern Hemisphere. Even today, full global coverage is not available.

The record, such as it is , does not indicate a continues world-wide warming. There was apparently a cooling period in the Northern Hemisphere from the 1940s into the early 1970s. This is inconsistent with the concept of greenhouse warming, but it may be due to other climate disturbances such as variations in the solar energy constant, or a change in the amount of volcanic discharge into the atmosphere, or it may simply represent internal variability within the system.

The overall warming for the past century is the right order of magnitude for the expected Greenhouse Effect. However, give the uncertainties about the actual temperature change, the climate feedback factor, the actual amount of CO_2 in the atmosphere in 1880, and the rate at which oceans absorb heat (which slows down the atmospheric warming), we cannot be more precise in determining what the expected warming would have been. Similarly, we cannot use the record to establish what the climate-feedbak factor really is.

Despite these qualifications, one aspect of the temperature record clearly stands out: during the past century, the four warmest years, globally, were all during the 1980s; this does not include 1988, which appears as if it will be the warmest year of all. This has occurred despite the eruption of the El Chicon volcano, putting additional dust into the air, and a decrease in the sun's energy output, both of which should have had a cooling effect. While modern temperature records may be contaminated to some extent by heat island effects which create warm areas in cities, the rapid rise of temperature during the 1980s is consistent with computer model projections. This suggests that the anticipated Greenhouse Effect changes may actually be appearing at this time.

Numerical models (called general circulation models) which simulate the known workings of the earth's climate system are used to calculate its response to increases in trace gases. The four models in current use all estimate that the doubled CO_2 climate will have a global average temperature some 4°C (7°F) warmer than today. They are thus all calculating similar climate feedback factors. However, even though many climate processes are handled similarly in the different models, their unanimity does not guarantee accuracy. For example, the treatment of cloud cover in all the models represents a major uncertainty. The models also differ to some extent as to the seasonal and latitudinal distributions of the calculated warming. It is thought unlikely that the models could be wrong by more than a factor of two, but this cannot be proven.

In addition, a climate change forecast should indicate when the warming would be expected to be evident, but only one model, the Goddard Institute of Space Studies (GISS) model, has been used to calculate the temperature increase over the next 50 years in response to a gradual change in greenhouse gas concentrations. Its results indicate substantial warming in the next decade. This calculation is affected to some extent by uncertainties in how much heat the oceans will absorb and the true climate feedback factor. Nonetheless, by providing an estimate of how much warming should be observed in the relatively near future, the model does give us a chance to test the accuracy of its projections.

It is estimated that the ice age climate was some 4°C cooler than today's. At that time (some 18,000 years ago), ice covered the area now occupied by New York City. Considering that the doubled CO_2 climate is estimated to be warmer to the same degree that the ice ages were cooler, large changes in the climate system may well be expected if this comes to pass. The GISS model's forecast for the next 50 years give changes of 2°C (3.6°F) by the year 2020, which would make the earth warmer than it is thought to have been at any point in historical time. Estimates for summer temperatures in the doubled CO_2 climate indicate that Washington, DC, which currently experiences 36 days of temperature above 90°F would routinely have 87 such days; Dallas would go from 19 days with temperatures above 100°F to 78 days.

Sea-level rise due to thermal expansion of the oceans would cause severe problems in many coastal cities, and this effect would be exacerbated if additional glacial melting occurred. Rainfall patterns would likely be substantially altered, posing the threat of large-scale disruption of agricultural and economic productivity, and water shortages in some areas.

We may start experiencing the effects of a chaining climate fairly soon. If we define a "hot" summer as the warmest one-third of the summers during the period 1950-1980, then, if the models are correct, during the 1990s we will experience "hot" summers twice as often, or two-thirds of the time. The summer of 1988 may be an all-too-tangible indication of how dire such changes in summertime climate can be.

The climate is being altered by the release of greenhouse gases due to fossil fuel consumption and industrial processes, and by deforestation. These factors are inherent in our current civilisation. It may be possible to limit specific trace gas increases (such as the CFCs) and slow down rates of increase of CO_2 through increased energy conservation. Our ability to manipulate the climate system deliberately, so as to offset the warming by some other process, is nonexistent. It is likely that the additional greenhouse gases which have been added to the atmosphere during the past 50 years have already built considerable warming into the system, which we have not yet experienced because of the slow warming response of the ocean.

The climate of the next century will very likely be substantially different from that to which we have become accustomed. Uncertainties in our knowledge of the true climate sensitivity prevent us from knowing exactly how different it will be. The consequences of the climate change that is currently being estimated would be enormous. With that in mind, it is worth-while for us to factor climactic change into decision-making processes related to our future, even though there are many uncer-tainties that still exist in our understanding of what may actually happen.

Increases may also occur in a gas we have not yet discussed in detail: carbon monoxide, which has the power to decrease the self-cleansing ability of the atmosphere. A rise in carbon monoxide concentrations is likely because its sources—fossil-fuel combustion, biomass burning and atmospheric reactions involving methane—are all expected to increase. On the other hand, a significant (but still not well-quantified) amount of the gas is formed in the atmosphere over the tropics from the breakdown of hydrocarbons emitted by vegetation, a source that human activities are removing. The future concentrations of carbon monoxide are therefore uncertain, although on balance many workers foresee a rise over the Northern Hemisphere.

Carbon monoxide undermines the self-cleansig ability of the atmosphere by lowering the concentration of the hydroxyl radical, which is an important "detergent" because it reacts with nearly every trace-gas molecule in the atmosphere, including substances that would otherwise be inert. Without hydroxyl, the concentrations of most trace gases would become much higher than those of today, and the atmosphere as a whole would have totally different chemical, physical and alimatic properties.

Our projections for the future are discouraging, then, if one assumes that human activities will continue to emit large quantities of undesirable trace gases into the atmosphere. Humanity's unremitting growth and development not only are changing the chemistry of the atmosphere but also are driving the earth rapidly toward a climatic warming of unprecedented magnitude. This climatic change, in combination with increased concentrations of various gases, constitutes a potentially hazardous experiment in which everyone on the earth is taking part.

What is particularly troubling is the possibility of unwelcome surprises, as human activities continue to tax an atmosphere whose inner workings and interactions with organisms and nonliving materials are incompletely understood. The Antarctic ozone hole is a particularly ominous example of the surprises that may be lurking ahead. Its unexpected severity has demonstrated beyong doubt that the atmospher can be exquisitely sensitive to what seem to be small chemical perturbations and that the manifestations of such perturbations can arise much after than even the most astute scientists could expect.

Nevertheless, some steps can be taken to counteract rapid atmospheric change, perhaps lessening the known and unknown threats. For example, evidence indicates that a major decrease in the rate of fossil-fuel combustion would slow the greenhouse warming, reduce smog, improve visibility and minimise acid deposition. Other steps could be targeted against particular gases, such as methane. Its emission could be reduced by instituting landfill operations that prevent its release and possibly by adopting less wasteful methods of fossil-fuel production. Methane emission from cattle might even be diminished by novel feeding procedures.

Perhaps more encouraging is the fact that many people and institutions are now aware that their actions can have not only local but also global consequences for the atmosphere and the habitability of the planet. A few recent events exemplify this awareness: in the Montreal protocol in 1987, dozens of nations agreed to halve their chlorofluorocarbon emissions by the end of the century, and several countries and the major chlorofluorocarbon manufacturers have more recently announced their intention to eliminate the chemicals by that deadline. Some of the same nations that have been involved in the Montreal protocol are now discussing the possibility of an international "law of the atmosphere." It would be directed at limiting the release of several greenhouse and chemical active trace gases, including carbon dioxide, methane and nitrous oxide, as well as sulfur dioxide and the NOx gases.

We and many others think the solution to the earth's environmental problems lies in a truly global effort, involving unprecedented collaboration by scientists, citizens and world leaders. The most technologically developed nations have to reduce their disproportionate use of the earth's resources. Moreover, the developing countries must be helped to adopt environmentally sound technologies and planning strategies as they elevate the standard of living for their populations, whose rapid growth and need for increased energy are a major cause for environmental concern. With proper attention devoted to maintaining the atmosphere's stability, perhaps the chemical changes that are now occurring can be kept within limits that will sustain the physical processes and the ecological balance of the planet.

Issues in Atmospheric Science

"Everybody talks about the whether, but nobody does anything about it." Atmospheric science includes understanding and predicting weather, of course, but it encompasses much more than that. Many chemical reaction in the atmosphere involve very small concentrations and occur at conditions of pressure and temperature that are difficult to obtain in laboratory measurements of rate constants. Coupled with complicated mass transport as well as photochemistry, this makes for an extraordinary set of problems, requiring both exotic laboratory procedures and great physical insight. *Science* regularly reports significant advances in all of the atmospheric sciences; the AAAS-Newcomb Cleveland prize for outstanding articles published in *Science* was awarded recently for two papers (one authority by Molina, Tso, Molina and Wang; the other by Tolbert, Rossi, Malhotra, and Golden) in the atmospheric sciences that address the problem of important reactions accruing on particulate matter, especially ices, in the stratosphere.

In spite of the rather fatalistic remarks about doing nothing about the weather, in recent years we have actually been doing quite a lot. And much of what we have been doing about the atmosphere and our climate is apparently undesirable. In this issue on atmospheric sciences we present overviews of four areas in which major advances have occurred in our understanding of the chemistry of the atmosphere and the impact that various human activities are having on it. The articles view aspects of atmospheric problems on scales that range from local urban areas to the entire global atmosphere. In this all-encompassing view, we see, for example, that ozone can be both good and bad. It acts as a significant pollutant locally, while at high altitudes it provides a valuable photochemical screen whose destruction plays an important role in overall global climate warming. Each of the authors addresses not only the causes and consequences of various activities, but also possible ways of attacking some of the problem.

Seinfeld discusses urban air pollution, a critical problem for everyone who lives in a large city. Seinfeld characterises the urban atmosphere as a giant chemical reactor in which a variety of undesirable materials are formed. A good deal of what is occurring is now well understood, but uncertainties remain.

Schwartz deals with a regional problem: acid deposition. In this arena of regional-scale air pollution, the difficulties are magnified because of the high density of emission sources and mutual zones of influence. Although much is now understood about the basic nature of the problem, it is not yet possible to put forward a credible source-receptor scientific model. It is clear that reduced emissions will, however, result in reduced deposition.

McElroy and Salawitch address the changing composition of the global stratosphere. Models show excellent agreement for many of the species. Ozone is a key compound here, and the effects of human activity, especially with regard to halocarbons, play an important role in the chemistry.

Finally, Schneider addresses global climate warming, the "greenhouse effect." Is there one? What are its consequences? It is clear from other work, some recently published in *Science,* that short-time climate problems, such as last year's drought, are a result of other effects. Schneider, however, makes the case that we are likely to see a rise in global temperature, and even a modest increase is predicted to have a substantial impact. Consequences and ways of dealing with the problems are presented.

Atmospheric science is an area in which scientific understanding is critical to our long-term quality of life. The progress that has been made recently has been remarkable. We need to respond now to what we know, while continuing to develop the knowledge base. Simply describing what is happening, however, should be only part of the story.

Environmental issues, more than anything else, are likely to bring the nations of the world together as the heat generated by these may well melt the tough attitudes of the West. Nature certainly recognises no national boundaries.

The June 20-29 meeting in London to consider amendemnts to the 1987 Montreal Protocol—which called for a 50 per cent reduction in ozone-depleting chlorofluorocarbons by 1988—was historic in ways more than one. Not only was the first step towards a consensus to tackle the serious issue of ozone depletion taken, but it was also an occasion when the developing world managed to score a victory. The United States had to climb down and face the harsh reality—that environmental issues cannot be dealt with effectively without the cooperation of the Third World, especially Indian and China, and that at least in this area the poor may be able to make the rich pay for their recklessnes.

More important, the London meeting laid the ground rules for the next big environmental event—the 1992 international conference in Brazil on global warming. Here too it was established that the Western nations had been spewing out greenhouse gases over the last century at a rate that threatened to upset completely the delicate balance in the carbon cycle and may, infact, result in runaway warming that could eventually kill all life. Scientists have predicted the meeting of the polar ice caps, an appreciable rise in the sea levels and vast climatic changes that could significantly affect agriculture world-wide. This scenario, they say, is not of the distant future, but could unfold in the next 30 to 50 years.

The Montreal Protocol was the first effort to correct the imbalances in nature, created by man's selfish quest for more and more comforts. But it did not work as it was not arrived at on the basis of consensus. More than half the world, including India and China, did not sign it. For a change, the Third World this time insisted that its viewpoint be considered before it takes on the obligations under the Protocol.

The Protocol, before the London amendments, did not pay attention to the fact that the technology to replace CFCs and halons—used in fire extinguishers—were with a few affluent countries, that the poor could not afford them and, more important, that it was the First World nations that had contributed to much of the ozone layer depletion. It was made clear that the Third World, especially India and China, would sign the Protocol only if it reflected these views. The London amendments do so and New Delhi and Beijing are now expected to sign the Protocol.

The US is the biggest offender, consuming some three lakh tonnes of CFCs per annum for a population of about 220 millions, which works out to 1.36 kg per capita. The entire developing world, including India and China, consumes less than 12 per cent of the world production of about 2 million tonnes of CFCs. While China consumes about 20,000 tonnes annually, India's production and consumption are around 5,000 tonnes, and for population of 800 millions it works out to just 0.01 kg per capita.

It is, therefore, only right that the developed world pays for the corrective steps necessary to repair the damage it has done. It should contribute to a fund to be used exclusively to meet the expenses of the developing world in switching over to an environment-friendly technology. This was the main demand of the group 77 developing nations, led by India. China threw its considerable weight behind this. Though some developing countries had signed the Montreal Protocol, India and China pointed out that these nations would not have been affected because they were importing CFCs. On the other hand, India is self-sufficient in them and has a fairly large indigenous refrigeration industry, the main user.

More than six months ago the developing nations had submitted their proposals to amend the Protocol. In April, India hosted a conference where the Third World's stand was harmonised. China was invited to the meeting and considerable effort was made to ensure that New Delhi's and Beijing's positions were identical. At the London meeting also the two delegations worked closely.

The Indian delegation to London was led by the Minister of State for Environment and Forests, Maneka Gandhi, and, at the official level, by the Secretary in the Ministry of Environment, Mahesh Prasad. Both were closely associated with the New Delhi meeting in April and had established a rapport with China and other Third World countries.

The amendments to the Protocol that were thrashed out have three main aspects. One, the setting up of a financial mechanism to meet the additional costs which the developing countries would incur in the switchover to a new technology and the crucial issue of ensuring its transfer. Two, the removal of certain glaring discriminatory clauses in the Protocol. And three, the tightening of the time schedule for a complete phase-out of the dangerous chemicals—a ratehr belated recognition of the fact that the already bedraggled ozone layer will not brook any further neglect.

To put it simply, the affluent nations had the unreal choice of driving a hard bargain only at the cost of facing, along with their poorer neighbours, the doomsday predicted by scientists. The deadly ultraviolet rays from the sun, reaching the earth through the holes in the ozone, layer, would not distinguish between the haves and the have-nots. These could play havoc with all life. Fortunately, the West was forced to see that its money and technology were not enough to ensure a healthy ozone layer and that without India and China the Protocol would not be worth the paper it was written on. Though the consumption of the ozone-depleting gases by he developing world is low, it is expected to increase dramatically in India and China even if they "developed" at a modest rate.

There is no doubt that the amended Protocol is more responsive to the urgent issue of ozone-depletion. But some enviornmentalists feel that this does not go far enough, fast enough. Jeremy Legget, Director of Science, Greenpeace, Britain, is sharp in citicising the amendments. In an article in *New Scientist* he says there is scientific evidence to suggest that the ozone layer has been depleted by 10 per cent since 1967. He points out that before the CFCs and halons are completely phased out, there will be a 30 per cent rise in the world production of these chemicals. Moreover, some of the substitutes themselves generate greenhouse gases, which will continue to add to the problem of global warming. He blames corporatioins such as ICI, the biggest producer of ozone-destroying chemicals in Europe, for lobbying to prevent a tougher Protocol.

For India, the amendments have been a gain. The Ministry of Environment and Forests some time ago had asked Touche Ross, a British consultancy firm, to estimate what the cost would be for India to phase out CFCs. An Indian firm was also involve in this exercise. It has ben put around $350 million over the next 10 years. From the interim fund, to be set up under the amended Protocol, India can hope to get straight way nearly $40 million in the next three years. Senior Ministry officials feel that money will not be a problem. The amended Protocol is clear on this.

The problem area could be the transfer of technology, for the Third World has not been able to get fully what it sought. The Protocol merely says that countries which have the new technology will take all practicable stamps to ensure its transfer. Western countries have apparently taken shelter under the plea that the technology is with private corporations over which they had no direct control.

There is no doubt that over the years the Indian consumer will also have to pay a price. The industry may still be able to get away lightly for it can hope to get funds from the Government, which in turn will get the money from the interim fund. Perhaps ways should be found to ensure that the consumer is not unfairly burdened.

The new Protocol is only the beginning. Drastic steps will have to be taken at the international level to tackle the threat man has himself created. The issue of global warming promises to be tougher to handle. For it is not restricted to a few industries but encompasses almost all industrial activity.

Every man, woman and child adds to the problem. The growing population puts pressure on land, and as more land is cleared of forests and vegetation, the process of global warming gets strengthened. This sets in a vicious circle. The threat of nuclear war, familiar in the Cold War era, has given way ironically to the threat posed by "development" by the very existence of man himself.

THE CRUCIAL CHANGES

At the London meeting in June to amend the 1987 Montreal Protocol, India certainly hogged the limelight as it led the Group of 77 developing nations in the campaign for major changes to accommodate the views of the Third World.

With China solidly behind this group, the West had to bend. It became obvious that the Protocol could not even present to be effective or meaningful if India and China stayed out. After 10 days of negotiations, which often stretched on to the night, an agreement was reached on major amendments after the United States was force to climb down from its earlier position. Thus the way was paved for India and China to become part of the first major effort to save the ozone layer by phasing out chemicals which have been assaulting it.

The important amendments can be classified under three heads. One, the financial mechanism to help developing countries in the switch-over to environment-friendly technology and the issue of its transfer. Two, the removal of disparities in conditions imposed on developed and developing countries. And three, the tightening of the time schedule to phase out the dangerous chemicals.

The Third World view was carried when the London meeting agreed to the setting up of a fund to help developing countries by meeting the incremental costs of the switch-over. Initially, an interim fund of $160 millions (about Rs. 272 crores) is to be set up. Another $80 millions (about Rs. 136 crores) will be added when India and China sign the Protocol.

The developing countries were also able to ensure a system of democratic control over the fund, an idea which was again opposed by the US initially. Developing and the developed countreis will have equal representation on the body controlling the fund. The first chairman will be from Finland, and his deputy from Mexico. These two posts will alternate between a developing and a developed country. The 14-member executive committee will have on it seven nations each from the two groups. Significantly, voting will be on a one country, one vote basis (the US had wanted weighted voting on the basis of contributions to the fund), and decisions would be arrived at by a two-thirds majority with at least a simple majority for the decision in each of the two blocks. This is to ensure that the developed world does not foist its will on the developing countries.

On the contentious issue of transfer of technology, India was not able to achieve completely what it had demanded. While the group of 77 and China wanted the ozone-friendly technology made available to the developing countries on a non-commercial and preferential basis, the other side argued that as the technology was with companies not controlled by governments, this was not possible. It was, however, agreed that the developed countreis would take "every practicable step" to ensure that "the best environmentally safe substitutes and related technology are expeditiously transferred to the developing countries and that these transfers occur under fair and most favourable conditions." The amended Protocol clearly says that the obligations of the developing countries to phase out the old technology will depend upon the implementation of the financial mechanism and the transfer of technology.

As regards the removal of disparties in the treatment of developed and developing nations, the success was 100 per cent as described officially in India. The Protocol, before amendment, allowed a developed country to withdraw from it and to export CFCs, while a developing nation was not allowed these privileges. But now nations from both groups can do so.

A major victory for environmentalists was the tightening of the time schedule for phasing out CFCs. Under the amendment, CFCs are to be phased out completely by the year 2010, with the developed nations doing it by the year 2000. For halons a similar phase-out schedule has been agreed upon with some exceptions such as in fire fighting where these are very effective. This is an important change because the 1987 document stipulated only a 50 per cent cut by the year 2000 for the developed worth, with a 10-year time lag for the Third World. The amendments have also added to the list of chemicals—carbon tetrachloride and methyl chloroform—that are to be eliminated.

THE INDIAN ANGLE

With the developed and the developing nations sorting out their differences over the Montreal Protocol at the June-end London meet, a decision was arrived at to phase out by the turn of the century the use of chemical products which have ozone-depletion potential. What does this imply for the industry engaged in this area in India?

The Montreal Protocol seeks to control the production of chlorofluorocarbons (CFCs); carbon tetrachloride (CTC), halons and methyl chloroform. Of these, CFCs are to be phased out compeltely while the halons and CTC are to be restricted to certain uses. The use of methyl chloroform is to be reduced by half.

India produces three CFC products—R11, R12 and R22—besides CTC and methyl chloroform. Halon production is at the pilot plant scale. Among CFCs, R22 is not covered by the Montreal Protocol. CFCs are used extensively as refrigerants, blowing agents in the production of foamed or extruded polymers, propellants in pressurised containers for aerosols, fire-extinguishing agents and solvents or cleaning agents in electronic industries. According to industry sources, the London decision is not expected to have an adverse impact in India. In fact, the production of CFCs is expected to go up in the next few years.

It is not that the industry is not concerned about the damage CFCs cause to the ozone, layer; its representatives have been interacting with the Union Ministry of Environment in the latter's efforts to facilitate the participation of developing countries in the Montreal Protocol. But considering the low base of production and use of CFCs, the projected growth in the industries using CFCs and the conditions necessary for a switchover to substitutes, there seem no immediate hurdles to expand the production of CFCs.

Comparatively a late entrant in this area, India set up its first CFC plant in 1968, three decades after the United States. Only in the last two years did three more plants come up., The four units, which have a total installed capacity of 17,500 tonnes per anum, are: Navin Fluorine Industries; SRF Limited; Gujarat Fluoro Chemicals; and Mettur Chemicals. But the current production is around 5,000 tonnes, the consumption level. According to R.N. Shende of SRF Limited, 70 per cent of the output goes for refrigeration and cooling, 20 per cent for aerosols, 5 per cent for foam-blowing and the rest is absorbed by others. This is in sharp contrast to the use pattern in the UK where 62 per cent finds application in aerosols, 18 per cent in foam-blowing, 12 per cent in miscellaneous uses and 5 per cent in refrigeration in and cooling. In the US 55 per cent finds use in the foam-blowing industry, 35 per cent in refrigeration, 23 per cent in miscellaneous and 7 per cent in aerosols.

The refrigeration and cooling industry, which absorbs the bulk of the CFC production in India, plays a vital part in food processing and preservation, drugs and pharmaceutical industry and defence. But till recently the growth rate in the industry was low because of which the consumption of CFCs had not picked up, the per capita use being around 3 grams against 1,700 grams in the US. This is changing fast and with industries using CFCs expected to grow at 15 per cent per annum, the installed capacity of 17,500 tonnes is expected to be achieved in the next five years and used in full. This still means, according to industry sources, a per capita consumption of not more than 30 grams by the year 2005 against the 300-gram limit set in the Montral Protocol. On the other hand in developed countries such as the US the per capita consumption is expected to be around 170 grams despite a 90 per cent cut.

"An analysis of the consumption of CFCs in developing countries shows that there are two categories of users. The first group includes those nations which have reached or are about to reach the consumption level of 300 grams per capita. The other group includes countries which shall never reach the level even by the year 2000. The second category can be given a further grace period of 20 or 30 years before being asked to comply with the Protocol," said Shende at the Indo-US workshop on CFC in New Delhi in April. This has been conceded and at London meeting it was decided to give the developing countries time till the year 2010 to phase out CFC use. Though Shende welcomes this, he hopes to get this extended by another five years and quotes the conclusions of a study by NASA (National Aeronautics and Space Agency) scientists that a 15-year lag by the developing countries in phasing out CFCs will delay only slightly the reaching of the 2 ppb (parts per billion) chlorine level in the stratosphere, the goal of the Montreal Protocol.

The Indian Industry will like the grace period to be long enough to synchronise with the development of substitutes with zero-depletion potential. The alternatives to CFCs, such as HCFCs (hydrochlorofluorocarbons), are not entirely free of elements that can damage the ozone layer; only their potential for damage is less. Even the zero-depletion-potential substitutes are said to have greenhouse effect, that can lead to global warming. Thus the CFC industry in the developing countries will first have to switch over to "transitionary" substitutes and then to permanent ones. Thus if the industry in the developing countries is spared the first switchover, to that extent it will save costs for the developed countreis which have to provide the finance and technology for the switchover. According to an estimate, the switchover will cost India around $2.5 billions.

Another arguement for a longer grace period is that the CFC substitutes are yet to be produced on a commercial scale. While semi-commercial plants are under construction, commercial plants will not be set up before 1993. The efficiency, power consumption and material compatibility of the substitutes are also yet to be established.

The National Chemical Laboratory at Pune has initiated a programme to assess the properties of the alternatives, for airconditioning and refrigeration. According to its Dr. S. Devotta, who presented a paper at the New Delhi workshop, the alternatives identified are R134a for R12 and R123 for R11. While the laboratory is to assess their properties, the Indian Institute of Chemical Technology in Hyderabad will synthesise and the Indian Toxicology Research Centre in Lucknow will study the toxicity. Then the product will be tried on compressors at the Ranchi station of the Central Mechanical Engineering Research Institute.

As regards halons, India imports all its requirements. As no substitute seems to be in sight, the Government has licensed indigenous production. If plants respond so well to additional carbon dioxide, then we would expect to see positive responses to the substantial increase in atmospheric CO_2 over recent decades. Several pieces of evidence suggest exactly such a response.

A fascinating report was published by scientists with the Finnish Forest Research Institute in the April 3, 1992 issue of *Science* magazine. The researchers reported a 25 to 30 per cent increase in the growing stock of forests in Austria, Finland, France, Sweden, Switzerland, and West Germany between 1971 and 1990, and attributed this growth in part to a 9 per cent increase in atmospheric carbon dioxide during the same period.

Hartwell Allen, a plant scientist with the US Department of Agriculture's Agricultural Research Service in Gainesville, Florida, has estimated that soybean yields have increased by 13 per cent because of a rise in global carbon dioxide concentration. Another source of dramatic evidence of rising biological productivity comes from the increasing amplitude of the atmosphere's CO_2 cycle.

The majority of the terrestrial vegetation of the earth is in the Northern Hemisphere, so more photosynthesis takes place in the spring and summer than in the fall and winter. The carbon dioxide level in the atmosphere begins to fall in the spring, and continues to fall through the summer months, as CO_2 is removed by the vegetative cover of the north. In late autumn, at least in northern latitudes, much of this vegetation dies and decomposes, and photosynthetic activity drops to a low level much of the carbon sequestered over the growing season is then returned to the air as CO_2.

This oscillation in the CO_2 level is called the "biosphere's breath," and its amplitude is measured by the Mauna Loa recording station in Hawaii. When the first records were made in 1959, there was a 6.5 ppm difference in CO_2 concentration between summer and winter. Last year, the difference was 7.5 ppm. What this means is astonishing: during their major growing seasons, the plants of the Northern Hemisphere have been able to sequester at least 15 per cent more carbon than they did 33 year ago.

Besides greater efficiency in photosynthesis and a reduction in water loss, higher levels of carbon dioxide provide other important benefits for plants. Rising levels of CO_2 compensate for teh deficiencies in light that frequently occur in the winter months in northern Europe, Canada, and the United States. Indeed, flowers and vegetables grown in CO_2-enriched greenhouses experience an even higher-per centage boost in plant productivity under very low light intensities than under normal light.

Enrichment of the air by carbon dioxide also appears to offer some protection to plants against both extremely hot and cold temperatures. There is also evidence that high atmospheric levels of CO_2 raise the optimal temperature for plant growth. The implication of this for the global warming debate is significant: if the higher-CO_2 world of the future leads to higher temperatures, plants will respond favourably both to increases in carbon dioxide and to the warmer conditions.

Plant responses to a higher carbon dioxide concentration do appear to be limited by deficiencies in nitrogen and other mineral nutrients. If plants are to take full advantage of future CO_2-enriched atmospheres, it may be necessary to apply more fertiliser in many parts of the world. Even so, higher CO_2 levels have a remarkably stimulatory effect on biological nitrogen fixation by legumes, such as soybeans. A classic study by Ralph Hardy and U.D. Havelka, published in *Science* in 1975, showed that a tripling of atmospheric CO_2 results in six-fold increase in biological nitrogen fixation—from 75 to 425 kilograms of nitrogen per hectare—by rhizobial bacteria in nodules attached to the roots of soybeans.

Elevated concentrations of CO_2 also offer protection against air pollutants. The partial closing of the stomata at higher CO_2 levels reduces the exposure of both C_3 and C_4 plants to ozone, sulphur dioxide, nitrous oxides, and other harmful substance in the air. The benefits are particularly pronounced for soybeans and other legumes that are especially sensitive to air pollutants.

Good News for the Planet

The benefits to plants that would result from a doubling of the carbon dioxide concentration do not necessarily mean that such a doubling is good for the planet. We do not know what the optimal level of atmospheric carbon dioxide should be. So many variables could be affected by a major increase in CO_2, including temperature and a redistribution of water resources, that the honest observer has to conclude he does not really know what will happen. Even so, the good news about plant growth makes it possible to project a number of features of the global ecosystem in the next century.

First, we can expect a rapid expansion of food production that may offset some of the presumed adverse climate effects. As crop yields rise with higher CO_2 levels, the amount of land devoted to agriculture can decline. It will be much easier to protect environmentally sensitive land areas from over-cultivation for crops.

Since C_3 plants will benefit somewhat more than C_4 plants from higher CO_2 levels, there will be some shift in the mix plants. Trees are C_3 plants, so we can expect more rapid reforestation and an enormous expansion in forest biomass. Of the 21 most important food crops, 17 have C_3 pathways. They include rice, wheat, barley, oats, rye, soybeans, field beans, hung beans, cowpeas, chickpeas, pigeonpeas, potatoes, sweet potatoes, cassava-yams, sugar beets, bananas, and coconuts. The exceptions are corn, sorghum, millet, and sugarcane, which have C_4 pathways, and which will probably decline in relative production. On the other hand, since 14 of the 18 most noxious weeds are C_4 plants, rising levels of atmospheric CO_2 will generally favour crop production over weeds.

Plants, directly or indirectly, provide 95 per cent of the total food of the earth. Since plants are at the bottom of the food chain, a boost in plant production should lead to major increases in bird, fish, and mammal populations as well.

The rising carbon dioxide concentrations in the atmosphere must be viewed with caution. But it is inappropriate for public discussion of the issue to focus only on the hypothetical dangers of global warming that might result from higher carbon dioxide levels. It is important to stress as well the known benefits of a higher carbon dioxide concentration for the productivity of food crops, trees, and other plants.

Scientists and Climate Change

Many scientists uphold the evaluations made in this number of All of us regards the modifications that might be taking place in the atmosphere and their subsequent effects on the climate. Others, in contrast, prefer not to commit themselves until more experimental evidence becomes available. There are yet others who play the issue down. These differences in criteria call for a socio-scientific analysis and a few reflections.

Science attempts to understand reality. With the experimental results it gathers, it constructs models and puts forward interpretations that change in the light of new results. The enormous influence science has attained in the last few decades has given it such prestige in the eyes of the majority that scientific models are often worshipped with dogmatic unction. This of course is wrong. The classical mechanics conceived by Newton in the seventeenth century was one of these models or approximations and allowed technological applications such as the steam engine or the internal combustion engine, but it proved clumsy and lacking in explaining the behaviour of particles moving at the speed of light. It took Plank and Einstein to develop the basis of quantum mechanics and relativity before the step could be made from the coffee grinder to the intersteller spaceship. Doubt and critical revision constantly preside over the work of the good scientist, while most of us prefer to entrench ourselves in the comfort of established truths.

Prefer to, or, to be more precise, need to. In fact, all of us need to, even scientists themselves when they are being ordinary citizens. It would be true to say that the climate has been going through uninterrupted variations ever since the world began, and also that there is no irrefutable evidence that present modifications have their origin in man. And whether they do or not, it is not entirely certain that the changes, if they take place, will be catastrophic for humans. Some scientists therefore prefer to reserve judgement or distrust the evil omens. The trouble is that this doesn't solve the problem, if there is one, which is why other scientists prefer to take a risk and make their point of view known, even if it is no more than a well founded hypothesis.

The least we can hope for from a provisional acceptance of these hypotheses would not be excessive caution, a good thing in a society that constantly errs in the opposite sense. Indeed, the same commendable rigour with which alarming hypotheses are criticised should have been applied when insufficiently tested products were admitted to circulation. The ecological equivalent of the juridical rule of presumed innocence ought to be the rule of contingent prevention. We cannot wait to know everything about reasonably suspected problems and then act scientifically. The most accurate medical diagnosis is always result of the autopsy, but this isn't appreciated by the patient, who would undoubtedly have preferred an incomplete diagnosis and a timely treatment: the wisdom of contingent prevention as opposed to the foolishness of complaining once it's too late. Taking a responsible risk is also a scientific approach. Perhaps, in all honesty, like no other.

CLIMATE DESTABILISATION

Between 1980 and 1990 we experienced that eight hottest years of the whole of the century. Does this mean that the earth's climate is changing?

In fact the climate undergoes both short-long and-termnatural variations. The shorter the time-scale, the smaller these natural climatic variations are. For example, the average surface temperature of the planet undergoes variations of between 5 and 6 degrees every 1,00,000 year. On the other hand, over period of the order of a few thousand years, these variations are of the order of one degree.

Carbon dioxide (CO_2), methane (CH_4), nitrogen dioxide (NO_2) and steam, amongst others, are molecules (some of them forming part of the natural composition of the atmosphere) which absorb long-wave or infra-red radiation. It is this inherent property of their structure that makes them greenhouse gases, since their natural concentration in the atmosphere means that not all the solar radiation reaching the planet is returned to our space, but is held back.

This blocking of part of the radiation (infra-red) and its reflection back towards the surface of the Earth is what known as the greenhouse effect. The effect makes the temperature on Earth suitable for life.

Every kWh produced from sun or wind avoids the introduction of 1 kg of CO_2 into the atmosphere

Human activities have always had an effect on the climate. But with the industrialisation of many parts of the planet, the effect have multiplied. Every year, industrial activities release almost 6,000 million tons of CO_2 into the atmosphere, along with hundreds of millions of tons of other greenhouses such as methane and the chlorofluoro-carbonates, so that their presence in the atmosphere is increased massively. Forest fires also contribute considerably to the emission of greenhouse gases.

Once released into the atmosphere, CO_2 remains there for a period of between 50 and 200 years. CH_4, on the other hand, remains for only 10 years, but can absorb between 20 and 30 times as much heat as CO_2.

All this makes the temperature of the Earth's crust increase above the limits of the natural variations. Detailed scientific analysis have shown that the average global temperature of the Earth's surface has increased by 0.5 degrees in the last 100 years.

The estimated contribution to the increase of the Earth's temperature by different sources of emissions are as follows: 50 per cent from energy sources (including transport), 20 per cent from chemical products (CFCs, halons), 15 per cent from tropical deforestation and 15 per cent greenhouse from agriculture and refuse.

If emissions of continue increasing at the current rate, then by the middle of the next century the temperature could reach the highest level in the last 20,000 years. This could have an unsettling effect on the climate of our planet.

AVOID IT!

CO_2 is the main factor

It is estimated that CO_2 is responsible for 50 per cent of the increase in the greenhouse effect. Most of the CO_2 released into the atmosphere is a result of burning fossil fuels, so that if we want to reduce emissions of this gas we must first reduce the amount of coal, oil and natural gas burn. It is also very important to burn these substances as efficiently as possible.

A large part of the energy consumed in the world today is not used to provide services from energy (heat, light, comfort, etc), but is simply degraded in the form of nonusable heat.

Stock-raising and refuse too

Methane gas (CH_4) is the second of the greenhouse gases in terms of the quantities released as a result of human activity. To contribute to its reduction, important changes must be introduced in many economic activities and consumer habits.

Methane is produced through the energy-related activities mentioned earlier (both during the extraction of fossil-fuels and during their distribution). Reducing energy consumption and improving efficiency is therefore doubly advantageous since it reduces emissions of methane as well as of carbon dioxide.

Stock-raising leads to considerable emissions of methane as a result of the products generated during the animals' digestive processes. The industrialised countries should aim to reduce their consumption of meat and increase their consumption of vegetable proteins. This would make for a reduction in the livestock population on a global level.

Rice cultivation also leads to important emissions of methane. In view of the food requirements of the human population the only way to reduce its effect is through the stabilisation of the population, which would have to take place respecting cultural diversity. Other forms of cultivaion that involve less emissions should also be investigated and implemented.

Urban refuse is an important source of methane. A greater effort must be made to reduce refuse production, introduce selective refuse collection and recycle. There are simple techniques for making compost and using micro-organisms that digest organic material. Their application should be extended.

Burning wood and stubble also contributes to methane gas in the atmosphere. This practice should therefore be avoided whenever possible, and when it is necessary (as carried out as efficiently as possible.

Combustion and fertilizers

Biomass combustion is responsible for most NO_2 emissions into the atmosphere- apart from volcanic eruptions which we humans do not control. This is another reason—besides the emission of methane— for avoiding the practice of burning organic material to no purpose.

Another important source of NO_2 emissions are combustion processes occurring at very high temperatures. One example is the internal combustion engine used in automobile transport. Reducing use of automobiles, therefore, reduces NO_2 emissions.

Fertilisers, whether chemical or mannure, also produce considerable quantities of NO_2. We should avoid the indiscriminate use of fertilisers and reduce intensive stockraising which causes concentrations of NO_2 production. At the same time, we should encourage the cultivation of the leguminous species which are nitrogen fixers, and the practices of permaculture and ecological agriculture which tend to respect the nitrogen cycle.

The climate change issue raises important questions of international equity. At present, the major proportion of greenhouse gases comes from the industrialised countries which contain only about 25 per cent of the world population. Agarwal and Narain (1991) have suggested that each person on Earth is entitled to an equal share of the capacity of the world's oceans and biosphere to absorb carbon dioxide. Others, including spokesmen from the developing world, insist that the developed countries must reduce their emissions so that more of the planet's capacity for assimilation of greenhouse gases will become available for the developing and newly industrialising countries and that the latter must be assisted through the transfer of adequate new and additional financial resources, through various forms of technical assistance, and through transferring to them environmentally-sound technologies in the broad sense of the term in order to pursue sustainable development pathways. This will be a major challenge in the coming decades.

An International Convention on Climate Change

Under the auspices of the United Nations, an Intergovernmental Negotiating Committee was established to negotiate a global convention on climate change.

The objective of the convention is to control and counter the adverse consequences of climate change.

The key elements in the negotiations are the commitments of the parties to:

- implement suitable measures to stablise the emission of greenhouse gases; starting with the industrialised countries;

- adopt policies and measures for reducing greenhouse gas emissions;
- update and report on national inventories of all sources and sinks of all greenhouse gases not controlled by the Montreal Protocol;
- protect existing and develop new sinks and reservoirs for greenhouse gases;
- co-operate in the development and dissemination of scientific, technical, socio-economic and legal knowledge relevant to climate change and potential responses thereto;
- co-operate in the transfer to developing countries of new technologies and techniques that consume less energy or produce less greenhouse gases;
- establish a fund for financing measures to counter the adverse consequences of climate change;
- establish a separate International Insurance Pool to provide financial insurance against the consequences of sea level rise.

The convention will also cover the institutional arrangements and procedures for implementation, review and further development of the convention, as well as managing the fund and insurance pool.

To support actions and policy analyses, a strong programme of studies, applications, research and monitoring is required. High priority items in addition to the above should include:

- increased understanding of the global cycles of the greenhouse gases;
- improved capability to produce climate change scenarios at the regional scale;
- strengthening of the Global Climate Observing System and the Global Oceanic Observing System;
- analysis of the underlying ethical questions of inter-regional and intergenerational equities;
- establishing the real costs of action and inaction in the area of climate change;
- establishing and/or application of economic incentives and disincentives for achieving stabilisation and reduction of CO_2 and other greenhouse gases not controlled by the Montreal Protocol;
- widespread and firm application of the 'Polluter Pays Principle' and the 'Precautionary Principle'.

ENSO and Climate Change

Variations in climate from year to year may appear at first glance to be random. However, careful examination of historical data reveals a coherent global pattern of oceanic and atmospheric fluctuations, called the Southern Oscillation. Extreme anomalies in this pattern involve dislocations of the rainfall distribution in the tropics, bringing drought to some regions and torrential rains to others. These anomalies last for many months. Related anomalies of the atmospheric circulation extend high into the atmosphere and polewards into the temperate zones. Some major changes in the ocean currents and temperatures are also related to fluctuations in the Southern Oscillation. The best known of these is the El Nino, a temperature increase that occurs every few years in the central and eastern equatorial Pacific with catastrophic effects on marine ecosystems. Because El Nino occurs when there is an extreme anomaly in the Southern Oscillation, the two phenomena are often referred to jointly as ENSO. The phase of the ENSO phenomenon when an El Nino occurs is also called an "ENSO event". The opposite phase, when the east Pacific is cool, is called an "anti-ENSO event" or "La Nina".

The distribution of anomalous rainfall associated with the two phases of ENSO (i.e. ENSO events and anti-ENSO events) has been well-documented. The majority of areas affected by the phenomenon are located in the Asia-Pacific region. Nicholls demonstrated that ENSO increases the variability of rainfall in the areas it affects (i.e. these areas suffer more intense droughts and well periods than in other areas). Changes in the behaviour of ENSO would, therefore, lead to changes in the variability of rainfall in the Asia-Pacific region. If ENSO stopped operating, for instance, climate variability in this region would be reduced. If it continued to operate but its behaviour changed (e.g. towards more frequent or stronger ENSO events and fewer or weaker anti-ENSOs) this would have significant effects on the frequency or intensity of droughts and floods in the Asia-Pacific region.

It is clearly important for the Asia-Pacific region to attempt to predict how ENSO might be affected by global warming. Several suggestions have been made about how this might be done. Most of these approaches are flawed and produce conflicting predictions. Therefore, no credible predictions of ENSOs likely reaction to global warming can be made at this stage. In the long run, a coupled ocean-atmosphere model capable of simulating the present-day behaviour of ENSO must be developed. The model could then be used to predict ENSO's reaction to global warming. Such a model is not yet available although several groups have developed rudimentry models which reproduce some aspects of ENSO's behaviours. These model "ENSO's" are very sensitive to the background model "climate". If the models adequately simulated ENSO's real behaviour this sensitivity would suggest that global warming may produce marked changes in its behaviour.

Until an adequate ENSO model is available perhaps the best approach to predicting ENSO's future behaviour is to examine its past behaviour. If we could be sure that its behaviour has not changed markedly in, say, the last 10,000 years, we might be more confident that it was robust with respect to climate changes. On the other hand, evidence of considerable variations in its past behaviour, such as frequency of occurrence, would indicate that it was not very robust with respect to climate change, as suggested by the rudimentary ENSO models presently available.

Large negative values of the SOI usually coincide with El Nino events; large positive values with La Nina events. The time-series shows little evidence of major changes or instability in behaviour over the past century, despite the 0.5^{0}C global warming observed over this period.

Unconventional data sources must be used to plot the behaviour of ENSO prior to the middle of last century, because of the paucity of appropriate meteorological data. Nicholls used documentary sources to demonstrate that ENSO was operating and affecting both sides of the Pacific in the first half of the 19th century. Quinn et al. documented strong and moderate El Nino events (using documentary records of heavy rainflls in northern Peru) from 1525, Enfield demonstrated that El Nino intensities and intervals between events seemed no different now than they were four centuries ago. Lough and Fritts (1989) reconstructed the SOI back to 1601 from tree-ring data and also found no clear trends in its behaviour. So ENSO Appears to have been robust at least with respect to climate changes of the magnitude of the "Little Ice Age" a period of globally lower temperatures that extended roughly from 1500 to 1850.

Whether ENSO exhibited its present-day behaviour earlier than this is difficult to say although several studies of proxy data such as ice-cores, coral-cores, sediments etc. suggest that at least some El Nino events occurred thousands of year ago. Enfield reviews some of this evidence. A different approach was suggested by Nicholls. As noted above ENSO causes highly-variable rainfall in the areas it affects. The wild-life in at least some of the areas affected by ENSO (e.g. Australia) is very well adapted to highly-variable rainfall implying that ENSO may have been operating in much the same way as today over evolutionary time-scales. If so it has survived major climate changes and might be expected to survive global warming of the magnitude predicted from the enhanced greenhouse effect.

From the evidence available at the moment there seems little reason to believe that major changes in ENSO will occur due to global warming in the next few decades. Certainly there is no way, yet, of predicting how ENSO might change. A definitive answer, however, requires the production of a model capable of simulating ENSO's present-day behaviour.

How ENSO Confounds the Monitoring of Global Warming

Even if ENSO does not change its behaviour as a result of global warming, its behaviour is still of relevance to a global change programme. Fluctuations on a decade time-scale can be observed in the numbers or intensity of ENSO events. For instance, between 1975 and 1987 several major ENSO events were observed, without any anti-ENSOs taking place. ENSO events tend to be followed by above average global temperatures, due to the warming of the tropical oceans and the eventual poleward transport of this warming. Anti-ENSOs tend to be followed by global cooling. Note that in the SOI values lead the global temperature variations by six months and some of these variations can, therefore, be attributed to ENSO. Thus some of the very recent warming might be due to a random fluctuation in ENSO. This can be examined by removing the influence of the SOI on global temperature.

The global temperature record for the past 40 years (provided by P.D. Jones of East Anglia University) is shown. Rapid warming has been observed during the past 15 years. The linear regression between the global temperature and the SOI was used to calculate a corrected global temperature with the effect of the SOI removed. Removal of the SOI influence has halved the apparent warming of the past 15 years. Some of the very warm years during the 1980s (which were associated with ENSO events) are cooled substantially when the SOI influence is removed. Some of the cool years in the mid-1970s (associated with anti-ENSOs) are warmed after removal of the SOI influence. Clearly some of the observed recent warming has been due to a random clustering of ENSO events in recent years, together with a cluster of anti-ENSOs (which led to cooling) in the mid-1970s. Note that the increase in ENSO events could not be due to the warming since he warming follows the SOI.

Warming is still evident in after the removal of the SOI influence. This residual warming may be due to an enhanced greenhouse effect although the simple method used here to remove the SOI's influence may not be optimal. A more sophisticated method may removed even more of the apparent warming. Other factors influencing global temperature (e.g. volcanic activity) might also be involved. In any case it is clear that random, short-term fluctuations in the behaviour of ENSO can produce short-term variations in global temperature that may be incorrectly interpreted as real trends. For instance, if the major anti-ENSO of heralds a period of several anti-ENSOs and no ENSOs we can expect global cooling. This cooling may be wrongly interpreted as evidence that the greenhouse effect is not real.

Changes in the mean climate expected from an enhanced greenhouse effect are difficult to predict. Predicting changes in phenomena such as tropical cyclones and ENSO is even harder. Even our present-day monitoring of these phenomena is far from optimal. Some comments about the magnitude of change that might be expected can, however, be made, even if the direction of change is not certain.

Tropical warming of the order of one degree might be expected from the enhanced greenhouse effect by 2030. This is of the order of observed interannual variations in tropical SSTs so changes in the behaviour of tropical cyclones and ENSO caused by this warming should only gradually leave the range of behaviour experienced at the moment. The warming of the tropical oceans should be gradual and so, therefore, should be any greenhouse-induced changes in tropical cyclone activity. It should be several decades before the changes in tropical cyclone numbers, location, and intensity clearly depart from the "envelope" of their recent behaviour.

Certainly, substantial "greenhouse-induced" changes in tropical cyclone behaviour should not yet be observable, since any tropical ocean warming should, so far, be very small. The decrease in tropical cyclone numbers and intensity in recent decades might be interpreted as evidence that predictions of more and stronger cyclones accompanying warmer SSTs are wrong. A different interpretion may be that the prediction of increased numbers, more polewards formation, longer seasons and greater intensity as the oceans warm may be right but that the predicted oceanic warming has not yet begun. Some of the apparent recent warming is due to a couple of recent, intense ENSOs. Some of it may also be attributed to changes in observing practices. The Indiana Ocean appears to have warmed about 0.5^{0}C in the past 30 years. This seems too large to be attributable to the greenhouse effect and may not be real. We may not yet have seen any general warming of the tropical oceans that might be expected to lead to increased tropical cyclone activity.

In summary, there seems little reason to expect, in the near future, rapid changes in the behaviour of tropical cyclones or ENSO, as a result of global warming. This warming should lead to a gradual increase in SSTs which, in turn, may lead to more cyclones, longer cyclone seasons, formation further polewards and stronger storms, although the evidence is not conclusive. For several decades it will be difficult, if not impossible, to separate these relatively small and gradual changes from the present-day interannual variability. Changes in ENSO seem even harder to predict although it appears to have been robust with respect to past climate changes. The immediate priority must be to improve monitoring, modelling, and understanding of tropical cyclones and ENSO so that future changes can be detected and better predictions made, and to improve our response to their present-day behaviour.

CLIMATE CHAOS

For the past several years scientists have issued ominous warnings about the future of the earth's climate. Predictions of dramatic global change arising from the continued dumping of industrial by-products into the atmosphere and forest loss of massive scale can no longer be ignored. Compelling scientific evidence now strongly suggest that world climate patterns, previously regarded as reliably stable, could be thrust into a state of turmoil. Emissions of natural and synthetic gases are increasing the heat-trapping capacity of the atmosphere through a phenomenon known as the greenhouse effect. The projected effects of this world-wide climatic disruption dwarf many of the environmental problems of the past and augur political, economic, and social disruptios on an enormous scale. Global warming could have catastrophic consequences for the habitability and productivity of the whole planet. The accompanying strain and upheaval on the international scene in turn could have serious foreign-policy consequences for all countries

Broad scientific agreement exists on the underlying theory of climate change, although the nature and magnitude of future effects from greenhouse warming as predicted by computer models remain in debate. Some of these, such as a rise in the sea level, have been established with greater certainty than others. Nontheless, the range of consequences is sufficiently clear and the magnitude of the resources at stake so enormous that policy action is required sooner rather than later. Once a crisis has been reached, it will be too late to act.

The international political and legal system remains ill-equipped to offer a solution that will assure the integrity of the earth's climate.

Although the greenhouse theory of warming has been accepted for about a century, policymakers have only recently become aware of its significance for the global environment. The international community cannot effort to continue to delay elevating the greenhouse effect to the top of the foreign-policy agenda. Arresting the impending climate instability will require a concerted international agenda and a reorientation of energy and development priorities in virtually all countries of the world. Heading this agenda for action should be a global multilateral agreement that sets strict, blinding standards for national emissions of greenhouse gases.

Human activities since the Industrial Revolution have dramatically altered the composition of the global atmosphere. A number of gases, emitted in small but significant amounts, absorb infrared radiation reflected from the surface of the earth. As the concentrations of these heat-absorbing gases increase, average global temperatures will rise.

Emissions of carbon dioxide (CO_2) are the single largest cause of elevated terrestrial temperatures from the greenhouse effect, accounting for approximately one-half of the problem. Concentrations of CO_2 inthe range of 280 parts per million (ppm), together with water vapour in the atmosphere, established the preindustrial equilibrium temperature of the planet. Since the middle of the 19th century atmospheric CO_2 levels have increased by about 25 per cent to approximately 4 per cent per year. Elevated CO_2 concentrations result primarily from the intensified burning of fossil fuels—coal, oil, and natural gas—which liberates the chemical in varying amounts. Coal burning releases the most CO_2 while the combustion of quantities of natural gas and oil needed to produce the same amounts of energy results in only about 57 per cent and 87 per cent as much CO_2 respectively.

The world's forest are vast storehouses or "sinks" for carbon. World-wide loss of forest cover, by releasing this vast stockpile of carbons into the atmosphere as CO_2 aggravates the greenhouse problem. Deforestation in Third World Countries is particularly severe, with the destruction of tropical forests in developing countries like Brazil and Indonesia exceeding 27 million acres annually from activities such as burning, logging, and conversion to agricultural and pasture land. Indeed, the release of CO_2 into the atmosphere as a result of deforestation amounts to 2-10 billion tonnes annually.

Concentrations of a second important greenhouse gas, nitrous oxide (N_2O), have also been rising, probably because of heavier fossil-fuel use, greater agricultural activity, and other ecological disturbances. Average global atmospheric levels of N_2O at the end of 1985 were approximately 300 parts per billion (ppb) and are increasing at an annual rate of .2 per cent. Both CO_2 and N_2O, unlike some conventional pollutants, are very stable compounds. CO_2 remains in the upper atmosphere for decades after its release and N_2O for considerably more than a century. Consequently, without major reductions in emissions of these gases with long atmospheric lifetimes, their concentrations will continue to grow.

A group of volatile chemicals known as chlorofluorocarbons (CFCs) is believed to be currently responsible for 15-20 per cent of the global warming trend. These chemicals, unlike CO_2 are strictly synthetic and are not known in nature. They have a number of uses as refrigerants, propellants, solvents, and thermal insolators. A related class of bromine-containing chemicals called "halons" are found in fire-extinguishing systems. Average global atmospheric concentrations of CFC-11 and CFC-12, two of the most commercially important chlorofluorocarbons, in 1985 were approximately .22 ppb and .38 ppb, respectively. Atmospheric concentrations of CFC-11 and CFC-12 are growing at a rate of more than 7 per cent annually as a result of increased world production in recent years.

Although their concentrations are small relative to that of CO_2, CFCs are up to 10,000 times more potent in absorbing infrared radiation. After release, CFCs and halons reside in the atmosphere for close to a century, or sometimes more, because of their great chemical stability at low altitudes. Consequently, an immediate 85 per cent reduction in emissioins of CFC-11 and CFC-12, for example, would be necessary merely to stabilise their atmospheric concentrations. With their long atmospheric lifetimes CFCs and halons eventually reach the upper atmosphere. There, they are the principal culprits in the world-wide loss of the protective stratospheric ozone layer, which shields life on earth from harmful levels of ultraviolet solar radiation.

Methane, the principal component of natural gas, is another significant climate-modifying chemical. It has an atmospheric residence time of about 11 years. Average global concentrations of methane were approximately 1,700 ppb at the end of 1985 and are increasing by about 1 per cent per year, the highest rate of any naturally occurring greenhouse gas, for reasons that are not now clear. Animal husbandry and rice cultivation have been identified as major sources of increased methane emissions. Coal mining and landfills are also significant sources, with large potential for rapid growth in the future.

Low-level ozone is another greenhouse gas. Although ozone in the atmosphere is beneficial, this highly unstable chemical is the leading component of photochemical smog pollution at the earth's surface.

While greenhouse gases are dispersed relatively quickly throughout the global atmosphere after release, industrial emissions of these heat-absorbing chemicals are highly concentrated in the developed world. In 1985, 23 per cent of total global CO_2 emissions of more than 20.5 billion tonnes of CO_2 originated in the United States—the single largest emitting country and the highest per capita contributor among industrial countries to the greenhouse problem. The second biggest contribution came from the Soviet Union, with 19 per cent of total CO_2 emissions. Western Europe emitted 15 per cent of the total, Japan 5 per cent, and the People's Republic of China 11 per cent. Other developing countries together accounted for only 20 per cent of total industrial CO_2 emissions.

Emissions of CFCs are even more strongly skewed. In 1980 the United States produced roughly 28 per cent of the global total of approximately 817,300 tonnes of CFC-11 and CFC-12, Western Europe produced about 30 per cent, industrialised Asian countries 12 per cent, and the East-bloc countries an estimated 14 per cent. The entire developing world accounted for just slightly more than 2 per cent of this amount.

Unfortunately, the very same characteristics that render chlorofluorocarbons rather inert enable them to reach the stratosphere unchanged. There they are exposed to strong ultraviolet radiation, which breaks them apart and liberates chlorine atoms that can destroy ozone by catalysing its conversion to molecular oxygen. (Catalysts accelerate chemical reactions but are freed unaltered at the end.) Indeed, every chlorine atom ultimately eliminates many thousands of ozone molecules. Primarily because of the emission of chlorofluorocarbons, the level of ozone-destroying chlorinated compounds in the stratosphere is now four to five times higher than normal and is increasing at a rate of approximately 5 per cent a year—developments that highlight the profound effect human activity can have on the stratosphere.

Stratospheric ozone (O_3) is formed initially when an oxygen molecule (O_2) absorbs shortwave radiation, which splits it into two oxygen with another oxygen molecule to form ozone. Usually, photochemical reactions catalysed by the NOx gases remove ozone at a rate equal to the rate of its formation. Chlorine catalytic cycles, which are growing rapidly in importance in the atmosphere, disturb this natural balance and result in a net ozone loss.

Particularly in Antarctica and to a lesser extent in the Arctic, frigid temperatures hasten the chlorine-catalytic cycles by removing nitrogen oxides, which strongly interfere with them. (Paradoxically, then although the NOx gases can destroy ozone, their presence in the stratosphere often ameliorates chlorine-catalised ozone destruction.) Together with water molecules, the NOx gases freeze to form particles making up what are known as polar stratospheric clouds. Worse yet, the cloud particles actually facilitate chemical reactions that release chlorine from compounds that do not themselves react with ozone, such as hydrochloric acid (HCl) and chlorine nitrate ($ClNO_3$).

Even if chlorofluorocarbon emissions stopped today, chemical reactions causing the destruction of stratospheric ozone would continue for at least a century. The reason is simple: the compounds remain that long in the atmosphere and would continue to diffuse into the stratosphere from the tropospheric reservoir long after emissions had ceased.

The depletion of global stratospheric ozone seems to be the handiwork primarily of a single class of industrial products—the chlorofluorocarbons-but several different emissions combine to raise the specter of a rapid greenhouse warming of the earth. Exactly how high global temperatures might climb in the years ahead is not yet clear. What is clear is that the levels of such infrared-absorbing trace gases as carbon dioxide, methane, the chlorofluorocarbons and nitrous oxide have mounted dramatically in the past decades, making added heating inevitable.

The trapping of heat near the surface of the planet by naturally emitted trace gases is a vital process: without it the planet would be too cold for habitation. Yet the prospect of a sudden temperature increase of even a few degrees Celsius is disquieting because no one can accurately predict its environmental effects, such as what the precise changes will be in precipitation around the world and in sea level. Any effects will probably be rapid, however, making it extremely difficult or impossible for the world's ecosystems and for human societies to adapt.

The extraordinary pace of the recent increases in greenhouse gases becomes strikingly evident when modern levels are compared with those of the distant past. Such comparisons have been made for several gases, including carbon dioxide—which alone accounts for more than half of the heat trapped by trace species—and methane, which is a more efficient infrared absorber than carbon dioxide but is significantly less abundant.

The histories of carbon dioxide and methane can be reconstructed on the basis of their concentrations in bubbles of air trapped in ice in such perpetually cold places as Antarctica and Greenland. Because the gases are long-lived and hence are spread fairly evenly throughout the atmosphere, the polar samples reveal the approximate global average concentrations of previous years.

Analyses of the bubbles in ice samples indicate that carbon dioxide and methane concentrations held stready from the end of the last ice age some 10,000 years ago, at close to 260 parts per million and 700 parts per billion, respectively. Some 300 years ago the methane levels began to climb, and roughly 100 years ago the levels of both gases began to soar to their current levels of 350 parts per million for carbon dioxide and 1,700 parts per billion for methane. Moreover, direct world-wide measurements made by several investigators during the past decade have shown that atmospheric methane levels are growing more rapidly than those of carbon dioxide, at the remarkably high rate of about 1 per cent a year.

The increases of both gases in the 20th century must be attributed in large part to the many expanding human influences on emissions. For carbon dioxide the sources are mainly fossil-fuel combustion and deforestation in the tropics; for methane, mainly rice cultivation, cattle breeding, biomass burning in tropical forests and savannas, microbial activity in municipal landfills and leakage of gas during the recovery and distribution of coal, oil and natural gas. As the world's population grows during the next century—and with it the demand for more energy, rice and meat products—the atmospheric concentration of methane could double. The climatic warming caused by methane and other trace gases could well approach that caused by carbon dioxide.

What are the expected trends for other trace gases? We as well as several other workers have extrapolated from the past and the present to make projections for the future, taking into account such factors are estimated increases both in population and in energy consumption. The estimates indicate that increases can be expected in the atmospheric concentrations of virtually all trace species in the next 100 years if new technologies and major energy-conservation efforts are not instituted to diminish the expected dependence on high-sulfur coal, an environmentally disadvantages fuel, as the world's major energy source.

For example, as part of a multicenter collaboration, we have looked at past sulfur dioxide concentrations over the Eastern US and Europe (estimated prior to the mid-1960's on the basis of emission rates) and have speculated about future levels there and over the little-industrialised Gangetic Plain of India.

The historical assessment for the U.S. shows a marked increase in sulfur dioxide concentrations between 1890 and 1940, paralleing the buildup of "smokestack" industries and the construction of many new power plants. The amount of sulfur dioxide then leveled off and decreased in the 1960's and early 1970's. To a great extent, the decrease reflects the increased exploitation of oil (which is low in sulfur) for energy as well as the success of clean-air legislation in curbing sulfur emissions.

The concentrations of sulfur dioxide increased over Europe between 1890 and the mid-1990's but then leveled off; they did not decline appreciably, because until recently emission-control effirts were less vigorous than in the U.S. For the Gangetic Plain, where industrialisation is rather recent, sulfur dioxide concentrations over some places have climbed from almost nothing in 1890 to velels that are now approaching those over the North-Eastern US.

The average sulfur dioxide concentrations over all three large regions are expected to increase, in part because low-sulfur fuels will probably become scarer (although extremely stringent emission controls could stabilise levels over the U.S. and Europe for a few decades). The increases could be most marked over India and other developing countries that have rapidly growing populations and access to abundant supplies of high-sulfur coal, which is relatively inexpensive. Clearly, major measures must be introduced in the energy sector to prevent sulfur dioxide concentrations from rising extremely high.

Atmospheric concentrations of carbon dioxide and other trace gases are increasing as a result of the combustion of fossil fuels, other industrial activities, and deforestation. These gases retain heat in the atmosphere that would otherwise be re-radiated into space, thus raising the temperature of the Earth and alterning major climatic patterns. State-of-the-art climate studies estimate that global average temperatures will increase between 2 and 5^0C with an equivalent doubling of the atmospheric carbon dioxide concentration, a concentration that will be reached in the first half of the next century given present rates of fossil-fuel use and deforestation. In the fall of 1987, climate scientists and policy-makers met in Villach, Austria under the auspices of the Bejer Institute, the World Climate Programme of the World Meteorological Organisation (WMO), the United Nations Environment Programme, and several other organisations to discuss plausibel future climatic impacts and possible policy responses.

The expected rise in temperature will be accompanied by many other climatic and geophysical effects, including changes in precipitation patterns, change in frequency and intensity of storms, and a rise in sea level due to both thermal expansion of the oceans and melting of land ice. Concerns have also been raised about the possibility of surprises—sudden abrupt shifts—in the way the oceans and atmosphere may respond to climate perturbations. State-of-the-art climate models can neither predict such surprises, nor rule out the possibility that they will occur.

Considerable uncertainties remain about the nature, timing, magnitude, and regional details of climatic changes. Although we could wait for more research and more detailed information on the environmental and economic impacts of climatic changes before taking actions, this approach has two serious flaws. First, the complexity of climate modelling means that the necessary research will be slow and difficult. Yet any delay in slowing the emissions of greenhouse gases commits the earth to greater and greater climatic changes. Second, any international agreement to prevent major climatic change may be complicated by a desire of certain actors (countries, subnational groups, corporations, alliances) to capitalise on perceived regional advantages in the midst of larger-scale negative effects. Those actors who perceive themselves to be beneficiaries of a warmer earth—even if that perception turns out later to be wrong—may choose not to cooperate in international negotiations to address in issue of climatic change.

More recently, attention has begun to focus on the rate of climatic change. In 1987, the Villach and Bellagio climate conferees concluded that slowing the rate at which temperature and sea level will rise is as important for preventing severe societal and ecological disruptions as in an absolute limit on climatic changes. This focus arises because of constraints on the ability of natural and societal systems to adapt. Of particular concern are limits to how quickly natural ecosystems can respond to climatic changes. Inland migration of wetlands is limited by sedimentation rates; upward growth of coral reef islands cannot exceed rates defined by temperature, tidal action, and nutrient availability; forest and plant communities move at rates constrained by soil conditions, water availability, and vegetation succession dynamics. Because of these concerns, attention is now turning to identifying critical rates of change in order to set long-term environmental targets that may have some use as regulatory guides. Such targets could be supplemented with absolute limits to climatic change, since unlimited warming must eventually be prevented.

Developed and Developing Countries: Vulnerabilities and Responsibilities

Future tensions over climate changes are likely to result from changes in superpower relations (East-West concerns) and from disparities between developed and developing countries (North-South concerns). Several fundamental factors suggest strongly that North-South tensions will be the most problematic. First, the industrialised countries are primarily responsible for the production of greenhouse gases. Second, the consequences of climatic changes will be far more widely distributed among both rich and poor countries. And third, developing countries have far fewer technical and economic resources at their disposal for adapting to or mitigating the impacts climatic changes than do the industrialised nations.

The industrialised world is primarily responsible for the production of greenhouse gases. Although rapid population growth in developing countries will lead to large increases in energy use, this disparity in per-capita production of greenhouse gases is likely to persist for a long time.

Health, Energy, and Greenhouse-Gas Impacts of Biomass Combustion

Poorly managed biomass combustion has three major environmental liabilities due in significant part to emissions of products of incomplete efficiency, with its attendant problems of onerous human labour requirements and pressure on biomass resources from harvesting. Second, some PIC are hazardous to *health* when bareathed in the concentrations comnoly found in homes using unvented biomass stoves,. Third, a different set of PIC are strong direct or indirect greenhouse gases, potentially contributing to global warming. To evaluate the impacts in these three areas, it is valuable to construct carbon balances of alternative stove/fuel systems in what can be called "triple carbon balance" analysis (TCBA). Here TCBA is applied to woodstoves in developing countries, and to one of the chief alternatives, kerosene. The tentative counter-intuitive result is that a switch to fossil fuels can sometime be justified on all three environmental grounds.

It has been said that wood is the fuel that heats you twice—once when you chop it and once again when you burn it. Like fossil fuels, however, biofuels also have the potential to heat you a third time as a results of enhanced greenhouse warming due to the gases released by combustion. It has generally been assumed that this potential is only realised when the biomass being burned is harvested on a nonsustainable basis. With sustainable harvesting, it is argued, an equivalent amount of carbon is recaptured by the regrowing biomass as released by combustion. Thus, the net greenhouse gas increment is zero.

Even when this is true with regard to the number of carbon atoms, it may not be with regard to their greenhouse equivalence. In particular, photosynthesis caputres only carbon dioxide (CO_2) from the atmosphrere, but actual biomass combustion emits other carbon-containing meterials with atmospheric warming impacts (Leving, 1991).

These products of incomplete combustion (PIC) are also of concern because of their effects on human health. Measurement so in village homes throughout the world has show that health-impairing concentrations of PIC are often encountered where people use wood or tower biomass for cooking or heating in poorly ventilated conditions (smith, 1987).

These same PIC also represent lost energy and contribute to the low engineering efficiency with which meals are cooked in much of the developing world (Baldwin, 1987). This, in turn, increases pressure on biomass resources, which, along with land clearing and other factors, has been associated with deforestation and accompanying environmental problems in some areas.

The apparent opportunity for decreasing forest-stressing biofuel demand as well as reducing health-threatening smoke exposures, has lured many local, national, and international organisations, both government and private, into programmes to disseminate improved biomass stoves in poor countries. Although there have been major successes, such as the Chineses national improved stoves programme (Smith et al., 1993a), which has reached more than one-half the nation's rural households, it is only in recent years that the per centage of success has been high (Barnes et al., 1994).

Recently, rising concerns about global warming from the buildup of CO_2, methane (CH_4), and other greenhouse gases in the atmosphere have focused attention on worldwide biomass combustion. Emitting 2100-4700 Tg carbon/year compared to 5700 Tg C/Y from fossil fuels, biomass burning plays important roles in the global carbon cycle (Crutzen & Andreas, 1990). Approaching 1000 Tg C/Y, household biofuel, in turn, accounts for a significant fraction of overall Biomass combustion (Meyers & Leach, 1989). A question is thus raised: "Would alterations in household biomass combustion such as might be brought about by improved stovoes, have significant implications for global warming?"

To explore these issues, it is essential to know the biomass stove emission factors for the varies airborne species that have significant implications for energy, health, and global warming. Some of this information is already available, for in the 1980s a number of studies were undertaken to examine the energy efficiency and health implications of bimass stoves, both indevleoped and developing country situations (Smith, 1987). Although a significant amount of greenhouse-gas research has gone into studies of biomass burning at large scale (forest fires, Sweden agriculture, savannah burning, etc.), relatively little has focused on the type of small-scale combustion found in household stoves of developing countries (Leving, 1991).

With such a focus, a pilot cookstove study was undertaken in Manila. Monitored were emissions of more than 80 greenhouse-related and health-related gases (total non-methane organic compounds, TNMOC) from traditional cookstoves burning wood, charcoal, kerosene, and liquefied petroleum, gas (LPG), which together account for the majority of all cooking in developing countries (Smith *et al.*, 1992, 1993b). Involving only a few stove/fuel combinations in each category, it is not possible to draw statistically valid global conclusions from this pilot study. Nevertheless, the measurements are quite suggestive and serve to illustrate how more detailed studies of this type could be useful. Table 1 summarises the results of the sampling and analyses in terms of emission factors, grams of pollutant per kilogram of fuel, for each of the major PIC.

A profitable way to compare the three major effects (energy, health, and global warming) of biomass stoves is to determine how carbon flows through each by means of a "Triple Carbon Balance" analysis (TCBA). TCBA derived for the composite wood-first cookstove in the Manila study. TCBA follows the typical fate of the 500 g of carbon contained in 1.0 kg of wood burned in such a stove. About 88 per cent of the carbon is emitted as CO_2 (weighing 1.6 kg) and the rest (60 g) is distributed as shown among several kinds of PIC, which together weigh about 130 g.

Energy To put the PIC in an energy context, each constituent needs to be weighted by its energy content, i.e., the additional energy that could have been released if it had been burned all the say to CO_2. The result is that the PIC contain about 12 per cent of the energy originally in the wood, i.e, the combustion efficency is about 88 per cent. In other words, compared to a stove with near 100 per cent combustion efficiency, this stoves requires about 14 per cent more fuel (1/0.88).

This inefficiency is part of the reason that traditional stoves use more fuel than it seems they should. The other major technical reason, of course, is low heat transfer efficiency (the fraction of heat released from the fuel that is taken into the cooking utensil).

As well as representing an energy loss, the 126 g of PIC represent the main health-damaging air pollutants from wood combustion. One way they can be aggregated and compared is by use of the Relative Hazard Index (RHI). This is simply the amount of air it would take to sufficiently dilue each pollutant until it reached the relevant health-based concentration standard (Smith, 1987). With US standards, the total RHI of the PIC is about 1,20,000 cubic meters. CO_2 is not much of a health hazard, as shown by the relatively small RHI, 1800m. (Obviously, application of standards from different countries would result in somewhat different relative weightings for the pollutants.)

Two of the PIC categories shown (TNMOC and RSP) are composites containing a vast array of, mostly, organic chemicals. Many of these individually are known to the health-threatening (e.g, benzene in TNMOC and ployaromatic by hydrocarbons in RSP). Thus, if the RHI included each separately, the total would be larger than that for the general categories.

Global Warming evaluates the same PIC in terms of their greenhouse-gas potential. To do this, it is necessary to apply some index so that the impacts of the different gases can be combined (Smith & Ahuja, 1990). This is so because the gases have different heat-trapping abilities, lifetimes, and interactions with other gases in the atmosphere. Here, I have used the Global Warming Potentials (GWP) developed by the Intergovernmental Penel on climate change (IPCC, 1990; Hayes & Smith, 1993). These are given as a ratio to CO_2 (either per molecule or per carbon atom), and thus can be interpreted as the degree to which the total warming of each compares to CO_2. Since the gases have different atmospheric lifetimes, the relative impact (GWP) depends on the chosen time horizon. Shown here are the results for time horizons of 20 and 100 years. In general, shorter time horizons make the non-CO_2 gases look more important relative to CO_2, since CO_2 is substantially longer lived.

The result is that, depending on the time horizon chosen, the non-CO_2 greenhouse gases, i.e., the PIC, have a total global-waming potential 20-10-% as much as the CO_2 itself. This implies that looking only at the CO_2 emissions of cookstoves may not give a good picture of their global-warming implications. It also implies that improvements in combustion efficiency could result in much larger reductions in total GWP than would be indicated simply by changes in CO_2 emissions.

Projecting emissions is an intricate exercise in social science. How much carbon dioxide humanity as a whole will be emitting in the future depends primarily on the global consumption of fossil fuels and the rate of deforestation (which accounts for perhaps half of the buildup since the year 1800 and 20 per cent of current emissions). Each factor in turn is affected by many others. Growth in fossil-fuel use, for example, will reflect population growth, the rate at which alternative energy sources and conservation measures are adopted and the state of the world economy. Typical projections assume that global fossil-fuel consumption will continue increasing at about its current pace-much slower than it grew before the energy crisis of the 1970s-yielding increases in carbon dioxide emissions of between .5 and 2 per cent a year for the next several decades at least.

Other greenhouse gases, such as methane, the CFC's, oxides of nitrogen and low-level ozone, together could contribute as much to global warming as carbon dioxide, even though they are emitted in much smaller quantities: they are much better at absorbing infrared radiation. But predicting future emissions for these gases is even more complicated than it is for carbon dioxide. The sources of some gases, such as methane, are not well understood; the production of other gases, such as the CFC's and low-level ozone, could rise or fall sharply depending on whether specific technological or policy steps are taken.

Given a plausible scenario for future carbon dioxide emissions, how fast will the atmospheric concentration increase in response? Atmospheric carbon dioxide is continuously being absorbed by green plants and by chemical and biological processes in the oceans. The rate of carbon dioxide uptake is likely to change as the atmospheric concentration changes; that is, feedback processes will enter the equation. Because carbon dioxide is a raw material of photosynthesis, an increased concentration might speed the uptake by plants, counter-acting some of the buildup. Similarly, because the carbon dioxide content of the oceans' surface waters stays roughly in equilibrium with that of the atmosphere, oceanic uptake will slow the buildup to some extent. (The slower the buildup is in the first place, the more effective, proportionally, oceanic uptake is likely to be).

It is also possible, however, that an increased concentration of carbon dioxide and other greenhouse gases will trigger positive feedbacks that would add to the atmospheric burden. Rapid change in climate could disrupt forests and other ecosystems, reducing their ability to draw carbon dioxide down from the atmosphere. Moreover, climatic earning could lead to rapid release of the vast amount of carbon held in the soil as dead organic matter. This stock of carbon-at least twice as much as is stored in the atmosphere-is continuously being decomposed into carbon dioxide and methane by the action of soil microbes. A warmer climate might speed their work, releasing additional carbon dioxide (from dry soils) and methane (from rice paddies, landfills and wetlands) that would enhance the warming. Large quantities of methane are also locked up in continental-shelf sediments and below arctic permafrost in the form of clathrates—molecular latices of methane and water. Warming of the shallow waters of the oceans and melting of the permafrost could release some of the methane.

In spite of all these uncertainties, many workers expect uptake by plants and by the oceans to moderate the carbon dioxide buildup, at least for the next 50 or 100 years. Typical estimates, based on current or slightly increased emission rates, put the fraction of newly injected carbon dioxide that will remain in the atmosphere at about one half. Under that assumption, the atmospheric concentration will reach 600 parts per million, or about twice the level of 1900, by sometime between the years 2030 and 2080. Some other greenhouse gases are expected to built up faster than carbon dioxide, however.

What effect will a doubling of atmospheric carbon dioxide have on climate? The historical record offers non clear quantitative guidance. Nor can climate—the product of complicated interactions involving the atmosphere, the oceans, the land surface, vegetation and polar ice—be physically reproduced in a laboratory experiment. In exploring the future of the earth's climate, my colleagues and I rely on mathematical climate models.

Recent investigations have also led to a better understanding of the effects produced by increased anthropogenic emissions. For example, the much studied phenomenon of "acid rain" by which we mean also acid snow, fog and dew) is now known to develop mainly as a by-product of atmospheric interactions involving the NOx gases and sulfur dioxide. Through various reactions, such as combination with the hydroxyl radical, these gases can be converted within days into nitric acid (HNO_3) and sulfuric acid (H_2SO_4), both of which are dissolved readily in water. When the acidified droplets fall to the earth's surface, they constitute acid rain.

Because water droplets are removed from the atmosphere rapidly, acid rain is a regional or continental, rather than global, phenomenon. In contrast, the atmospheric lifetimes of several other trace gases, including methane, carbon dioxide, the chlorofluorocarbons and nitrous oxide, are much longer and so the gases spread rather evenly throughout the atmosphere, causing global effects.

Since the beginning of the Industrial revolution in the mid-18th century, the acidity of precipitation (as measured by the concentration of hydrogen ions) has increased in many places. For example, it has roughly quadrupled in the North-Eastern US since 1900, paralleling increased emissions of sulfur dioxide and the NOx gases. Similar increases have been found elsewhere in the industrialised parts of the world. Acid rain has also been detected in the virtually unindustrialised tropics, where it stems mainly from the release of the NOx gases and hydrocarbons by biomass burning.

Wet deposition is not the only way sulfuric and nitric acids in the troposphere find their way to the earth's surface. The acids can also be deposited "dry," as gases or as constituents of microscopic particles. Indeed, a growing body of evidence indicates that dry deposition can cause the same environmental problems as the wet form.

Acid deposition clearly places severe stress on many ecosystems. Although the specific interactions of such deposition with lake fauna, soils and different vegetation types are still incompletely understood, acid deposition is known to have strongly increased the acidity of lakes in Scandinavia, the North-Eastern US and South-Eastern canada, thereby leading to reductions in the size and diversity of fish populations. Such deposition also appears to play some role in the forest damage that has been discovered in parts of the North-Eastern U.S. and Europe.

There is little doubt that acids deposited from the troposphere also contribute to the corrosion of outdoor equipment, buildings and works of art, particularly in urban areas—costing tens of billions of dollars each year for repairs and equipment replacement in the U.S. alone. Particles containing sulfate (SO_4) have other effects as well. By scattering light efficiently, they decrease visibility; by influencing cloud albedo, they may have important implications for climate.

In and around cities photochemical smog is another negative consequence of modern life. The term technically refers to the undesirable mixture of gases formed in the lower troposphere when solar radiation acts on anthropogenic emissions (particularly the NOx gases and hydrocarbons from vehicle exhaust) to produce reactive gases that can be destructive to living organisms.

Ozone is a major product of such photochemical reactions and is itself the main cause of smog-induced eye irritation, impaired lung function and damage to trees and crops. The severity of smog is therefore generally assessed on the basis of ground-level ozone concentrations. In other words, the same three-oxygen molecule that is critically important for absorbing ultraviolet radiation in the stratosphere, where some 90 per cent of atmospheric ozone is concentrated, is a problem when it accumulates in excess near the earth's surface.

Investigators have measured ozone levels in the atmosphere since the late 19th century, first from the ground and then within the atmosphere, aided by sophisticated airbone devices. Some of the earliest data showed that the "natural" level of ozone close to the ground at one measuring post in Europe roughly a century ago was about 10 parts per billion. Today the typical ground-level concentrations in Western Europe are from two to four times higher. Abundances more than 10 times higher than the natural level are now often recorded in Western Europe, California, the Eastern US and Australia.

Photochemical smog is also appearing in broad regions of the tropics and subtropics, particularly because of the periodic burning of savanna grasses; the same territories may be set afire as often as once a year. This practice releases large amounts of precursors to smog. Because solar radiation is plentiful and strong in those regions and photochemical reactions occur quickly, ozone levels can readily climb to perhaps five times higher than normal. As the populations in the tropics and subtropics grow, unhealthy air should become even more widespread there. Such a prospect is particularly worrisome because the properties of the soil in those regions may make the ecosystems there more vulnerable than the ecosystems in the middle latitudes to smog's effects.

Although a decrease in ozone near the ground would benefit polluted regions, any decrease in stratospheric ozone undisturbed, because the resulting increase in ultraviolet radiation reaching the earth could have many serious effects. It could elevate the incidence of skin cancer and cataracts in human beings, and it might damage crops and phytoplankton, the microscopic plants that are the basis of the food chain in the ocean.

So far, the extent of stratospheric ozone depletion has been most dramatic over Antarctica, where an ozone "hole," a region of increasingly severe ozone loss, has appeared each southern spring since about 1975. In the past decade springtime ozone levels over Antarctica have diminished by about 50 per cent. A more global assessment of the stratospheric ozone layer is still in a preliminary stage, but in the past 20 years depletions of from 2 to 10 per cent have apparently begun to occur during the winter and early spring in the middle-to-high latitudes of the Northern hemisphere, with the greatest declines in the higher latitudes.

Is it now quite evident that chlorofluorocarbons, particularly CFC-11 ($CFCl_3$) and CFC-12 (CF_2CL_2), are the major culprits responsible for ozone depletion. These anthropogenic chemicals, whose emissions and atmospheric concentrations have grown rapidly since their introduction several decades ago, are widespread as refrigerants, aerosol propellants, solvents and blowing agents for foam production, in part because they have what initially seemed to be a wonderful property: they are virtually unreactive in the lower atmosphere, and so they pose no direct toxic threat to living organisms.

Thousands of scientific experiments have been conducted to measure the effects of carbon dioxide enrichment on specific plants. In most green plants, productivity continues to rise up to CO_2 concentrations of

1,000 ppm and above. For rice, the optimal CO_2 level is between 1,500 and 2,000 ppm. For unicellular algae, the optimal level is 10,000 to 50,000 ppm.

Bruce Kimball, a research leader of the Water Conservation Laboratory of the US Department of Agriculture in Phoenix, Arizona, has pulled together nearly 800 scientific observations from around the world measuring the response of food and flower crops to elevated CO_2 concentrations. The mean (average) response to a doubling of the CO_2 concentration from its current level of 360 ppm is a 32 per cent improvement is plant productivity, with varied manifestations in different species.

Greenhouse-grown vegetables, including tomatoes, cucumbers, and lettuce, show earlier maturity, larger fruit size, greater numbers of fruit, a reduction in cropping time, and yield increases ranging from 10 to 70 per cent, averaging 20 to 50 per cent.

Greenhouse-grown flower crops, including roses, carnations, and chrysanthemums, grow to earlier maturity, and have longer stems and larger, longer-lived more colourful flowers. Yield increases range from 9 to 15 per cent, with a mean of 12 per cent. Flowers and ornamental plants propagated by cuttings, such as geraniums and a number of herbaceous and woody species, show faster and more extensive rooting, with greater plant heights and dry weight. There are also significant reductions in the time needed to grow a marketable product.

Cereal grains with C_3 metabolism, including rice, wheat, barley, oats, and rye, show yield increases ranging from 25 to 64 per cent, resulting from a rise in carbon fixation and reduction in photo-respiration. Flag leaves, the ones closest to grain panicles or heads, show a 60 per cent increase in photosynthetic rates.

Food crops with C_4 metabolism, including corn, sorghum millet, and sugarcane, show yield increases ranging from 10 to 55 per cent, resulting primarily from superior efficiency in water use.

Tuber and root crops, including potatoes and sweet potatoes, show dramatic increase in tuberisation (potatoes) and growth of roots (sweet potatoes). Yield increases range from 18 to 75 per cent.

Legumes, including peas, beans, and soybeans, show yield increases of 28 to 46 per cent. For soybeans, frequently planted not only for their food value but because they naturally fertilise the soil, there is a spectacular increase in biological nitrogen fixation, as will be shown below.

Aquatic plants are commonly limited in their growth by a shortage of carbon dioxide. CO_2 enrichment induce three-to five-fold increases in algal biomass when light and mineral nutrients, especially phosphorus, are plentiful. In lakes and ponds with severe pollution, a higher CO_2 concentration may therefore accentuate the danger of oxygen depletion through a buildup in algae population. This could lead to the death of fish, as happened in Lake Erie until there was a major reduction in pollutants, including phosphates. Aquatic species with foliage above the water, such as the water hyacinth, produce about 40 per cent more biomass when the CO_2 concentration is doubled. Such an increase could aggravate an already serious weed problem in fresh water lakes and streams.

Flower Power

The benefits of carbon dioxide enrichment are wellknown among commercial growers of flowers and vegetables. Paul Nielsen, a rose grower in Santa Barbara, California, puts carbon dioxide into his greenhouse every morning before the vents are opened. "We find that roses respond well to carbon dioxide levels of 1,000 to 1,100 ppm," says Nielsen. "The stems are larger and thicker, and there are more stems per plant."

Dick Oglevee, of Connellsville, Pennsylvania, is America's largest grower of geraniums. From the beginning of September through the middle of May, he burns natural gas in his greenhouses to keep the CO_2 concentration at 1,000 ppm. "More carbon dioxide gives us stronger stalks, better breaking, and more cutting per week," says Oglevee. "I wish we could do this in the summer too, but it's too hot to keep the vents closed."

Richard Gerhart, a cucumber and tomato grower in North Ridgeville, Ohio, is also a big fan of CO_2 enrichment: "Half the dry matter in a tomato or cucumber is carbon, and the only place that comes from is carbon dioxide in the air." Gerhart maintains a 600 ppm concentration in his greenhouses, except on cloudy days, when he gets even better results by raising the concentration to 800-1000 ppm.

American commercial greenhouses have used carbon dioxide fertilisation for tomatoes, lettuce, cucumbers, flower and foliage plants, and bedding plants for at least 30 years. The benefits of this enrichment were first discovered by nurserymen in Germany 100 year ago, and the practice is widely used in Sweden. Denmark, Holland, Germany, Australia, and Japan, as well as the United States and Canada. Carbon dioxide enrichment is economical when greenhouse vents can be closed. It is therefore used most often in winter in northern areas and in the southerly latitudes of the Southern Hemisphere.

It is also standard practice for laboratory scientists working with algae cultures to conduct their research in CO_2 enriched environments. "Most experiments with algae are conducted at CO_2 concentrations of up to 20,000 ppm," says N.E. Tolbert, professor emeritus of biochemistry at Michigan State University. "Cultures that can be grown in three days at high levels of CO_2 would require 10 to 14 days at the normal atmospheric concentration.

Phenomenal Response in Trees

Some of the most convincing evidence that the rising level of atmospheric carbon dioxide is good for plants comes from the response measurement of individual trees and overall forest growth. Forests cover approximately one-third of the earth's land area, and account for -two-third of global photosynthesis. They have C_3 metabolism, and, like other C_3 plants, respond favourably to higher concentrations of carbon dioxide.

Trees and their seedings grown under controlled environments or in open top chambers simulating the outdoors have shown remarkable growth responses to elevated levels of CO_2. Practically every species evaluated thus far in the seedling stage has shown a positive response. Addition of carbon dioxide to black walnut seedlings—at concentration of 1,000 to 2,000 ppm for three months—increases dry weight by 80 per cent, height by 96 per cent, and leaf area by 79 per cent. Similar results have been obtained for sugar maple, oak, ash, sweet gum, pine, and eucalyptus. The forestry department at Michigan State University has produced plantable trees in months, rather than years, by subjecting seedlings to 1,000 ppm CO_2 concentrations under optimal conditions of light, temperature, day length, and nutrients. The Water Conservation Laboratory of the US Department of Agriculture has compared the growth of orange trees under the current atmospheric CO_2 concentration of 360 ppm, and a concentration of 650 ppm. The trees at the elevated levels have accumulated 2.8 times more biomass in five years, and in their first two years of production produced 10 times more oranges.

The models, which have been built at Princeton University's—Geo-physical Fluid Dynamics Laboratory, the Godard Institute for Space Studies, here at NCAR and elsewhere, consist of expressions for the interacting components of the ocean-atmosphere system and equations representing the basic physical laws governing their behaviour, such as the ideal gas laws and the conservation of mass, momentum and energy. Given values for, say, the input of energy from the sun and the composition of the atmosphere, a model calculates "climate"—temperature land, in sophisticated models, pressure, wind speed, humidity, soil moisture and other variables.

To keep the task computationally manageable, the calculations are down at discrete points in a simplified version of the real world. In the most complicated models—global-circulation models (GCM's), which were first developed for long-term weather forecasts—the atmosphere is represented as a three-dimensional grid with an average horizontal spacing of several hundred kilometers and an average vertical spacing of several kilometres; climate is calculated only at the intersections of the grid lines. In spite of the simplification, running such a GCM for only one simulated year can take many hours on the fastest available supercomputers. To study the effect of a trace-gas buildup, a modeler simply specifies the projected amount of greenhouse gases and compares the model results with a control simulation of the existing climate, based on the present atmospheric composition. The results of the most recent GCM's are in rough agreement: a doubling of carbon dioxide, or an equivalent increase in other trace gases, would warm the earth's average surface temperature by between 3.0 and 5.5 degrees C. Such a change would be unprecedented in human history; it would match the five-degree warming since the peak of the last ice age 18,000 years ago but would take effect between 10 and 100 times faster.

The shortcomings of computer models limit the reliability of such forecasts. Many processes that affect global climate are simply too small to be seen at the coarse resolution of a model. Such climatically important processes as atmospheric turbulence, precipitation and cloud formation take place on a scale not of hundreds of kilometres (the scale of the grid in a GCM) but of a few kilometres or less. Since such processes cannot be simulated directly, modelers must find a way of relating them to variables that can be simulated on the model's coarse scale. They do so by developing a parameter—a proportionality coefficient—that relates, say, the average cloudiness within a grid cell to the average humidity and temperature (something the model can calculate).

This strategy, known as parameterisation, has the effect of aggregating small-scale phenomena that could act as feedbacks on climatic change, either amplifying or moderating it. Clouds, for example, reflect sunlight back to outer space (tending to cool the climate) and also absorb infrared radiation from the earth (tending to warm it). Which effect dominates depends on the clouds' brightness, height, distribution and extent. Recent satellite measurements have confirmed two-decade-old calculations showing that clouds currently have a net cooling effect; the earth as a whole would be much warmer under cloudless skies. But climatic change might cause incremental changes in cloud characteristics, altering the nature and amount of the feedback. Present models, crudely reproducing only average cloudiness, can say little that is reliable about cloud feedback—or about the many other feedbacks that depend on parameterised processes.

Another shortcoming of present models is their crude treatment of the oceans. The oceans exert potent effects on the present climate and will surely influence climates to come. Their enormous thermal mass will act as a "thermal sponge," slowing any initial increase in global temperature while the oceans themselves warm up. The magnitude of the effect will depend on ocean circulation, which in turn may change as the earth warms. In principle, a climate model should couple a simulated atmosphere with oceans whose dynamics are simulated in equal detail. The computational challenge is staggering, however, and in most GCM's applied to greenhouse warming the dynamics of the oceans are simplified, treated at coarse resolution or left out.

In addition to limiting the reliability of global forecasts, the simplified treatment of the oceans also prevents the models from giving a definitive picture of how climate will change over time in specific regions. Ideally one would like to know not only how much the world as a whole will warm but also whether it will, say, get drier in Iowa, wetter in India or more humid in New York City. Yet, as long as the oceans are out of equilibrium with the atmosphere, their thermal effects will be felt differently at different places. An area in which there is little mixing between surface waters and cold, deep waters might warm quickly; high-latitude regions where deep water is mixed up to the surface might warm more slowly. These thermal effects could in turn affect wind patterns, thereby altering other regional variables, including humidity and rainfall. (Regional forecasts are also compromised in many models by simplified representations of vegetation, which ignore climatically important processes such as the release of water vapour by plants and their effect on surface albedo, or reflectiveness).

Nevertheless, climatologists have grounds for considerable confidence in their models' forecasts of global surface-temperature change. Individual model elements can be verified by comparing them with the results of a more detailed submodel—a smaller, finer-scale simulation—or with real data. Cloud parameterisations, for example, can be tested against actual measurements of the relation of temperature and humidity to cloudiness within an area corresponding to a cell in the model.

The skill of a model as a whole, and in particular its ability to account for relatively fast processes, such as changes in atmospheric circulation or average cloudiness, can be verified by checking its ability to reproduce the seasonal cycle—a twice-yearly change in hemispheric climate that is larger than any projected greenhouse warming. In spite of parameterisation, most GCM's map the seasonal cycle of surface temperature quite well, but their ability to simulate seasonal changes in other climatic variables, including precipitation and relative humidity, has not been studied as thoroughly.

Perhaps the most interesting aspect of clouds is that they appear to function as the key element of a powerful global thermostat that is biologically modulated. This intriguing aspect of climate regulation only has been eluvidated within the past few years and is well outside the domain of current GCM simulations, which presently deal only with physical and chemical processes.

The biological modulation of climate begins with an initial impetus for warming, such as would be provided by an increase in the CO_2 content of the atmosphere. As warming is experienced by the microscopic algae or phytoplankton that live in the surface waters of the world's oceans, they begin to grow faster. As they phtosynthesise at increasingly greater rates, they produce, as a by-product of their metabolism, increasingly greater quantities of a substance called dimethylsulfonio propionate, which ultimately is released to the sea, when they either die or are eaten by zooplankton. There, a portion of it decomposes to form a compound known as dimethylsulfide that diffuses into the atmosphere, where it is oxidised and converted into sulphuric and methanesulphonic acid particles, which then function as cloud condensation nuclei.

Some of these nuclei produce new clouds, while others increase the droplet number concentrations of pre-existent clouds, making them more highly reflective of incoming solar radiation. Both of these effects tend to cool the planet dramatically, acting as a brake on the initial impetus for warming. It has been estimated that this negative feedback is of about the same strength as the primary CO_2 greenhouse effect.

The primary climate expectancy of the CO_2 greenhouse effect is, thus, a small upward trend in the mean surface air temperature of the globe. However, the concomitant increase expected in low-level cloud cover greatly should modify the primary impetus for warming. Specifically, more low-level could cover should reduce, negate, or possibly even overwhelm any tendency for daytime warming, but will augment the greenhouse effect at night. Hence, it is likely that greenhouse warming will be marked by a decrease in the amplitude of the mean daily temperature wave. Such a trend has been observed in the historical climate records of many parts of the Northern Hemisphere. Climatologist Tom Karl has noted that this decreased temperature range is "one of the few pieces of evidence. . . that is consistent with the potential impacts of increased greenhouse gases."

The only reason rising temperatures are believed to be detrimental to the biosphere is that they are assumed to increase the severity and frequency of maximum temperature events which define the upper thermal limits of various plants' and animals' life zones. Hence, if almost all of the CO_2-induced temperature rise is confined to minimum temperatures, there will be little change in stress conditions at the high end of the thermal adaptability spectrum, but a great reduction in stress conditions at the low end. In addition, several studies have demonstrated that the direct biological benefits of atmospheric CO_2 enrichment grow ever larger as daily mean temperatures rise. Both of these consequences are desirable.

In light of these several observations, I suggest that, rather than posing a clear and present danger to the inhabitants of the globe, the CO_2 greenhouse effect significantly will enhance the direct biological benefits of atmospheric CO_2 enrichment and thereby help to thrust the planet into a new era of enhanced biospheric productivity, creating a clear and present *benefit* for all the many species with which we share the Earth.

What does the future hold?

As the CO_2 content of Earth's atmosphere continues its return to the high levels characteristic of past ages of biological prominence, the biosphere gradually should awaken from the great lethargy it has experienced over the past two millennia of near-CO_2-starvation. In so doing, it should reverse the long-term trend of species extinction brought about by the gradual lowering of the air's CO_2 content over the geological history of the planet. It also should provide the help we so desperately need to feed and clothe the burgeoning world population adequately, alleviating the severe pressures humanity currently is exerting on natural ecosystems the world over. Furthermore, what little greenhosue warming may occur could be just what is necessary to forestall the next "scheduled" glaciation of the global, which, geologically speaking, lies just around the corner.

We thus find ourselves at a significant cross roads in Earth's history, with an important choice to make. On the one hand, we seem poised on the threshold of a veritable rebirth of the biosphere. On the other, in many parts of the world, we appear to be balanced on the brink of ecological collapse. We owe it to ourselves and to generations yet unborn, as well as to all the other species with which we share the planet, carefully to investigate the many and varied aspects of atmospheric CO_2 enrichment before we cast our vote in the developing debate on global warming. It well could be that the rising CO_2 content of Earth's atmosphere is actually a blessing in disguise and one of the better things that could happend to mankind and nature.

Special Characteristics

The first is that global warming epitomises the idea that everything relates to everything else. The issue arises from the growth of human population and wealth. It is rooted in industrialisation, technological change, dependence on energy, striving for economic growth, and the many other aspects of a human population that aspires to improve its quality of life. The buildup of greenhouse gases is a result of normal, not aberrant, human behaviour and is a product of innumerable independent decisions by individuals, industries, and governments in daily life and over the globe.

Societies are increasingly familiar with the need to regulate and influence decision making dispersed among individuals and institutions; the control of greenhouse gases—or of their effects—is not different in kind from other policy issues in the management of risks and control of extrenalities of technology. In the climate debate, however, the nature of the interactions, the number of relevant actors, and the scale involved all mean that core economic and political interests within and among countries will be directly involved in measures to limit emissions, as well as in the effects of warming that do take place. Agriculture, energy, industry, water, ecology, population, Third World development, and specific political commitments—such as tax policy—are only the moist obvious interests at stake.

The breadth of such interests greatly complicates bureaucratic politics within governments and among international organisations. Virtually all government ministries and international organisations have a stake in global warming, either in its mitigation or in its possible effects. Overlapping jurisdiction, differing agendas and priorities, varying knowledge and influence, and competition for budgets and power all conspire to complicate tradeoffs and make concerted policy difficult to achieve. Moreover, the interaction of so many interests and issues means that policies will probably have significant unexpected consequences, and that developments in subjects far afield, such as political change in Eastern Europe or China, will affect the climate issue.

A second major characteristic of the issue is the interaction of two vast and complex systems, the planet's ecosystem and the human socioeconomic system. Large systems usually have large inertia, making change slow and difficult. Natural changes in the ecosystem occur over periods normally measured in centuries, which has made a reasonably consistent and predictable climate a staple of the development of civilisation. The evolution of the socioeconomic system, once also measured in centuries, is now clearly moving more rapidly, both in growth and in internal structural modification. Nevertheless, such major components of the overall system as energy technologies and the location of urban population can still be changed only slowly.

The fact that climate change involves the interplay of large human and physical systems has several consequences. The most obvious are that the time horizon of policy intervention must be very long and that changes in either of the fundamental systems cannot be achieved quickly, even if there were an agreed will to do so. The momentum in the systems as a whole already guarantees some atmospheric change, however great that change turns out to be and whatever policies are put in place. But this momentum also means that delaying action will extend the effects further into the future and with greater magnitude. Effective irreversibility becomes an inherent and unavoidable aspect of the problem.

This latter characteristic makes it particularly difficult to evaluate policies intended to reduce emissions. The present value of benefits that will be realised decades from now is small. Even if direct comparison of the future with the present were possible, the costs and benefits of ill-defined future consequences would be much more uncertain than the near-term costs and benefits of specific actions today. Thus, specific policies are likely to be assessed principally for their short-term impact and for their costs that would deny funding for other needs. The more theoretical and controversial long-term consequences are likely to receive short shrift. The less certain the future costs, the more likely the future will carry less weight in comparison to the present.

Ironically, the complexity of these large systems also suggests they could be vulnerable to dislocation in unexpected ways. Complex systems that are incompletely understood may have unsuspected nonlinear responses when subjected to unusual strees. Major war is an example of stress that can produce large discontinuities in the socioeconomic system. Although the ecosystem is resilient and self-correcting under a variety of disruptive events, it is now being taxed by human activities at a greater rate than ever before. The ecosystem's responses to that stress are uncertain, making confident prediction of stability hazardous. Surprises such as the Antarctic ozone hole cannot be ruled out on the basis of current knowledge.

A third characteristic of the issue is the importance of uncertainty. In the case of climate change, the uncertainties are particularly large, contentious, and slow to be reduced. The policy process needs information first about the phenomenon itself and then about its causes. Yet that information is of little policy interest without knowing the effects of global warming on precipitaion, sea level, crops, ice cover, frequency and violence of storms, and other consequences. At the next level, the effects of the phenomenon on agriculture, coastal zones, economic development, energy demand, international status, population migration, prevalence of disease and pests, and urban areas become important. At this level policymakers need analyses of alternatives for moderating the rate and extent of warming and the possibilities for adapting to warming. At each successive level, the uncertainties are likely to be greater, as the variables and possible future effects become less well-defined.

Removing or reducing all these uncertainties is not essential; policy is often made without definitive "downstream" analysis. However, if policies are especially costly, affected interests are apt to use the absence of a definitive analysis of effects as an arguments to defer or prevent action.

A fourth major characteristic of the climate issue is the global nature of its causes, effects, and the actions and policies required to modify it. No country can solve the problem on its own, and the actions of one can be negated if others fail to act. That does not mean unilateral action is irrelevant, only that it is not a solution.

Universal action clearly seems required. An appreciation of that fact may be valuable in promoting a global approach, but it can also deter national action if it proves difficult to achieve common policies, or if a nation tries to act as a free rider, benefiting from the actions of others without contributing. The complicated bargains over policy that must be struck within nations will be rendered more difficult by international negotiation over issues that are not likely to parallel those domestic bargains. Perhaps more significant in the long run is the absence of any institution in the international system with the authority to dictate or enforce trade-offs. For the time being, national attitudes are not likely to accept such a dramatic ceding of power to international institutions.

Further complicating the global nature of climate change are the profound differences between developing and industrialised countries in both causes and consequences. Today, the developing world (including China with its large coal reserves) accounts for about one-third of the emissions of greenhouse gases, but its contribution will grow rapidly as population and development increase. Developing countries are not likely to compromise their objective of economic growth in response to a problem created by rich countries. As developing nations are fond of pointing out, with justification, it is the industrialised countries that have benefited from using the global atmosphere as a sink for their carbon by products and that now propose limits on the use of that shared resource. Curbing greenhouse gases is not the highest priority of developing countries; and in any case, their response would require in immediate infusion of money, skilled labour, and technology.

To include the developing countries in an attempt to cap or reduce emissions would require global bargains between the North and the South including a large expansion of development assistance. A new global fund of "only" a few billion dollars would be on the wrong scale, no matter how useful such a fund might be in its own right. In addition, some countries may see themselves as winners if warming occurs, reasoning that they will benefit from longer, more productive growing conditions or from a more hospitable climate. Or they might calculate that the costs of warming would not be significant to them and would be lower than the costs they are being asked to bear to prevent it. Such nations may be lukewarm participants at best in negotiations toward common international action. Ironically, reducing uncertainty might show that some states would benefit, thereby dimming their interest in an international agreement.

The global character of the issue also has important implications for international cooperation, first for reaching agreements and later for the oversight and management of commonly agreed rules, resource transfer, research, monitoring, and dispute settlement. International organisations perform all these functions today within their areas of responsibility, and do so more effectiely than is often believed. Nevertheless, national governments are not about to delegate significance authority over matters central to their political and economic interests to an external body. Negotiation and implementation of agreements will be carried out within the basic nation-state system. The contending national policy processes striving for agreement will ultimately determine what is done.

A fifth dominant characteristic of the global warming issue is its dependence on the theories, research, and calculations of scientists even to know there is a problem. Only uncertain and conflicting data indicate a warming trend consistent with the predictions. There is no tangible evidence yet—nothing that can be seen or felt. Although the word of scientists alone is enough for some things to be judged dangerous—the thinning of the Earth's ozone shield and the unseen relationship between a chemical and cancer, among others—observable evidence of some kind is usually necessary to catalyze corrective action. For climate change, there is no demonstrable effect. Indeed, comprehensive data showing a sustained buildup of CO_2 in the global atmosphere began to be collected only in the 1950s.

The evidence of such a buildup is growing and becoming more precise, and imaginative paleoclimatic research is uncovering a relationship (not necessarily causal) between the atmosphere's greenhouse gas content and swings in the planet's average temperature. However, estimates of future temperatures and other effects rely on complex computer models of the Earth's atmosphere, designed and steadily improved by a relatively small group of scientists. These scientists would be the first to concede the imperfecions of their models and the incompleteness of their data, especially with regard to the oceans and their interaction with the atmosphere. Potentially significant feed-back effects (such as clouds or changes in absorption of CO_2 by the ocean) are not adequately accounted for, and computers are still incapable of performing computations on the needed scale. Many scientists would also admit to a considerable degree of incestuousness within the community: Common paradigms are reinforced through intensive interactions among a small group of researchers.

Nevertheless, a consensus among many researchers (though by no means all) holds that the consistency in the predictions of various model, along with corollary evidence, provides a reasonable basis for anticipating the indicated temperature rise. No scientist asserts that the effects *will* definitely occur in a particular time frame, but many leave no doubt that they accept the general consensus.

In essence, elements of the scientific community have reached a judgment based on limited evidence and imperfect models that has massive implications for the health of the eco-system and for the fate of people and of nations. The image of an inverted pyramid comes to mind, of a steadily broadening body of implications that rests ultimately on the point of a relatively small band of dedicated scientists who recognise the uncertainties of their work.

The dependence of the public and policy-makers on scientists for basic information on global warming has few parallels—weapons issues may be the closest example. The uncertainties, however, make it inevitable that respected scientists will be found on all sides of the issue; judgments will differ about the importance of the shortcomings in the data, the models, the interpretations, and the expectations of what future research will show. When it comes to difficult, costly, detailed economic and political choices, scientific experts will be testifying in opposition to one another.

The press plays in important role here, too. Much of the public concern about global warming is undoubtedly a result of the media focus on the more dramatic predictions, since apocalypse makes for good copy. However, on detailed questions media coverage that strives to present all sides tends to portray differing scientific viewpoints as equal in merit, whatever the actual balance is among the scientists.

In fact, scientists face serious dilemmas in presenting theri evidence, in characterising uncertainty, in steering a course between objectivity and advocacy, and even in handling potential conflicts of interest as they seek research funding. Brutal candor about the imperfections of data and models may deter appropriate political action, just as subtle advocacy may lead to unjustified action. The importance of effective science policy mechanisms in governments thus becomes clear: They must be able to sort out conflicting views and relate knowledge and uncertainty to the demand of the policy process. Such mechanisms cannot resolve uncertainty or differing scientific judgments, but they can clarify policy choices and their implications.

Finally, the very planetary characteristic of the issue may come into play. Climate change is seen as a threat to the entire planet in a way only previously encountered in the destruction of the ozone layer. There is, or may be, a visceral public response to that kind of threat that goes beyond other environmental concerns. Is this response significant in the ecological consciousness being expressed on this issue? It is an open but possibly critical question.

These considerations lead to a conclusion that may be deeply troubling to some but is nevertheless compelling: Unless overwhelmed by a strong and enduring public consensus or by political leadership not yet in evidence, the political processes within a among nations are not likely to bring forth substantial policy action until the uncertainties surrounding climate changes are greatly reduced, and probably not until evidence of warming is palpable. This conclusion holds, even though the Earth's climate may sustain irreversible changes if some of the forecasts are correct.

Among industrial countries, the United States may prove to be the most resistant to formulating comprehensive policy and to taking early action, especially in the absence of clear leadership. The divided American governmental structure—with conflicting bureaucratic and legislative goals and overlapping jurisdictions and agendas—makes agreement difficult even on less demanding issues. Virtually every federal agency has a legitimate interest in the climate issue; correspondingly, most congressional comittees will be involved in the debate, each with different turf to defend or expand and each with a limited vision of the whole. Conflict between the overlapping jurisdiction of House and Senate authorising committees and between the Congress and a fragmented executive branch further complicate the matter. All of this will militate against policy that would levy serious costs on powerful segments of society or that conflict with deeply held political commitments to such subjects as taxes or the appropriate scope of government. The central role assigned to the president's science adviser, Allan Bromley, to lead policy formulation in the executive branch on the climate issue is an encouraging step, though the obstacles to coherent policy are clearly formidable.

Significantly, the inherent uncertainties of the issue extend to the costs of adaptation to the warming. The costs of adaptation for some countries, especially the richer countries, may not be very large even when they are compared directly—without any discounting for the future—to the costs of prevention. Thus it becomes quite unlikely that actions with substantial short-term costs will be taken, at least until much more is known.

The growing strength of public support for environmental protection and the accompanying growth in mobilisation and influence of national and international environmental movements will become more important. But it remains doubtful that these movements can move public and national policy processes sufficiently in light of the uncertainties of the phenomenon itself and the more certain costs of action.

The successful conclusion of the Montreal Protocol in 1987 to cut production and use of CFCs might indicate otherwise. The agreement was an important precedent for international action before evidence of damage was in hand. Yet it also underscored the difficulties: The ability to reach agreement was aided by dramatic evidence of the unpredicted ozone hole over the Antarctic, while the immediate economic costs of regulation were relatively minor. Even so, agreement was difficult and resulted only in a commitment to cut CFCs by 50 per cent, though many industrialised countries subsequently called for a total phase out of CFCs by the end of the century.

The experience of the CFCs negotiation is relevant to the global warming issue, but it hardly compares in complexity and economic implications.

The presence of hard evidence is particularly important to the staying power of public attitudes. The North American heat and drought of 1988, heralded by a prominent scientist in congressional testimony as the arrival of the greenhouse effect, helped catapult the issue high on the environmental agenda, at least in the United States. Although subsequent analysis did not show a direct relationship between that summer's weather and global warming, the heat and drought nonetheless served to focus attention and opinion. Undoubtedly, several more summers of that kind—whether or not a causal tie to temperature rise exists—will continue to fuel the public pressure necessary to override obstacles in the policy process. On the other hand, a series of cooler-than-normal years might lessen both public interest and the ability to mobilise political action on the issue.

This confusion of appearance and reality can be avoided only through the ability to convincingly relate gradual climate changes or other environmental effects to changing average temperatures. One principle seems clear: The risk of capricious public opinion cannot be managed by dramatic pleas that are later shown to be wrong. The credibility of science and government is a powerful card, but one easily damaged.

In 1957 Roger Revelle and Hans E. Suess of the Scripps Institution of Oceanography observed that humanity is performing a "great geophysical experiment," not in a laboratory, not in a computer, but on our own planet. The outcome of the experiment should be clear within decades, but it essentially began at the start of the Industrial Revolution. Since then human beings have increased the atmospheric content of carbon dioxide by about 25 per cent by burning coal, soul and other fossil fuels and by clearing forests, which releases carbon dioxide as the liter is burned or decays.

Carbon dioxide makes up only a thirtieth of 1 per cent of the atmosphere, but together with water vapour and other gases present in much smaller quantities, such as methane and the chlorofluorocarbons, it plays a major role in determining the earth's climate. As early as the 19th century it was recognised that carbon dioxide in the atmosphere gives rise to a greenhouse effect. The glass of a greenhouse allows sunlight to stream in freely but blocks hear from escaping, mainly by preventing the warm air inside the greenhouse from mixing with outside air. Similarly, carbon dioxide and other greenhouse gases are relatively transparent to sunshine but trap heat by more efficiently absorbing the longer-wavelength infrared radiation released by the earth.

By now the atmosphere's heat-trapping ability has been well established. For example, as seen from space, the earth radiates energy at wavelengths and intensities characteristic of a body at ⁻18 degrees Calsius. Yet the average temperature at the surface is some 33 degrees higher: heat is trapped between the surface and the level, high in the atmosphere, from which radiation escapes. There is virtually no doubt among atmospheric scientists that increasing the concentration of carbon dioxide and other gases will increase the heat trapping and warm the climate.

What, then, is the question that the ongoing geophysical experiment will settle? Even though there is virtually no debate among scientists about the greehouse effect as a scientific proposition, there is controversy? Will the increase take 50, 100 or 150 years Will it be drier in Iowa or wetter in India? There is still more controversy when it comes to policy: Should steps be taken to reduce the greenhouse warming or to anticipate its effects? What steps, and when? In the face of so much controversy, an understanding of what is well known, known slighly and not known at all about the greenhouse warming is essential.

Circumstantial evidence from the geologic and historical past bears out a link between climatic change and fluctuations in greenhouse gases. Between 3.5 and four billion years ago the sun is thought to have been about 30 per cent fainter than it is today. Yet life evolved and sedimentary rock formed under the faint young sun: at least some of the earth's surface was above the freezing point of water. Some workers have proposed that the early atmosphere contained as much as 1,000 times today's level of carbon dioxide, which compensated for the sun's feeble radiation by its heat-trapping effect. Later an enhanced greenhouse effect may have been partly responsible for the warmth of the Mesozoic era—the age of the dinosaurs—which fossil evidence suggests was perhaps 10 or 15 degrees C warmer than today. At the time, 100 million years ago and more, the continents occupied different positions than they do now, altering the circulation of the oceans and perhaps increasing the transport of heat from the Tropics of high latitudes. Yet calculations by Eric J. Barron, now at Pennsylvania state University, and others suggest that paleocontinental geography can explain no more than half to the mesozoic warming.

Increased carbon dioxide can readily explain the extra heating, as Aleksandr B. Ronov and Mikhail I. Budyko of the Leningrad State Hydrological Institute first proposed and a as Barron, Starley L. Thomspon of the National Centre for Atmospheric Research (NCAR) and I have calculated. A geo-chemical model constructed by Robert A. Berner and Antonio C. Lasaga of Yale University and the late Robert M. Garrels of the University of South Florida suggests that the carbon dioxide may have been released by unusually heavy volcanic activity on the midocean ridges, where new ocean floor is created by upwelling magma.

Direct evidence linking greenhouse gases with the dramatic climatic changes of the ice ages comes from bubbles of air trapped in the Antarctic ice sheet by the ancient snowfalls that built up to form the ice. A team headed by claude Lirous of the Laboratory of Glaciology and Geo-physics of the Environment, near Grenoble, examined more than 2,000 meters of ice cores—a 1,60,000-year record—recovered by a Russian drilling project at the Vostok Station in Antarctica. Laboratory analysis of the gases trapped in the core showed that carbon dioxide and methane levels in the ancient atmosphere varied in step with each other and, more important, with the average local temperature (determined from the ratio between hydrogen isotopes in the water molecules of the ice).

During the current interglacial period (the past 10,000 years) and the pervious one, a 10,000-year period around 1,30,000 years ago, the ice recorded a local temperature about 10 degrees C warmer than at the height of the ice ages (At the same time, the atmosphere contained about 25 per cent more carbon dioxide and 100 per cent more methane than during the glacial periods. It is not clear whether the greenhouse-gas variations caused the climatic changes or *vice versa*. My guess is that the ice ages were paced by other factors, such as changes in the earth's orbital parameters and the dynamics of ice buildup and retreat, but biological changes and shifts in ocean circulation in turn affected the atmosphere's trace-gas content, amplifying the climatic swings.

A still more detailed record of greenhosue gases and climate comes from the past 100 years, which have seen a further 25 per cent increase in carbon dioxide above the interglacial level and another doubling of atmospheric methane. Two groups, one led by James E. Hansen at the National Aeronautics and Space Administration's Goddard Institute for Space Studies and the other by T.M.L. Wigley at the Climatic research Unit of the University of East Anglia, have constructed records of global average surface temperature for the past century.

The workers drew on data from many of the same recording stations around the globe (the Climatic Research Unit also included readings made at sea), but they had different techniques for analysing the records and compensating for their shortcomings. Certain recording stations were moved over the course of the century, for example, and readings from city centres may have been skewed by heat released by machinery or stored by buildings and pavement.

This "urban heat island" effect is likely to have been disproportionately large in developed countries much as the U.S., but even when the same correction calculated for the U.S. data (by Thomas R. Karl of the National Climatic Data Center in Asheville, N.C., and P. D. Jones of East Aglia) is applied to the global data set, about half a degree C of unexplained "real" warming over the past 100 years remains in both records. In keeping with the trend, the 1980's appear to be the warmest decade on record and 1988, 1987 and 1981 the warmest years, in that order.

Is this the signal of the greenhouse warming? It is tempting to accept it as such, but the evidence is not definitive. For one thing, instead of the steady warming one might expect from a steady buildup of greenhouse gases, the record shows rapid warming until the end of World War II, a slight cooling through the mid-1970's and a second period of rapid warming since then.

What trajectory will the temperature curve follow now? Three basic questions must be answered in forecasts of the climatic future: How much carbon dioxide and other greenhouse gases will be emitted? By how much will atmospheric levels of the gases increase in response to the emissions? What climatic effects will be resulting buildups have, after natural and human factors that might mitigate or amplify those effects are taken into account?

2. Global Warming and Greenhouse Policy Issues

"IF THE Lord Alimighty had consulted me before embarking on the Creation, I would have recommended something simpler," wrote Alfonso of Castille, a medieval patron of astronomy. For better or worse, the Lord did not, and several centuries later we are becoming even more aware of the earth's complexity. Scientists still understand little of the workings of the atmosphere and the oceans, of their interactions that drive the planet's climate machine, and of the biogeochemical cycles of carbon, nitrogen, sulphur, and the few other elements essential to life. And while nature is complex and dynamic enough in itself, natural forces are now being changed by man's own activities, whose effects reach from the oceans to the top of the stratosphere. In generating a global "greenhouse" warming, man has embarked on a vast, unplanned planetary experiment that poses unprecedented challenges to his wisdom, forsight, and scientific capacity.

The theory of a greenhouse effect was first proposed almost a century ago. The earth's atmosphere, it was pointed out, is largely transparent to incoming solar radiatioin but absorbs the lower energy (longer-wavelength) radiation that is reradiated by the earth. The extra energy that remains in the atmosphere warms it in a manner akin to the warming of a greenhouse. The major atmospheric gas responsible for this effect is carbon dioxide, though several other gases are also involved.

The greenhouse theory is supported by many lines of evidence. Climatologists know that without the natural greenhouse effect, the planet would be permanently covered with ice and, presumably, lifeless. Astronomers have discovered that Venus, which has greenhouse gases in its atmosphere, is much warmer than one would expect on the basis of its distance from the sun. The atmospheres and climates of the other close-in-planets are also consistent with the theory. More recently, computer climate models of increasing sophistication have yielded consistent estimates of the expected warming of the earth's climate over the coming decades from rising concentrations of greenhouse gases. Finally, within the past few months, the strongest empirical evidence obtained to date that a greenhouse warming is actually under way has become available.

This evidence comes from three sources. First, researchers at the University of East Anglia in Great Britain combined temperature readings obtained over the land and in the oceans during the past 130 years to produce the first comprehensive estimates of global mean temperature. The readings show a marked warming trend beginning about 1890. The three warmest years of the 134-year record are 1980, 1981 and 1983, and five of the nine warmest years have occurred since 1978. The overall warming of about 0.5°C is consistent in magnitude with what is predicted by the most sophisticated greenhouse projections. Second, temperature readings taken in the Arctic permafrost show that the temperature there has risen about 2 to 4^0C over the past several decades to a century. This larger temperature rise in the Arctic is also consistent with theoretical predictions, which show a warming at high latitudes several times greater than the global average. A third line of evidence is the sea-level of 2.0 to 2.5 millimetres per year measured over the last 50 years, which is in part due to thermal expansion of the water.

There are five principal greenhouse gases. Carbon dioxide, the best known, accounts for about half of the current effect. Its atmospheric concentration has risen by 25 per cent over the past 100 years. Recently, the rate of growth has accelerated; the concentration of carbon dioxide has increased 10 per cent in the past 30 years. The principal cause of this increase is the combustion of fossil fuels. An additional, though uncertain, component of the rising concentration is tropical deforestation. The other four greenhouse gases—methane, nitrous oxide, tropospheric (lower atmosphere) ozone, and a collection of closely related compounds known as chlorofluorocarbons (CFCs)—together account for the other half of the current effect. Although present in very small concentrations, each is a far more potent greenhouse gas than is carbon dioxide, so that rising concentrations have a correspondingly greater effect.

Analysis of air bubbles trapped in global ice has made possible the measurement of long-term trends in atmospheric methane, which apparently began increasing about 300 years ago. Very recently—during the past 20 years or so—the rate has increased sharply to about 1 per cent per year. The reasons for this increase, however, are unknown. Among the many suspected causes are larger cattle populations and growing areas of rice paddies, both of which release methane. Nitrous oxide is also increasing steadily, though much more slowly. The causes are also unclear because its natural cycle is little understood. However, increasing fertiliser use and combustion of fossil and biomass fuels are suspected. The tropospheric ozone concentration may be on the rise by as much as 1 per cent per year. Ozone is produce in the presence of sunlight from hydrocarbons and nitrogen oxides, the concentration of which are also increasing due to greater fossil fuel combustion and industrial activity. Only in the case of the CFCs is the reason for the increasing concentration absolutely clear, since all CFCs are manmade. The atmospheric concentration of the principal CFCs has recently been growing very rapidly, at an annual rate of about 7 per cent.

What is important to note from this brief reviews is that, with the exception of CFCs, the sources of all the greenhouse gases are either uncertain, diffuse (and therefore difficult to control), or principally the result of energy use. Therefore, if or when control of emissions of greenhouse gases is required, the options are principally two: limit or eliminate CFC use and control energy use, particularly fossil-fuel burning. Slowing or reversing deforestation may also help.

How likely is it that such actions will be needed? Despite many uncertainties about the detailed impacts of a greenhouse warming, a broad consensus exists on many of its features. A greenhouse effect equivalent to that which would be caused by a doubling of the preindustrial level of carbon dioxide in the atmosphere is expected to cause a rise in average global surface temperature of 1.5 to 4.5^0C (or about 3 to 8^0F), with the more recent projections near the high end of this range. A 3^0C rise would put the global climate beyond the range of human experience, in the same neighbourhood as the climate that prevailed 13 million years ago. An atmospheric commitment to such an increase is now expected as early as the 2030s.

The expected warming is projected by the date of its "atmospheric commitment" rather than its actual appearance because of the substantial lag between the time higher greenhouse gas concentrations occur and the time the global average temperature reaches its new (warmer) equilibrium. The length of the time lag, which is influenced by the moderating effect of the oceans, is not known but is believed to be a few decades. Although the location is awkward, the distinction in the way warming is measured is important. This measure of how much warming the atmosphere is committed to is the meaningful one, of course, because once the greenhouse gases are emitted, the eventual effects are inevitable.

In considering appropriate policy responses to global warming, it is necessary to keep in mind two additional aspects of the problem: the unpredictable discontinuity of climatic changes and the way in which they magnify the likelihood of so-called 'natural' disasters. For example, commenting on the greenhouse effect, Broecker makes the case that changes in the Earth's climate is more likely to be sudden rather than gradual:

Earth's climate does not respond to forcing in a smooth and gradual way. Rather, it responds insharp jumps which involve large-scale reorganisation of Earth's systems. If this reading of the natural record is correct, then we must consider the possibility that the main responses of the system to our provocation of the atmosphere will come in jumps whose timing and magnitude are unpredictable. Coping with this type of change is clearly a far more serious matter than coping with a gradual warming.

Similarly, 'in terms of environmental and socio-economic impacts, the slow changes in mean climate or sea-level may often be manifested as large changes in the risks of extreme events'. For example, 'threats from future sea-level rise and salt water intrusion are linked to extreme drought occurrence, because existing water resource supply systems are, based on past experience, already adjusted to only comparatively moderate fluctuations in precipitation and resultant salt intrusion'. Both those aspect of climatic change need to be carefully considered in the design of policy measures to counter the greenhouse effect.

The likely range of policy responses to global warming and their appropriate stage of intervention in the greenhouse effect 'causal chain'. Note that the essential trade-off is between doing nothing now—which obviously involves little current cost, but obviously may incur greater future costs in terms of reducing vulnerability and modifying effects—and investing heavily in the control and reduction of fossil-fuel emissions, or in more sustainable management of tropical forests, in order to avoid the anticipated future costs.

We are only at the beginning stages of understanding the nature of the policy options available to us in a global warming scenario; a full assessment of the costs and benefits of each option is not yet possible. Nevertheless, several points crucial to this analysis are worth highlighting.

- There is clearly a waiting cost for doing nothing; the best guess is that we are already committed to some global warming in the near future (e.g. 1.5^0-4.5^0C). Thus the longer we do nothing, the higher will be the future costs of forced adaption and residual effects. The crucial questions are how large is this waiting cost and how fast is it growing over time?
- There is also a high cost attached to an extreme (i.e. surprise) event suddenly occurring. As noted above, both the unpredictability surrounding the impact of the greenhouse effect and its tendency to increase the risk of extreme events suggests caution in understanding the probability of a 'surprise' event occurring or in being too inflexible in the assumptions made concerning the timing of possible events.
- As a result, conventional discounting

assumptions may need to be modified in a analysing the policy responses to the greenhouse effect. For example, using a positive discount rate in project or policy analysis virtually eliminates the net present value of effects occurring 30 or more years in the future. Policy analysis must not be biased by such a standard result because it leads automatically to the rejection of options which involve future costs and benefits. Rather, in addition to using sensitivity analysis with different discount rate assumptions, there should be modification of the benefits and costs under various contingencies concerning risk. There should be as well a full assessment of the costs of maintaining the 'sustainability' of current economic activities and essential natural resources.

- Analysts of policy options will need to be aware of both monetised and unmonetised costs, and not to be automatically biased to consider only the former. For example forced adaption would involve both monetised costs (e.g. costs of rebuilding a flooded village) and costs that are more difficult to monetise (e.g. environmental damage, human mortality and morbidity). Residual costs will also be extremely difficult to monetise. A fair assessment of the trade-offs among policy options will therefore require a thorough analysis of both monetised costs and those that are less easily monetised.

In sum, the purpose of such an analysis should be to determine whether it is worth investing today in a number of important anticipatory adaptive and preventive measures for controlling a global warming and limiting the impacts of its effects.

Adaptive measures would include:

- the development of a global food security system while Northern Hemisphere countries are still able to generate an agricultural surplus at relatively low costs;
- the provision of additional financial assistance to developing countries, especially those in semi-arid and flood-prone regions, to help them establish a self-supporting agricultural infrastructure, appropriate techniques and improved water management and control;
- the allocation of development funds to Third World countries in order to expand food production and to develop a sustainable agriculture;
- a greater international efforts to halt the trend of accelerating global desertification; and
- a commitment to humane methods of curbing population growth where in inhibits development.

These adaptive measures would to some extent lessen the more severe welfare impacts of the economic and environmental disruptions accompanying any global greenhouse effect. They also involve economic costs that would mean the diversion of resources away from other economic activities. On the other hand, some of these options, such as the commitment to sustainable agricultural development in the Third World, have long been advocated in their own right; the greenhouse effect just adds force to arguments in their favour. In any case, the total economic costs of all these options are one estimate of the welfare consequences of climatic instability and change. The alternative to paying these costs now may be to face the full brunt of their consequences at a later date.

Preventive measures would include:

- reducing burning of fossil-fuel, particularly through the development of non-fossil-fuel energy sources and through improvements in the efficiency of extracting useful energy from fossil-fuels;

- reducing the emission of trace gases from other anthropogenic sources;
- increasing pollution control and developing technological processes to 'scrub', recover and recycle the carbon and other trace residuals emitted when fossil-fuels are burned;
- halting unnecessary tropical deforestation through alternative development strategies and incentives; and
- increasing the rate of replanting in deforested areas, encouraging afforestation and improving forest management.

To make any major impact on reducing carbon and trace gas emissions quickly would require rapidly implementing the above measures and should necessitate diverting resources from economic activity. This could result in some decrease in short-term economic growth. However, this might not necessarily be the case, because measures that improve the efficiency of fossil-fuel consumption could simultaneously result in overall savings in energy costs.

Effective implementation of these preventive measures will eventually require their acceptance by developing countries, whose consumption of fossil-fuels is continually increasing. However, those countries cannot be expected to invest in such measures without assistance from more advanced industrialised countries, either in the form of specific investment flows or by making the appropriate emission-reduction and non-fossil-fuel technology available. With the possible exception of investments to improve energy efficiency that directly influence energy costs, one cannot expect developing countries to 'go it alone' with implementing such preventive measures. Indeed, without a substantial commitment to apply such measures in their own countries, the advanced industrialised nations have little moral authority with which to persuade the developing and newly industrialising countries to adopt such measures. On the positive side, these same preventive investments are linked to controlling other global and transnational problems, such as 'acid rain' and the impacts of CFCs on the ozone layer to which public opinion and policy-makers in the advanced industrialised countries, are beginning to react.

Finally, with such a potentially vital future problem, which is plagued with uncertainties over the estimates of climate change and sea-level rise and uncertainties over the sustainability of our current pattern of global economic activity, there is a mandatory call for further research. In the case of the greenhouse effect, the need for research is essential. Fortunately, the research effort into this effect is growing. The next and urgent step, however, is more rigorous analysis of the policy options currently available. This is turn requires the continuation of three major thrusts of international cooperation in future research: monitoring of climatic changes, ocean and atmospheric circulations, biogeochemical and other ecological processes, changes in the sea-level and ice caps; research and development of warning systems for predicting sudden disasters from climatic and sea-level changes; and analysis of the costs and benefits of both the impacts of global warming and the adaptive and preventive measures necessary to ameliorate them.

Climate change has become a staple of national and international politics. It is by far the most dramatic environmental issue yet to receive world-wide attention, as human activities pose unprecedented threats to the planet's environment. The threat has spawned widespread proposals for treaties, international institutions, and major new policies, many of draconian dimensions. Even British Prime Minister Margaret Thatcher, who so assiduously defends her country's sovereignty in the European Community, has called for binding international protocols to control greenhouse gas emissions, with supervision and monitoring by international organisations. Analysis of the threat has also led to assertions that the climate will not change in the dimensions claimed, or that the proposed cures would be more costly in lives and prosperity than the climate change itself.

This often frenetic activity has legitimate roots. Responsible forecasts have predicted that the growing concentration of certain gases in the atmosphere could lead to climate change with major effects on health and living standards. The economic and political affairs of nations could be substantially altered and international relationships significantly changed. The forecasts raise many profound questions of—equity, Third World development, conflict over food and land, and the appropriate limits of human intervention in the natural environment—now measured on a planetary scale.

The many proposals for urgent action, however, rarely reflect the limits of policy processes, or what would be necessary to lead to major new policies. The question is not whether to take action now, as it appeared to be in the April 1990 White House conference on global climate change. Rather, it is to recognise what is feasible and what will fail, given the state of knowledge and the unique characteristics of the subject. Some proposals are wildly unrealistic even for an era of aroused environmental consciouness, such as those calling for the establishment of a new supranational institution with enforcement powers to supervise mandatory reductions in greenhouse gas production. Even more modest proposals to encourage energy efficiency and conservation have proved difficult or impossible.

Significantly, discussion of the proposals often neglects the real costs of premature action: the resources denied to other crucial needs and the effects of what might prove to be costly policy errors. If pressure from the scientific community and environmental movements leads to costly measures that later prove to have been unwise or unnecessary, the public may lose the trust in scientific analyses essential to sustaining long-term environmental protection.

It is important to understand the sometimes unique difficulties surrounding this subject. Without that understanding, there can be little clarity about what is possible and on what time scale, and what will ultimately determine the boundaries within which effective action can be taken.

The central problem is that outside the security sector, policy processes confronting issues with substantial uncertainty do not normally yield policy that has high economic or political costs. This is especially true when the uncertainty extends not only to the issues themselves, but also to the measures to avert them or deal with their consequences. The responses of different political systems will, of course, vary, but that initial judgment is valued for most.

The climate change issue illustrates—in fact exaggerates—all the elements of this central problem. Indeed, no major action is likely to be taken until those uncertainties are substantially reduced, and probably not before evidence of warming and its effects are actually visible.

Unfortunately, any increase in temperature will be irreversible by the time the danger becomes obvious enough to permit political action.

The only real prospect for a different policy outcome in the near future would be if public consensus and international negotiations over-came the stubborn nature of the policy processes of governments. Certainly the rising concern about the environment is much in evidence, as growing political power of the Greens in many countries makes clear. And the escalating number of international meetings and calls for action puts pressure on governments to formulate new policies. It is not inconceivable—though still improbable—that policies and agreements could result that can not be expected on the basis of economic or policy analysis alone. The "framework" treaty expected to be negotiated by 1992 is likely to be an empty she for many years. The barriers to action and agreement are high, and the public's commitment has not yet been seriously tested on international issues that are both so costly and so intangible.

The global warming phenomenon is by now well-known. The basis greenhouse effect itself is neither a new concept, nor in doubt. In fact, it is essential for life on Earth. Without the warming action of the atmospheric blanket, the surface would be approximately 35°C (63°F) cooler and unable to support life as we know it, or possibly any life at all. Earth's atmosphere is largely transparent to short-wave energy from the sun. Some of the energy is reflected back into space, but a portion is absorbed by the surface and clouds and reradiated as long-wave energy—that is , as heat. Several gases absorb energy at these wavelengths and, in turn, warm the atmosphere.

The gases contributing most to the greenhouse effect include water vapor, carbon dioxide (CO_2) methane, and several others, the most important of which include the manmade chlorofluorocarbons (CFCs). Several of these greenhouse gases have been increasing rapidly as a result of human activities—on a scale sufficient perhaps to alter significantly the determinants of climate and weather. The concentration CO_2 the gas that accounts for approximately half of the human-caused warming effect, has increased some 25 per cent in a century, largely as a result of the burning of fossil fuel and the destruction of forests. Other greenhouse gases come from agriculture, mining, automobiles and industry.

This part of the story is straightforward and generally not in dispute. Estimates of the effects, however, are less definite. They come from theoretical analysis, calculation using complex models of the atmosphere and oceans, and the interpretation of a limited but growing body of measurements. Important uncertainties are introduced, and controversies over the models and their interpretation necessarily arise. The more prominent projections by such authoritative scientific bodies as the National Academy of Sciences coalesce around the estimate that if greenhouse gases increase to the equivalent of a doubling of the atmospheric concentration of CO_2, average temperatures would rise by .15°-4.5° C (3°-8°F)—less at the equator but up to twice the average at the poles. This would lead to average surface temperature higher than any recorded in human history. When a temperature rise would occur depends on the rate at which gases are emitted and on complex feedback effects of clouds and oceans that are not now adequately understood. In the absence of controls on emissions, and with of surges or large dips in economic growth, typical analyses predict it would take place around the middle of the twenty-first century.

The surface effects of such a temperature rise are not predictable in geographic details; they would vary regionally and even locally. Many estimates, however, predict reduced availability of fresh water during summer growing seasons in inner areas of the continents, migration of agriculture toward the poles, increased of agriculture toward the poles, increased desertification, and more weather extremes. Probably the most publicised prediction is a sea level rise by one-third to one-half a meter by the middle of the next century. Low-lying areas would be particularly vulnerable to storm surges. If the forecasts are correct, all of this will happen more rapidly than has virtually any ecosystem change in the last 160,000 years, raising questions of such unanticipated and possibly serious, nonlinear effects as shifts of major ocean currents.

Not only are these forecasts uncertain and controversial, they depend on future developments in the economy, in technology, and in policy, as well as on the result of research. And not all the effects will necessarily be damaging; some activities will be enhanced and some localities and nations may benefit by the changes, at least in relation to others.

The greenhouse effect and its possible physical and socioeconomic effects are frequetly presented in cataclysmic terms; certainly the visions of catastrophe have stimulated a variety of proposals for the United States and other countries to reduce CO_2 and other greenhouse gas emissions to "stabilise" the atmosphere. The subject is rich in complexity and detail, but there are several characteristics that make it different from or larger than other environmental issues.

In the light of the global survival imperative which these climate changes now impose, equity is simply the trans-temporal sharing of risks, responsibilities, rights, rewards, responses and restraints in a balanced way-or fairly. This equity-survival premise is axiomatic and empirical. It is both moral and expedient. We can choose which way to view it, but rejecting this linkage, as some (like the UK government) still do, is rooted in a failure to understand the universal nature of this crisis. Such an attitude can only precipitate deepening conflict with each and with nature.

Understanding this requires understanding the complex system of feedbacks in the bio-sphere. Such an understanding may deliver the new synthesis upon which to model and protect the future prospects of us all. However, as we shall argue, this will need to embrace dealing more with the causes of the problems we face, rather than ambulance-chasing their effects. Because of growth in all sectors of the global community, these sectors are now impacting both upon each other and upon the biosphere. Consequently all are now vulnerable to the adverse consequences of their own, each other's overall continuing growth. Traditionally we have all been directly vulnerable to each other in the context of aggressive behaviour. From this, conventional 'attack and defend' security strategies evolved. However, we are now also indirectly vulnerable to each other through our respective (and even ostensibly non-violent) impacts on the bio-sphere. Survival is now a collective enterprise and security is now fundamentally redefined because of this. If historically victims could pay the collateral costs of violence, the new reality is that from now on-with violence or even merely with inappropriate development—we can only make collective victims of ourselves. Conversly only with collective and right action can we hope to survive. This article attempts to point up some of the right actions and to lay out some of the preconditions for that right action becoming collective.

A Dynamic Trans-Temporal Model Pointing to a New Strategic Imperative

Graphic shows a model of the bio-sphere as a closed' system. All feedbacks in the system are self-internalising and result in an overall condition of benign evolutionary self-regulation. Our bio-sphere loop is looped upon itself with three cross-overpoints. These give rise to the appearance of three sub-loops or 'petals' all sharing a common space at the centre of the 'knot'. The common space at the centre especially represent the geo-sphere/biosphere upon which all human activities depend. This is only initially unaffected by human activities. The petals we shall assign to represent population, production and pollution and growth of these.

It is a trans-temporal progression based on this loop. The progression takes the loop through stages of transformation and (in human terms) so-called 'growth'. It shows, the growth of population, production and pollution over time and should suggest the interaction of these. It shows how these are now impacting on each other whilst also closing on the biosphere like a tightening noose. It is this growing prevalence and proximity of human impacts in diminishing ecological space which redefines conventional security and engages the equity-survival linkage as a strategic imperative in the politics of the global commons.

It is imperative we recognise that when resources (and global commons resources such as a stable global climate system) diminish, we have only limited choices of response and access. We can either auction them, or fight over them, pursuing variations of a conventional property-based approach. Alternatively, and since the critical resources are now obviously systemic and part of the global commons, we can ration them or their use. This could be done at least until the crisis is substantially reversed. In our view, auctioning will not work. Allocating fundamental rights (such as the right to survive, and the right to providence by income is obviously grossly inequitable. But it also compounds positive feedback, and this is counter-productive at a time when providing negative feedback is what is required. Fighting has never diminished environmental impact. In the context of global warming and the management of climate change, fighting is an unrealistic proposition as any 'victory' would be Pyrrhic. Rationing is the only potentially viable option. This is now the guts of our global dilemma. The post-war period saw the dilemma of 'mutually-assured-destruction' (MAD). This evolved on an East/West axis. It was a situation primarily characterised by the direct and mutual vulnerability to each other's use of weapons of mass destruction. However, whilst there were various strategic initiatives to circumvent the progressive 'lose-lose' certainties of nuclear war, nuclear disarmament—or the negative growth of armaments—was only ever the truly rational way out of that dilemma. To some extent that rationale has now conditionally begun to catch on and there is a relevant lesson in this.

However, this debate still continues somewhat blind to the new and more urgent security dilemma of global climate change and creeping ecocide. Our new strategic dilemma is total and cultural rather than just global and military/political. It is on a more North/South axis. Because of long-term and intricate feedbacks, it is much more complex and deeply-rooted and deals also with much higher degrees of uncertainty, risk and momentum. To face it, all conventional economic development assumptions must be radically challenged.

(a) at any one time so far, it has only been a minority of people on the planet who have been engaged in the high-intensity development activities which have led in turn-albeit unintentionally-to global scale geo-sphere/bio-sphere changes of an adverse nature;

(b) the generation of these changes has occurred over large time frames with much actual (if acknowledged) and further potential for direct and indirect damages to first, second and third parties and the biosphere itself;

(c) in fact, largely undetectable thresholds of biosphere viability are being crossed, with the potential for irreversible changes and even catastrophic feedbacks being increased;

(d) this uniquely heralds the bio-sphere and its protection as 'mediator' in the pragmatic merging of the interests of first, second and third parties as the development of policy responses to this crisis unfolds;

(e) it is this set of factors in which the military connotations of security become progressively less relevant since global development cannot be restrained by military means and neither can environmental protection be delivered by military means;

(f) it is also this set of factors in which the survival/equity linkage becomes evident, because 'self-interest' is gradually being understood as surrendering to the "universal interest' rather than clinging to 'vested interest';

(g) so far however there has been only the so-called 'no-regrets' policy response where, faced with all the above, status-quo development praxis is merely being modified in the direction of 'efficiency', and then only to the extent that the isolated efficiencies so gained are deemed commercially advantageous; all the efficiencies so gained are still being lost to the net growth of environmental and social impact;

(h) it is this set of factors in which the survival/equity linkage and the need for surrendering to the universal interest is vainly being ignored or rebutted by vested interest;

(i) this 'postponement' is conveniently assisted by the belief in 'detached' or 'value-free' science or the insufficiently acknowledged scientific dilemma about 'objective proof' where humans are part of the evidence, (or humans are both observer and observed). Proof sought for these global changes, their causation and the validity of the damage prognosis, is deemed dependent on what are regarded as 'objective; externalities. However, because of the global scale of human impacts, these externalities are already being influenced by 'us; and have become therefore, part of the dynamics of subjectivity. Humans and human development culture are now critically part of the overall feedback process. This makes "value-judgements; and 'bias' unavoidable and also radically challenges the ethos of laissez faire.

The preceived impacts of a particular environmental problem can change as the problem involves increasingly larger and more complex ecological and political systems. For example, at the local level, a problem such as deforestation primarily affects slash and burn cultivators and peasants attempting to establish stationary farms. But the same problem, multiplied by thousands of small farms and combined with extensive commercial logging, can exacerbate the global-level issue of accelerated atmospheric warming. As an environmental problem shifts from a subnational to an international concern, diverse political systems must be co-ordinated before workable solutions can be implemented.

International development organisations such as the World Bank are able to influence local perceptions about environmental problems through their financial assistance programmes. Poorly managed development programmes may result in serious local conflicts without warranting much attention from national policymakers more concerned with promoting economic growth. However, intrenational development agencies, increasingly concerned about long-term natural resource management, may push national and, subsequently, local policymakers toward more environmentally sound land-use practices by tying economic assistance packages to the results of environmental impact evaluations.

Environmental politics can be viewed in a hierarchical framework with five layers that represent local, subnational, national, international, and global eco-political scales or stages. These layers form an upside-down pyramid of eco-politics with four expanding sides. Each side represents an attribute of a given environmental problem or issue: the number of actors, the number of political jurisdictions, the complexity of the ecological cause and effect relationships, and the institutional obstacles to enforcement. From one level to the next, the variables become more numerous and their interactions more complex. Consequently, the perceived threat to human survival increases, as does the level of political intransigence that impedes effective action. The jump from legislating national policies to negotiating international and global policies may be the most challenging of all. At the global level, for example, issues such as climate change involve most, if not all, national governments as well as international NGOs and supranational agencies.

At the bottom of the inverted pyramid are subnational concerns that embrace their own hierarchy of political interactions—from the neighborhood level up through the first-order administrative units; provinces, states, and republics. The cause and effect relationships that govern local environmental problems are relatively clear and the roles of various actors relatively well defined. Subnational environmental politics, however, can be quite contentious with overlapping agencies, competing economic interests, ambiguous regulations, and numerous special-interest groups.

Environmental politics at the national level takes place among state or provincial governments, central government agencies with constituencies affected by the environment, and the three different branches of government. The role of the national government in mediating between the competing concerns of subnational polities and special interest groups can be particularly difficult when the burden of resolving an environmental problem, be it acid rain or deforestation, is not equally shared among subnational regions. NGOs have effectively lobbied national policy-makers and politicians, particularly during election years, on issues affecting both domestic and foreign relations.

Environmental regulations have been controversial issues in most countries that have attempted to formulate and implement them. In the international arena, environmental agreements are even more contentious because of problems with national jurisdictions, division of responsibility, unequal economic support and technical resources, and resistance to the creation of a supranational authority to decide and enforce the agreed standards and restrictions.

Marine pollution is a good example of the promise and problems of international environmental agreements: International marine pollution agreements, including over 70 multilateral instruments and the comprehensive 1982 UN Convention on the Law of the Sea, have a mixed record. Although some multilateral accords have helped establish protocols for monitoring and responding to such problems as oil spills, attempts at reaching global agreements covering a broad range of marine environmental issues have been bogged down by lengthy negotiations, questionable economic and technological assumptions, and ratification problems. World-wide environmental problems such as global warming potentially involve all national governments as well as some key supranational organisations.

Unlike multilateral agreements involving several countries, global programmes require the participation of a large number of countries, many of which may consider the issue a low priority or primarily another country's concern. The political processes required for effective environmental action become much more complicated with many governments and national development agendas are involved. Co-operation between rich and poor countries will be greatly affected by wide-ranging differences over what the major problems are, who caused them, and who will pay to resolve them.

North-'South' Conflicts

The archetype of southern politics-UNCTAD—was built on polarised political blocs organised around adversarial and conflict-prone economic relations between the poor and wealthy states. In the 1980, however, the unity of the South's Groups of 77 (G77) had been severly stressed already by the rise of the newly industrialised countries in Asia which heraded the end of the 'third world'. The end of the second Cold War combined with the steady decline of the 'South' and unified entity may make it easier to delink old debates from new agendas such as climate change .

Thus, even the concepts of 'South' and 'North-South' conflict may be obsolete and irrelevant in the climate change arena. Indeed, since late 1991, negotiations over climate change have resulted in a fractured and pragmatic set of political axes. Completely new alignments of co-operation and conflict emerged that are still fluid but no longer mirror the old North-South cleavage. Thus, in 1991 Anne Kristin Sydnes identified at least five groups from the 'South' in the climate change negotiations. These were:

1. radical, like-minded activist states (e.g. Bangladesh and Maldives) which view climate change as a major threat to their national existence;
2. potential, like-minded activist states (e.g. Maxico) which see climate change as a good way to extract additional concessional aid;
3. energy-consuming, hardliner states (e.g. Brazil, China, and India) which support more research but object to costly commitments and actions;
4. energy-exporting hardliner states (OPEC states led by Saudi Arabia and supported by Australia) which object to potential market shrinkage and trade impacts;
5. unpredictable 'transition' states (e.g. Taiwan, South Korea, and some Eastern European states) which are already ambivalent as to the North-South cleavage given their position in the world economy and international hierorchy of states.

By December 1991, the G77's unity had virtually collapsed at the greenhouse negotiations in Geneva. A breakaway Group of 24 (G24) proposed that developing countries consider acting on greenhouse issues while awaiting action by the OECD. Two other southern strains also emerged in addition to the centrist G24, namely a groups of energy exporters which backed the United States in stalling agreement; and the AOSIS island states which joined the European Community, Austalia, Canada, and New Zealand in calling for a strong convention. The AOSIS states believed that the dilatory and ideological stance of the hardline, big poor states in G24 jeopardised their chance of obtaining any resources from the OECD. A fourth group of still uncommitted countries emerged, including Argentina and Mexico. These divisions continued up to the signing of the Climate Change Convention in June 1992 and are reflected in its text. It is difficult to believe that the state elites of the G77 can reconstruct their solidarity while negotiating protocols to the Convention now that they have discovered that their interests diverge fundamentally in relation to climate change.

The fragmentation of the South places developing countries in a weak bargaining position on the centre; issues of financing, technology transfer, and compensation payments that are still to addressed in the Convention. The greenhouse issue exemplifies a general dilemma that developing countries face in global environmental politics. Global and regional environmental predicaments present them with new demands on scarce resources for regime and national survival as well as new bargaining opportunities with the OECD states. It remains to be seen exactly how southern elites will respond to these pitfalls and opportunities in the greenhouse arena.

On the negative or threatening side of their security, environmental problems could shift the priorities of wealthy trade and aid partners away from political and social stability in the South to global dilemmas of less concern to the southern elites. They also confront new and unruly domestic, social movements often aligned transnationally with powerful counterparts in OECD states. Vulnerable states could launch ideological campaigns against environmental issues in an attempt to wrest the political initiative away from the OECD states in the international area. Polarisation around issues such as climate change between the big, poor states and the big rich states block rather than foster international co-operation. Conflicts at a global and regional level on environmental issues could spill over into geo-economic and geo-political dimensions of interstate relations salient to climate change, thereby gridlocking ongoing negotiations.

Big, poor states may also use environmental issues to extract concessions from the OECD states in long-standing geo-economic and geopolitical areas. The greenhouse issue is unique in that the South influences a global asset that is greatly valued by the North: Earth's climate. Negotiations to date have been stalled by ideologies transposed from prior North-South conflicts into the greenhouse arena. But the elites of big, poor states have also tried to play a climate distruction card in a slow motion game of global climate change poker.

This strategy may fail, however. As Dallas Burtraw and Michael Toman have explained, most negotiations have two phase. The first phase is the bargaining over terms and content of agreement, which was partly completed at Rio. The second phase now underway is concerned with ratifying and implementing the agreement. 'Any proposed outcome that cannot be credibly implemented in the second phase of the game,' they note, 'cannot be credible in the first phase of the game.'

Due to weak administrative and market institutions, states such as India, China, or Brazil may be unable to abate in accordance with global commitments to reduce greenhouse gas emissions—with or without transfers of resources from the wealthy nations. In addition, poor states are more vulnerable to the economic and social impacts of climate change than wealthy states. For these reasons, it is lilely that the leaders of the OECD will be unmoved by implicit threats form the developing countries of destroy the earth's climate system unless the rules of international commerce are reformed.

Ironically, demands by big, poor states for massive resource transfers undermines the credibility of such threats for two reasons. First, it suggests to donors that the problems are so large that aid recipients may not be able to deliver the abatement, even if they receive additional support on a large scale. Second, to the extent that large-scale resource transfer achieved abatement and stimulated development, it sould increase the dependency of southern elites on this source of external support. Recipients who defected from the regime would be sawing off one of the branches supporting them. It is not surprising therefore that the elites of the big, poor states are not persuaded that co-operating in the climate change arena helps rather than harms their prospects of staying in power.

A New Power Game

The potential economic impacts of mitigation and adaptation strategies have elevated environmental concerns from low to high politics, on a par with traditional economic and military preoccupations of the great powers. In contrast to the nuclear arms race, for example, no single state or groups of states so predominates in emissions or abatement capability that it can impose an internnational regime on everyone else. The potential candidate—the United States—abdicated from its potential hegemonic role in this ragard at Rio by refusing to commit itself to reduction targets. Moreover, unlike the geo-political and geo-economic domains of interstate relations, there are as yet no widely accepted ideologies that frame ecological issues such as climate change.

Consequently, ecological alignments in international relations remain fluid and unpredicatable. No single state or group of states can lead or coerce other states to join a greenhouse regime or to build an oppositional grouping. It remains an open question whether governments will construct a meaningful greenhouse gas regime. Faced with this agnostic prognosis, there are three reasons to be optimistic about the medium- and long-term future of a greenhouse regim: technological innovation, the contribution of scientists to elite and popular understanding of climate change issues, and social movements.

Technological Innovation

Rising energy efficiency is closely associated with technological dynamism, in turn an attribute of competitive firms and economies. Domestic and international competition drives technological innovation that will reduce the cost of greenhouse gas reductions, even at high levels of abatement. Governments can impede or encourage this phenomenon, but they can't stop it in the long run.

Nonetheless, an international climate change agreement that sets the ground rules for investors and states would enhance this phenomenon. Relatedly, bilateral and regional initiatives to demonstrate the technological feasibility and economic attractiveness of greenhouse gas reduction will be an important immediate step toward a greenhouse regime. It is crucial to identify the costs of abatement above the 20-30 per cent level of reduction for which data is available today. Only when the true cost of preserving the world's climate is known will political leaders be able to respond meaningfully.

Scientific Research

Scientists will continue to develop a common stock of scientific knowledge on greenhouse issues out of which political elites can forge consensus on policy issues over time. Environmental regimes reflect no only interests, including the influence of domestic stakeholders in international affairs, and legal authority flowing from or ceded to international institutions, but also different world views. The smaller the common understanding and commitment to shared values, the weaker the regime. Thus, consensual knowledge is critical to overcoming the divisions of interest, authority, and belief systems that militate against international agreement.

Political scientist Peter Haas calls this influence 'epistemic' because scientists have been able to shape the images held by politicians and diplomats as to what is at stake in negotiations. He notes that scientists have already provided an ecological basis to international agreements in the Mediaterranean Action Plan and the Vienna Convention to stop ozone depletion. There is little doubt that the Intergovernmental Panel on Climate Change has and will continue to play exactly this role in goading governments to grapple with the problem of global climate change.

Social Movements

The inability of governments to concur in immediate stringent greenhouse gas reductions under the Convention may stimulate even greater efforts by social movements to address the issue of climate change on a global basis. In the developed countries, these organisations prefigure emerging social trends, are paradoxical in that they represent contradictions in the social order, and often transform the status quo. In many developing countries, non-governmental organisations are among the few well springs of social activity that are not nominated by govoernment or administrative structures. Non-governmental organisations are able to pioneer creative and innovative solutions to many problems that stymie governments.

Although local citizen groups are the bedrock of social movements aimed at increasing accountability and participation in decision making of governments, they have invented new ways of communicating across cultural and institutional barriers, both within and between countries. Citizen groups provide a unique interface at the intermediate level of society to link national governmental policies and programmes with local realities via a host of social, economic and political organisations at the provincial and district level, including federations of co-operatives, trade unions and business, institutes, and churches. In many developing countries, this level of civil society is weak and must be strengthend to complement efforts to decentralise national public bureaucracies on to local, autonomous governmental institutions.

Another hallmark of environmental politics is the role of strong national, regional, and transnational social movements concerned with environment, development and social justice. The Climate Action Network exemplifies this trend. Established first in Western Europe, North America and Australia, the Network now includes vibrant and self-reliant regional networks in South-East Asia, South Asia, Latin America, and Eastern Europe. The Network marshalls scientific information to present policy-makers with strong recommendations and backsup these positions with strong political pressure on national governments and in the course of climate change negotiations.

Many citizen groups are starting carbon abatement projects without waiting for governments to reach international agreement. They are the key to reaching the millions of decision makers and billions of people who must change their daily routines if greenhouse gases are to be reduced to ecologically acceptable levels. They can inspire, complement and (when necessary) circumvent governments to initiate shifts in popular and elite world views. Non-governmental networks increasingly cross national boundaries to generate common positions on issues that divide their respective governments, including the old North-South divide. They can also monitor the implementation of the agreement and trumpet loudly when governments fail to meet their commitments. The production of independent inventories of greenhouse gas emissions is an important contribution in this regard which has already had an impact on international negotiations on climate change.

An important component of an international greenhouse strategy in the short-to medium-run is to increase the participation of non-governmental organisations in private and public international financing of energy and environmental investment projects. The participation of non-governmental organisations in the project cycle of the World Bank's Global Environment Facility (GEF) is an important first step in this direction. In Mexico, the EMIR *(Efficientacion Mexicana de illuminacion Residencial)* project began in 1991 when the International Institute for Energy Conservation working with US scientists and the *Comision Federal de Electricidad* (the Maxican electricity utility) proposed to the GEF that in lend $10 million to Mexico to improve the efficiency of residential lighting by promoting CF (compact fluorescent) lamps. Over its life, each 16 watt CF lamp that costs $10 to install will save about $33 of electricity and $9 of incandescent lamps, and will reduce greenhouse gas emissions by about one fifth of a tonne of carbon.

Citizen groups are also pushing for a direct role in the funding decisions of multilateral banks. At the NGO Global Fortum that was held at the same time as the governmental 1992 Earth Summit, they committed themselves to urge the governments of developed countries to provide adequate, new and additional funds on concessional terms to developing countries, and to ensure that they participate in expenditure decisions and implementation of funded projects to ensure that these resources are well spent.

The creation of an NGO consultative committee by the GEF is an important step in this direction, and one that other organisations, especially the regional developments banks, should emulate. Indeed, by December 1992 citizen groups had produced already a positive, even visionary reform agenda for the GEF.

Market-driven technological innovation, increased popular participation in decision-making, non-governmental mobilization for sustainable development, and the role of the scientific community in policy formation may impel governments to overcome all the barriers to agreement. The first steps toward creating a global greenhouse regime are likely to be small rather than large, bilateral rather than multilateral, and regional rather than global. The transfer of resource is likely to be pragmatic, linked closely to abatement activity, and largely additional to existing aid flows.

Initially, therefore, a greenhouse regime must be flexible enough to demonstrate what is possible rather than to strive for final policy commitments that are simply ignored. The low (or possibly negative) net cost of abatement for the next one or two decades will grant the world a breathing space in which to explore the frontiers to social and technological possibility.

As Ralph Buultjens has written, no other issue has the ability to bring together so many people and nations as does climate change. The negotiations to create a global greenhouse regime are a rare opportunity to form a global coalition of interests that transcends national boundaries and historical antagonisms. It is perhaps the first time in history that the poor in the developing countries have a powerful ally among influential citizen groups and even some governments in the developed world. Thus, Greenpeace International's greenhouse gas scenario uses a development scenario that favours the poor countries rather than simply projecting the unequal global status quo into the future, as did the Intergovernmental Panel on Climate Change.

It is conceivable, therefore, that humanity will not march over the precipice of climate change, but will stop, look down, and will head instead toward a sustainable future.

Developing countries are subsidising unsustainable energy practices in the North to the tune of $3.4 trillion a year, an environment research body said here.

A document by the London-based Global Commons Institute (GCI) calculated just how much industrial countries, or "energy debitors", owe developing countries, or "energy creditors".

The document has been presented to the Intergovernmental Negotiating Committee for a Framework Convention on Climate change (INC/FCC), which is meeting at U.N. headquarters this week to discuss funding for the concention.

Anandi Sharan of GCI explained the links between gross domestic product (GDP) and carbon dioxide emissions, noting that the "two go up and down together," so that "the higher the GDP, the greater the carbon dioxide output."

An Intergovernmental Panel on Climate change (IPCC) has said a minimum of 60 per cent cuts in carbon dioxide emissions were needed to slow further warming and curb adverse changes in the climate system.

"At present there are countries which are massively over their allocated quota limit and countries which are well under that limit," Sharan said.

She said that currently total global GDP amounted to some 20 trillion dollars per year, but that based on the IPCC recommended cut in greenhouse gas emissions, only seven trillion dollars worth per year is sustainable. And of the global 20 trillion dollar GDP, industrialised countries account for 19.2 trillion dollars. But on the basis of their population, industrialised countries are only allocated $2.7 trillion worth of global GDP, "so they're actually appropriating, or in debt to the tune of $16.5 trillion annually based on 1990 figures," she said.

Based on IPCC the assumption of a 60 per cent cut in greenhouse gas emissions, the GCI calculates that environmentally non-damaging carbon dioxide output per person annually would amount to some 0.46 metric tonnes of carbon. But it says that today, the United States alone emits between seven and eight metric tonnes of carbon dioxide per person annually. "When you look at the per capita consumption figures, we find that India and China can triple and quadruple their emissions without getting anywhere near the present levels of developed countries' emissions," Sharan said.

Commenting on the INC talks so far Sharan said, "the biggest scandal is that the only accountability that seems to be being discussed here is the accountability of developing countries through the Global Environmental Facility (GEF)."

The GEF is the interim financial mechanism mandated to provide the resources on a grant or concessional basis to developing countries to help them implement the commitments of the Climate Convention. The convention, signed by 160 countries at the Earth Summit in Rio de Janeiro last June, invites countries to return to their 1990 levels of greenhouse gas emissions as soon as possible.

"There is no formal accountability, as yet at all, by the emitter countries-there's an 'intention' to cut emissions, but the stated commitment is to 'strong and sustainable economic growth'." Sharan said. She noted that through the energy subsidy the South pays the North, the industrialised nations can run a system, "which allows them to dump their unsustainable technologies and energy systems on the world at prices that drive out sustainable technologies like renewables and reproductive holistic systems."

"It is of paramount importance that we stop talking about developing countries at all in the context of climate change, and that we concentrate whole-heartedly on getting eco-restructuring in the North," she said.

And Jeremy Leggett, Scientific Director of Greenpeace International's Climate campaign warned Wednesday that, "time is everything in this game."

He said, "as every month goes by, we learn depressing news out there in the natural world-it looks less and less likely that this series of record-breaking storms around the world, is not at least, in part, getting its excess energy from the known heat-trapping ability of greenhouse gases." He warned that insurance companies in the industrialised countries, on whose health successful economies depend, could soon be ruined by the avalanche of recent windstorms.

According to Greenpeace, between 1966 and 1987, there was no windstorm anywhere in the world which cost more than billion dollars in insured losses. But is says that during the period 1987-1992, thee have been at least ten such catastrophes.

Let us look at some of the key feed backs in a bio-system where humans now influence both the temperature and the climate of the biosphere. We shall concentrate on the relationships between;

(1) carbon dioxide (CO_2) and temperature

(2) . . . and population

(3) . . . and production

(4) . . . and industrialisation

(5) . . . and emissions from fossil fuel burning.

(6) . . . and gross domestic product or GDP

(7) ....and carbon dioxide and temperature and so on.

The comparative values for atmospheric concentrations of atmospheric CO_2 temperature over a period of 160,000 years up to the present. The correspondence between the two is considerable. They tend to go up and down together. The period embraces much of the ice-ages. During these times the total global population of homo sapiens was thought to have been around one million. During this period, varying CO_2 concentrations in the atmosphere could not have been driven upward by humans even though they had mastered the use of fire, as there weren't enough of them and atmospheric CO_2 concentrations were failing. However, the temperature started to warm from around 20000 BC. CO_2 concentrations started to rise and population started to increase.

Positive feedback between these elements would be difficult to assert at this time. But, from around 8000 BC onwards, agriculture started to replace hunting-gathering, CO_2 concentrations rose further to nearly pre ice-age levels and by 1000 AD world human population had risen to a figure in excess of two hundred million. CO_2 emissions from forest-land clearance, primarily in the Northern hemisphere, almost certainly helps to explain the increased CO_2 is the principal greenhouse (or heat-trapping gas). If this is so, already at this stage there is a potential example of overall positive feedback, or a self-reinforcing loop of climate-change causation in which the growth of humans and the growth of their activities are implicated.

However, it is with the onset of the industrial revolution, that this kind of positive feedback becomes more strikingly plausible. At the start of the industrial revolution, atmospheric CO_2 concentrations were at 280 parts per million by volume (ppmv). With the burning of fossil fuels (such as coal) for energy, concentrations have risen by 25 per cent to the current all-time high level of 353 ppmv. Global mean temperature has risen the best part of one degree Celsius during this period and population has risen from a figure of below one billion to the current figure of around 5 billion. In the context of growing numbers globally, population in Europe retained a clear lead density through this period.

It is this industrialising phase that a proper evaluation of geo-sphere, bio-sphere and socio-sphere feedbacks and 'what is causing what', is essential to a proper evaluation of humanity's prospects and policy requirements. This is a complex task, not least because the system is becoming more and more unstable and the empirical problems can only increase as human activities leading to this instability increase.

The global industrial CO_2 output as a percentage curve from 1860-1990 and global gross domestic product (GDP) from 1960-1990. (In both cases 1985 is given the value of 100 per cent). In the case of the GDP (or production curve), it has also been exponentially regressed to 1860. This does not mean that GDP was being used as an economic indicator from 1860 forward, but it does reasonably reflect the amount of industrial production and so-called 'wealth-creation' which was in progress throughout.

In essence the consolidation of Western economic, political and military dominance throughout the globe occurred during this period. In the light of this, it is predictable to find that 20 per cent of world population living in the industrialised countries receive more than 80 per cent of global income and the 80 per cent of world population living in developing countries receive 20 per cent of global income. In fact this 6:1 income split is more like 16:1 if atmospheric CO_2 stabilisation factors are included in the calculations. This version has two thirds of world population living at zero to negative global climate impact levels, or on 5 per cent or less per capita of the average fossil fuel per capita income level in the OECD countries. On 1990 global fossil fuel GDP: CO_2 values, this provides an energy subsidy to the North worth $3.4 trillion annually.

The global mean temperature variation during this same period covering a range of one degree Celsius with an average mean rise of 0.6 of a degree and with a sequence of three hottest years on record at the end of the 1980's. It was during this period that the atmospheric concentration of CO_2 rose to the all time high level of 353 ppmv at and a rate more quickly than any on the record.

In response to this, the Intergovernmental Panel on Climate Change (IPCC) announced in 1990 their finding that if atmospheric concentrations of CO_2 were to be stabilised at 1990 levels, immediate 60-80 per cent cuts would be required in the emissions of CO_2 from human sources. Also directly in response to these dangerous trends, the UN negotiations for a convention on Climate change began in 1991. However the politics of vested interests intervened and the treaty text signed at the Earth Summit in Rio in 1992 made a legally-binding commitment to "strong and sustainable economic growth" whilst merely outlining a non-legally-binding option of OECD countries to restrain CO_2 emissions to no more than 1990 levels by the year 2000. This latter point reflects nonetheless less a general (if caveat-ridden) political acknowledgement of the pre-emptive timing and character of Western industrial development.

Gross global industrial CO_2 emissions 1995-1990. It also shows global CO_2 and global GDP per cent rates of growth 1955-1990. The variations in each of the two growth rates are very strongly correlated. The chart also shows the projected gross emissions growth until 2025. It also shows within this, the 1990 levels cut by the IPCC 60 per cent. It also shows the maximum impact of emissions stabilisation in OECD by 2000 AD (white segments after 1990) should this optional undertaking be fulfilled by OECD. In fact the ratio of the OECD self-restraint segments in this graphic is generous. It was calculated on the basis that all OECD countries stabilised CO_2 emissions output as of 1990, rather than trying to guess the rates at which these countries might respond. Even in spite of this, it is obvious that the impact of these cuts on the overall trend of emissions growth is slight. The convention's commitment to economic growth explains this and shows the perceived (and irresponsible) nature of the political priorities at this time.

Even more intractable than the positive feedback between GDP: CO_2: temperature, is the positive feedback between population growth and the above. Population growth and economic growth certainly feed each other. However as this feedback has evolved to an intensity of global significance, the lead element has become economic growth. Demographic transition is utterly contradicted by this fact. Demographic transition theory settled for the view that beyond certain levels of industrial development and per capita wealth attainment, population growth rates stabilise because people desire—or need—children less. But this is only a sub-set phenomenon applying to a minority. Industrial wealth creation has also been a process of global emiseration, and poverty is ultimately the key reason why people have more children. The Rio Earth Summit identified the need to address mass poverty for both environmental and development reasons. It also identified the need to address 'over-consumption' for the same reasons, particularly in the industrialised countries. This was thus, an acknowledgement of industrial-caused emiseration globally. However, what was not made explicit was the underlying feedback between the growth of global GDP and population growth. The basic 'cure' of further economic growth everywhere was prescribed. This can only draw our difficulties nearer and the noose of equity and survival tighter.

The conclusion to draw from all of this has comparatively little to do with the marginal—and highly conditional—North to South redistribution schemes which were mooted at Rio on the back of this further global economic growth. Far more conclusive at this stage, would be an unconditional commitment to eco-restructuring in the industrial countries. At the heart of this must be the overall restructuring of their system of global economic exploitation. Since industrial countries have caused the lion's share of global damage and developing countries have provided the lion's share of global benefit by having very low consumption rates, promoting both equitable and sustainable consumption ratios globally should be the immediate priority of the industrial countries. If this does not happen and it is not founded in self-restraint, overtime developing countries will simply mimic the politics of the West and impose restraint or retaliation. Promoting this parity at sustainable levels will for example have little to do with CO_2 emissions trading—or the 'auction' approach to the management of climate change—suggested by the World Bank and others. Currently these seek to award environmental rights by income. Perpetuating the right to pollute and exploit in the direction of these who can 'afford' to, is a pseudo-efficiency argument won at peoples' expense and will only result in further emiseration and overall positive feedback. For these reasons GDP growth itself, the underlying causes of growth, the structural consolidation of the growth momentum in the financial system, even the fundamental desire for growth must be challenged, if we are interested in the survival of our species. This means, both for local and for global reasons, looking for where best to apply negative feedbacks. In the industrial countries this means self restraint and if necessary, developing countries should force the issue on the West by imposing tariffs pro-rata on imports from countries receiving the fossil fuel subsidy.

Once all this is confronted, the full tension of the global equity' survival linkage is revealed. Fighting for survival may be how some will attempt to type-cast our options and prospects. Politically we are still being led by the blind. But as a viable global survival strategy, it is not in the model. There are no exits from this system if we want to survive. Reality can't be avoided for ever-equity and survival are inextricably linked.

Lifestyle is the Issue

At the UN Conference on Environment and Development (UNCED), the so-called *population question* was reconstruct in chapter five of the Agenda 21 as *demographic dynamics and sustainability*. This shift is more than semantic. While *population question* is a formulation indicative of a quantity-based thinking, *demographic dynamics and sustainability* is clearly and attempt to grapple with issue of quality, loss of quality and disbeneficiaries. It is primarily open-ended and industrially-driven economic growth which is responsible for this quality loss. As Agenda 21 states, *"unsustainable consumption patterns... place increasingly severe stress on the life-supporting capacities of our planet."*

Southern countries resisted stereotype formulation of population as sheer numbers of people and their growth rates per country. Their partly successful efforts was at reformulating the debate, making the *rising rates of resource use per capita* the meaningful measure. This concentrates not so much on people per se, as on people as units of consumption and their comparative rates of resource consumption per country over time.

These are rate/intensity measurements and this approach is a move towards comparative environmental impact assessment of peoples' lifestyles. This is a necessary development towards quality and loss-of-quality-based measurement in a world of shrinking economic and ecological space. Vulnerable to this, the countries of the North-but particularly George Bush's America-resisted with remarks such as *"the American way of life is not up for negotiation"* and *"we believe in the principle of consumer sovereignty"*. They did this because such an approach clearly turns the tables on the North. Although it has yet to be properly recognised, this call for accountability over lifestyle impact was the beginning of the end for consumerism and the political economy of growth.

In what follows we survey various positions, recent and current, and thereafter go into a retrospective assessment of the historical changes from which the present tensions have developed. We will also explore some of the philosophical contradiction which lie at that the very heart of the crisis. Our interest is to present an interpretation of the energies and events which exposes the fundamental delusion behind the 'growth optimist' rationale of the economists and commercial society. This growth-optimist perspective has been and remains the immediate, the proximate and the long-term cause of the cultural. Political and ecological distortions which comprise the global crisis.

In the immediate preamble to Rio, the Economist Magazine published an irritable editorial about global warming and the extent to which this issue had distracted attention from, "The Question Rio forgets". *"Population growth"*, they insisted, was the most crucial issue, grumbling about the extent to which climate change and global warming had come to dominate the agenda. The whole article was very typical of the disarray of the laissez-faire economic growth-optimists position and the real tension was revealed out-right,-"*...some environmentalists appear to see the global warming the revenge they have long expected nature to take on humanity for economic growth and technology... now, argue gleeful greens, the limits to growth are at last in sight."*

They went on-in the face of powerful evidence to the contrary—to claim that *"...much of the damage that could result from climate change in the next century is occurring right now, and it has nothing to do with carbon dioxide... the root cause of this damage is a combination of population growth and poverty in the third world."* It is especially the last point here with which it is such folly to concur.

We can see quite clearly here how the population question has been deployed in an overtly political way to distract attention from the issue of human-induced global climate change. The industrial CO_2 output linked to economic growth has an established record spanning at least one hundred and thirty years.

As demonstrates, the accumulated output separated by region, has the industrialised countries of the North responsible for 83 per cent of CO_2 output since 1860. As demonstrates, the linkage of global industrial CO_2 output and global industrial CO_2 output and Global Domestic Product (GDP) is correlated to a very high degree. As goes on to reflect, the

Intergovermental Panel on Climate Change (IPCC) made a statement of scientific consensus in 1990 in the run-up to UNCED, declaring that immediate 60 per cent cuts in the human output of CO_2 were necessary if the rising atmospheric concentrations of this principle greenhouse gas were to be stablised at 1990 levels. In their view, the planet was warming and action for restraint was needed because the probability of drastic ecosystem failure was increasing.

This re-introduces us to the realities of the human condition. With this there is the beginning of a global environmental measurement of the diminution of human benefit (or the global disbenefits of industrial growth) as a consequence of the pursuit of open-ended economic growth. It is measurable not so much as impact on specific individuals or on social groups, but rather as impact on the biosphere on which we all interdepend. Measurement of who caused this is reflected has irresistible political content.

Malthus (at the beginning of the industrial revolution) published his population catastrophe theory in response to the egalitarian and pro-equity arguments in William Godwin's Enquiry Concerning Political Justice. Present day economists (at the beginning of the de-industrial revolution) attempt to diffuse the equity implications of global climate change by raising the same spectre. They do this for the same reasons. The Economist again:—"*By the middle of the next century, world population will almost certainly have doubled. Before it eventually levels off, it may have quadrupled. Many of these extra people will be crowded into the third world, mainly in the countries that already find it hardest to provide food, water, jobs and health care for their people. By 2050 Bangladesh, which now has 113 million people, may have 245 million. Its population density, already almost double that of Holland, would then be 1,700 people per square kilometre (4,400 per square mile). Nigeria, with roughly the same population as Bangladesh today may have about 440 million people by mid century.*" This description not only suggests the Malthusian stereotype, it critically reinforces the wrong analysis from which such stereotypes derive. Prior to the UNCED, the British Royal Society and the American National Academy of Sciences published similar observations.

This may seem a bleak description of the probable response to the threat, perhaps bleaker than it actually will be, given the many activities within nations, at the United Nations, and in international scientific circles. Certainly policy attention to the problem has spawned a growth industry of conferences, seminars, analyses, studies, and policy research. It would be ungenerous to wonder whether this frenetic activity will ultimately be productive; even if it may be inefficient, greater understanding of the issue is essential for policy formation.

While scores of policy proposals have already been advanced, a few general prescriptions seem in order here. Perhaps the most important, though the most difficult to accept, is to recognise that the uncertainties surrounding climate change are genuine. A premature commitment to action can pose serious dangers of error and backlash and can incur costs that would affect a wide variety of interests. Not all these interests are narrow; indeed, one is the economic growth necessary to have resources adequate for later needs as they emerge. Measures that reduce emissions and contribute to other goals, such as improved energy efficiency, are numerous; but they must be justifiable on their own merits. If not, and if they increase costs, the achievement of other major social objectives will be unnecessarily harmed. Still, the risk of global change with serious consequences is sufficiently great that heroic measures are justified to understand the true situation better and to prepare policy responses that could prove necessary.

Several steps should follow. The first is the most compelling, even if it is the most obvious: Adequate funds must be made available for the research necessary to reduce uncertainty. It is difficult to imagine an issue for which that argument has more validity, no matter how banal the recommendation. The $1 billion proposed in the United States for the president's fiscal year 1991 budget tour research on global change is substantial, but it includes several expensive space projects. The remainder appears small in comparison to the scale and danger of the problem. Expanded US funding might also encourage additional funding in other countries.

Much policy research is also necessary. In the future it will be critical to understand some of the interactions discussed here. These include careful assessments of the real costs of adaptation, of the tradeoffs among energy resources, of international negotiation processes, and of a host of issues associated with international organisations and Third World development. The fervor with which universities and a wide variety of other institutions, in the United States at least, have embraced the issue is a bit frightening, and in some cases self-serving. On the whole it is encouraging, however, because ideas and policy approaches will begin to be sorted out.

The appropriate international steps are more difficult to assess. Current discussions in the UN and other international forums are raising the issue's visibility among governments and publics. These negotiations are creating political momentum that will lead to some form of global negotiations. By September 1990 the Intergovernmental Panel on Climate Change set up by the UN is scheduled to offer its assessments of the scientific evidence and the socioeconomic impact, and to suggest response strategies. In addition, a negotiating body will be established to draft a climate-stabilisation treaty for the 1992 UN. Conference on Environment and Development. There is a good chance such a treaty will be attained at that conference.

However, in spite of the time and attention the international community is devoting to the subject, such a treaty and supplemental binding agreements to restrict emissions are not likely to have substantial reach, or to be accepted if they do, for all the reasons outlined above. The United States, the USSR, and Japan have made clear their skepticism about the adequacy of the knowledge base for taking strong measures. The need to develop policy for the international discussions will, however, be a useful spur for increasing understanding of the issues, for reducing the uncertainties, and for increasing the pressure on governments to focus attention and resources.

The international focus on global warming may also lead to two other important results. The first is that prospects may improve for measures, unilateral or international, that offer many benefits but do not find their primary justification in combating global warming. The complete cessation of CFC production or steps to increase energy efficiency and conservation are two broad examples of such measures. The second may be the development of contingency policies and a framework of international agreements if action proves necessary. Stringent policies to cut emissions may be politically impossible or even inappropriate today; but if they prove to be justified in the future, it would be of enormous value to have a clearer idea of the issues at stake, the policy alternatives, and a process for rapid response.

The role of the United States will continue to be central in the climate issue. As the biggest source of CO_2 and the nation with the largest scientific enterprise relevant to the subject, the United States will strongly influence the roles and attitudes of others. So far, the Bush administration has followed a low-key path, both domestically and internationally, and has given no impression of advancing the issue more than the minimum required. The question now is not what specific positions the administration takes on controlling greenhouses gases, but rather what determination it shows, including through funding, to bring the matter into focus and to take a leading international role in spuring collaborative research and the development of contingency policies.

The climate issue cries out for imaginative marshaling of political resources and for leadership from Washington, but not for questionable actions launched prematurely. The issue deserves a concerted commitment of domestic funds and personnel, imaginative international participation, and something more than the past penurious support for the UN. The world is waiting for America to show that it understands this broader meaning of the role of power.

Equity is on the Agenda

During the UNCED, Maurice Strong-- Secretary General of the UNCED—made more progressive comment. His audience was more international than that of the Economist, and the comment reflects the precarious political balance he had been seeking to strike. *The gross imbalances that have been created by concentration of economic growth in the industrial countries and population growth in developing countries is at the centre of the current dilemma. Redressing these imbalances will be the key to the future security of our planet—in environmental and economic as well as traditional security terms. This will require fundamental changes both in our economic behaviour and our international relations.*

Maurice Strong's remarks was reflecting the inverse distribution of power and poverty on the planet. His remark skates lightly on the thin-ice of international diplomacy. However it still treads heavily on the disbeneficiaries of the development process and is not specific about the cause of population growth. Remembering the link between GDP and industrial CO_2 pollution, the inverse symmetry between pollution and population in China and the United States over the past one hundred and thirty years. Both countries have an area of approximately one billion hectares. China has been responsible for only 3 per cent of the global accumulated CO_2 output next to 33 per cent from the USA.

Whilst our pairing of the USA and China suggests the inverse symmetry between pollution and people (and power and poverty), we need to go to the global picture of industrially driven global economic growth to identify causation. The pursuit of this *pre-emptively* created disbenefit and disbeneficiaries in the form of poverty and ecological dysfunction. The political consequence of this is that equity is on the agenda again, as Mr Strong implicitly recognises by referring to redress and security. Security is on the agenda because the industrial growth process is raising pollution (as the IPCC defined) to levels which threaten global survival. Redress is needed because we are all (whether powerful or poor) directly vulnerable to our impacts upon each other over time, and indirectly vulnerable to each other through our direct impacts upon the biosphere.

"Impeccable economic logic extended..."

It is well-known that economists are not keen on retrospective accounting. This is understandable as it is complex to the point of being baffling and up against a world of 'fait accompli'. However, in response to the evidence of massive pollution output differentials between industrialised countries and developing countries, the senior economists at the World Bank, Mr Lawrence summers, circulated a memo in the run-up to UNCED talking about the *"implicable economic logic"* of encouraging for example African countries to commercially exploit their *"under-pollutendess"* by selling pollution space to the over poluted countries of the North. The memo drew a howl of rage and he was obliged to publish a defence and an apology. What was not recognised was the opportunity value this approach represented for the deconstruction of the dominant paradigm of economic growth optimism. It is possible to extend his impeccable economic logic to the US/China example taking in population as well. Obviously of China is under-polluted, the US is underpolulated.

In Search of Sustainability

The historical accounting typified by Mr Summers and his profession, does not acknowledge the man----. Nor, even accepting Mr Summer's logic and the hypothetical extension of this, does retrospective snap-shot symmetry explain the problem. The point is the evolution of political imbalance, and the extent to which the industrial development process rooted in the North, has entrenched ancient quarrels. Industrial capitalism has been pre-emptively stealing a march on shared global interests and security through a massive campaign of resource appropriation and the creation of disbeneficiaries.

Much of this was already recognised as an environmental problem twenty years ago. At the time of the Stockhokm UN Conference on *"The Human Environment"*-U Thant, who was then Secretary-General of United Nations, stated:—*"I do not wish to seem overdramatic, but I can only conclude from the information that is available to me as Secretary-General, that the Members of the United Nations have perhaps ten years left in which to subordinate their ancient quarrels and launch a global partnership to curb the arms race, to improve the human environment, to defuse the population explosion, and to supply the required momentum to development efforts. If such a global partnership is not forged within the next decade, then I very much fear that the problems I have mentioned will have reached such staggering proportions that they will be beyond out capacity to control."* Ten years after the decade prescribed by U Thant for reconciliation, UNCED contributors could do little but acknowledge the lack of such partnership and the loss of control.

In 1992 the newly formed Business Council for Sustainable Development (BCSD)—the elite of transnational capitalism in their Manifesto "Changing Course" retained a position in favour of universal economic growth. They came nonetheless to an important recognition:—*"Environmental arguments for slowing human reproduction rates in the south pale before the truth that most environmental offences with global impacts are committed by rich minorities with hazardous production patterns, energy use, and consumption styles."*

Nearer the nub of the argument, Akio Morishima, Dean and Professor of Law at Ngoya University, Japan has asked the following question:—*"I want to stress that we must not demand that the developing countries alone, one-sidely, must do something (about population growth). We have to be constantly asking ourselves if our economic welfare in developed countries is forcing sacrifices from the people living in poverty in developing countries; if our over-abundant consumption of energy and affluence, enjoyment of rich food, are not creating their poverty. We must reflect on this matter constantly. I believe this is the starting point of environmental ethics."* This question is embarassing. It acknowledges disbeneficiaries whilst suggesting the source of causation. It is a reminder of the traditional Japanese inclination to view things in paradoxical terms.

Fritjof Capra deepens the embarassment.—*"The present global population crisis is due to the rapid increase of population in the Third World, and this increase continues because the conditions for the second phase of the demographic transition have not been met. During their colonial past the Third World countries experienced an improvement in living conditions that was sufficient to reduce death rates and thus initiated population growth. But the rise of living standards did not continue, because the wealth generated in the colonies was diverted to the developed countries, where it helped their populations to become balanced. This process continues today as many Third World countries remain colonised in this sense. This exploitation continues to increase the affluence of the colonisers and prevents Third World populations from reaching the standard of living conductive to a reduction in their rate of growth."* Here too, not only the creation of poverty but also population growth is traced to the system of wealth creation and its global impacts in modern times. Capra ascribes the cause of population growth very specifically to a failure to achieve demographic transition in developing countries because of international exploitation. This is an explanation which has been re-itereatd many times throughout the decade. He goes on to make a further point about the connection between global ecological balance and social justice and this is a key theme which has come more and more to the fore in the statements of the green-left and social-ecology groupings North and South throughout this decade as well. *"The world population crisis is an effect of international exploitation, a consequence of the fundamental inter-relatedness of the global ecosystem in which every exploitation eventually comes back to haunt the exploiters. From this point of view it becomes quite apparent that ecological balance also requires social justice."*

We must remember that the world is made up of nation-states concerned about their security in the anarchical international society. And though there is some recognition that all countries of the world share a 'common future', political realities dictate that competition does still exist among states. A government is expected not only to further the interest of its own people (hence the recognition, in the previous section, of the role that economic costs play), but it is also not expected to afford other states a disproportionate share of any benefit mutually achieved in an international dealing. If states are to pursue cooperative arrangements, their leaders will expect not only to achieve an absolute gain, but also to avoid a relative loss.

Therefore a second major precondition for cooperation is the presence of fairness, in a system-wide sense. It has been recognised in the literature that in order for a state to enter into an agreement, the terms of the cooperative venture must be seen, by the state, to be equitable (for example, Young, 1989: 368-369). Such 'equity' will not only be based upon a comparative analysis of costs and benefits, but it will also be based upon less tangible factors, such as history and ideology.

This issue of fairness has played, and will continue to play, a significant role in the pursuit of a cooperative agreement on global warming. Individual governments have already hinted at reasons why they should be given special consideration in any venture. Japan might argue that restrictions would only be fair if they were seen to encourage high GNP-per-unit of fuel consumed, thereby rewarding its efficient industries. Similarly, the United States might want some concession for its large landmass—stating that a 'fair' agreement would accept higher American demands owing to its reliance upon a high energy-consuming transportation system (Grubb, 1990: 75-76). Other countries will continue to argue their own 'special cases' and this will undoubtedly affect cooperative efforts. But, regardless of the merits of the claims made by these states, the largest problem of fairness is, and will continue to be, raised by North-South relations.

Industrialisation in the developed world has resulted in two important consequences: a higher standard of living for these states and an increasingly-fragile environment, world-wide. The developed world is now asking the developing world to join it in a global effort to stabilise the atmosphere. But the developing world has not significantly contributed to the build-up of greenhouse gases in the atmosphere and therefore does not feel a responsibility to bear the burden of helping to solve this problem. This burden would involve sacrificing the use of coal-burning facilities (which are the most appropriate means, practically and economically, to meet the developing world's energy demands) and thereby hampering their attempts to progress toward a higher level of industrialisation. It seems that the full cooperation of these states will depend upon a satisfactory response, from the developed world, to their demand for justice.

This demand for justice has been articulated in the South's call for a general climate fund to be established in order to help the developing world pursue sustainable methods of industrialising. A number of proposals has been put forward; most include the transfer of technological and financial resources, free of charge, to the developing world. But the propositions remain theoretical since clear responses from the developed world have, so far, been nonexistent.

Within the debate around the global warming issue, justice and fairness will be subjectively defined by every different actor. However, the dominant consideration in this controversy will continue to be between the developed and the developing worlds. Therefore, if international cooperation is to be achieved on global warming, justice as it applies to North-South relations will have to be agreed.

With regard to dilemmas of collective action, the attainment of the benefit is necessarily dependent upon the action of others. Though cooperation may yield the most desirable result for society, each state is also subjected to the temptation to cheat and therefore free ride—that is, not pay the cost, but perhaps receive the benefit nevertheless; and therefore realise a higher net gain. Economic, political, social or even militarily coercive pressure from certain sectors of society may force a government to choose this action, and therefore forbid the state from making any cooperation arrangements. Such pressure will be even greater if there is a feeling that others will cheat because, for a variety of reasons, states would not want to shoulder an inappropriate share of the burden.

Thus, another precondition for international cooperation involves expectations—i.e., the extent to which participants believe that the developing arrangements are viable and that they will be adhered to. In the theoretical literature, it has been shown that the realisation of mutual trust, and subsequent compliance to any agreement, will be aided by the availability of reliable information on the actions of other players.

With regard to the global warming issue, this suggests that effective cooperation will require states to publish production and emission figures for a wide variety of substances and industrial processes. Monitoring and verification will therefore become important, because no state will not want to be making sacrifices while others 'cheat'. What may seem to be a straightforward consideration is hampered by the important matter of data sovereignty. Keohane has stated that information can be regarded as 'a significant variable in world politics' (1984: 245).

Indeed, data can be manipulated so to be used as powerful instruments of coercion. Thus, many states, especially the weak and the suspicious, may think that the global warming data will be used for other ends—goals which do not bring about world-wide collective benefits, but instead entrench some sort of neo-imperialist dependent relationship. Therefore it would seem that some sort of atmosphere of trust is a precondition for satisfactory expectation, which in turn, is a precondition for cooperation.

Role of Efficiency

Energy efficiency gains made over the past 15 years, now save industrialised countries hundreds of billions of dollars per year and made their economies more efficient and competitive. Between 1973 and 1985, oil use in OECD nations declined by 15 per cent while GDP rose by 21 per cent. Over the same period, overall Japanese energy-use declined by 6 per cent while per capita GDP rose by 21 per cent. Energy intensity (energy consumed per unit of output) declined more rapidly for Japanese industry than for the rest of the economy, dropping 4.7 per cent year.

Japanese corporations and entire industrial sectors have made remarkable gains in reducing energy waste. For example, the Japanese steel industry cut energy required per tonne of steel produced by more than three quarters between 1973 and 1986. In fact, Japan has made a focused national effort to improve energy efficiency, thereby reducing oil imports, driving down manufacturing costs and gaining a cost advantage in international markets. The sustained Japanese economic success demonstrates, in part, that increasing energy efficiency—by permitting a shifting of expenditure from energy consumption to productive investment—is a demonstrably successful national competitive strategy that other nations would do well to emulate.

Despite mounting evidence that efficiency is a cost effective investment—not just in industrialised nations—but also in developing countries, OECD development assistance has done little to promote energy efficiency in the developing world. Between 1972 and 1980, while more than 90 per cent of energy financing by multilateral and bilateral development assistance agencies went to large-scale electricity generation projects, less than 1 per cent went to promoting efficiency. And despite increased interest on the part of developing nations and agencies in the potential of efficiency, funding for end-use efficiency remains negligible.

In part, the lack of funding for energy efficiency is due to the decentralised nature of efficiency projects, which makes it difficult to identify investment possibilities and to implement project control. A second reason is that some development assistance funding especially from bilateral aid agencies is intended to serve the dual purpose of both providing assistance and expanding exports from donor nations. These two objectives are expanding an inefficiently-used power supply is a wasteful application of scarce resources.

A 7.7 per cent/year growth in electricity generation in the developing world from 1988 to 2008 would create a potential market worth US$ 908 billion for power equipment and services. However, if electricity demand was held at 4.9 per cent/year the export market for power equipment and services would be cut by 60 per cent. Despite the potential loss in exports for domestic manufacturers, capital shortages and environmental problems are prompting development agencies to turn to energy efficiency as a way to allow economic growth at much lower financial and environmental costs. It isn't news that developing nations are suffering a financial crisis. They face huge interest obligations on mounting debts, reduced borrowing capacity and limited foreign exchange earnings. External borrowing to expand energy consumption accounts for more than a quarter of total indebtedness in many developing nations. Even if these countries could support the debt and interest burdens, external funding to finance sustained rapid expansion of energy consumption is unavailable.

The IMF has estimated that net external borrowing available to developing countries in 1989 and 1990 will be about $45 billion, less than half the level in the early 1980s. A 6 per cent annual growth in power supply (slightly below the average rate experienced through the 1980s), would require an average annual investment of $125 billion/year, of which about $56 billion would have to be in the form of foreign exchange. In other words, foreign exchange costs of the projected power supply expansion in developing nations exceeds the projected available external financing for at least the next few years.

The balance of payments on goods, services and private transfers in developing countries is projected to worsen considerably, from $14.6 billion in 1987 to about $45 billion in 1990. Deepening financial shortages in developing nations make continued rapid expansion of energy supply in most developing nations increasingly difficult, if not impossible, and has begun to force an evaluation of energy efficiency as a more cost-effective solution to energy needs in developing nations.

Potential Role of Energy Efficiency in Developing Countries

A 1988 report to the US Congress on power shortages in developing countries summarised the problem this way:

If current trends of energy demand continue, over $125 billion per year in capital requirements would be needed over the next twenty years. Since developing countries already spend on the average of 25 per cent of their public development budgets on power, such capital expenditures are simply not available.

After concluding that a sustained rapid expansion of electricity supply to developing countries is financially infeasible, the study—by the US Agency for International Development (USAID)—urged an alternative energy investment strategy in which slightly more than half (53.3 per cent) of projected electricity service requirements are supplied by improving energy efficiency rather than by expanding power supply. The financial benefits of emphasising efficiency are substantial. A mixed efficiency-power expansion strategy would cut required capital expenditure on the power sector in developing nations because cutting energy waste costs much less than building and operating new electricity generating plants. According to USAID, this shift toward efficiency would slash required investment in energy in developing nations by $55 billion/year, reducing it from $125 billion to $70 billion, while foreign exchange requirements would be cut from $56 billion to $32 billion.

The benefits of shifting to an energy efficiency strategy alone the lines proposed by USAID are enormous. Efficiency would permit a sharp cut in required capital expenditures, would drive down energy cost inputs in manufacturing, and/or would lessen taxes or national debt burdens. As in the case of Japan, improved energy efficiency would make developing nation exports more competitive in international markets. Major gains in energy efficiency would also translate directly into large cuts in the growth of CO_2 emissions, thereby slowing global warming.

USAID proposed to achieve these gains through a two pronged approach: reducing losses during generation, distribution and transmission, and increasing efficiency of the motors, appliances, etc., that use the electricity. Specifically, USAID envisages a 10 per cent to 20 per cent improvement in plant capacity and a reduction in transmission and distribution losses from 25 per cent currently down to 15 per cent to 20 per cent. These measures would allow projected required electricity expansion to be cut by about one quarter at a cost of between $200 to $300 Kw. This is about one seventh of the $1 830 kW cost estimated for building new power plants.

USAID efficiency costs in production, transmission and distribution are based on experience in developing countries. In contrast, because there has been so little work to improve the efficiency of energy use in developing nations, USAID drew on experience in the USA to estimate a cost of about $300/kW for improving developing country energy efficiency. This raises several questions.

First, given the relatively low level of energy use in developing nations and in view of existing institutional barriers, such as energy subsidies, is it realistic to hope for large gains in end-use efficiency? Are these gains in energy efficiency significantly cheaper than expanding electricity supply? And finally, could energy efficiency have a significant impact on limiting CO_2 emissions in developing countries? This paper seeks to point to answers by reviewing five recent studies on energy efficiency in developing nations. Before turning to energy efficiency potential, this paper briefly reviews two widely discussed electricity generating options: nuclear power and efficient gas turbines.

Rising levels of a range of greenhouse gases, largely produced by human activity, are altering the thermal structure of the earth's atmosphere, and threatening to cause significant, irreversible changes to the world climate. The problem of 'global warming' has come to be accepted as the most serious environmental issue facing the world community today.

The latest model-based projections suggest that the combined effect of anthropogenic greenhouse gases will be equivalent to a doubling of atmospheric CO_2 levels between pre-industrial times and about 2030: after they have remained virtually unchanged for millions of years. The world's average surface temperature, which has altered by no more than one or two degrees since the last ice age, may rise by up to 4.5^0C as a result.

It must be emphasised that these figures, although increasingly familiar, are no more than estimates derived from computer simulations. They may exaggerate the problem, but are equally likely to underestimate it. Climate change is, perhaps, the first completely global phenomenon to confront the world's decisionmakers: it is impossible to identify a limited region in which it has already occurred, to test theories about its possible implications. While past emissions may have already committed the world to a warming of between 1 and 2^0C by 2030, actual changes to date in key observable parameters directly related to climate (global mean surface temperature, mean sea level) have been within the bounds of natural fluctuations. Yet given the thermal inertia of the atmosphere-ocean system (which causes a delay of the order of 50 years between a certain level of warming becoming inevitable and it actually being observed), the world community cannot afford to wait for a more conclusive signal before taking action.

Some warming is already inevitable, and may have already begun. The consequences cannot be predicted with any degree of certainty, but may include major changes in regional patterns of rainfall, and consequent shifts in global ecological zones. The ability of the societies of the world to adapt, particularly the agricultural societies which support the majority of the world's poor, will largely depend on the rate at which changes occur. New crop strains for different rainfall conditions take time to be developed and introduced; irrigation systems take time to adapt and so forth. Large-scale social and economic disruption will only be avoided if changes are relatively small and take place sufficiently slowly.

Thus it is imperative to keep the pace and magnitude of climate change within manageable bounds, and the only available means of doing so is through reducing the rate of greenhouse gas accumulation. Although much research is still required to determine what the limits are, what rate of greenhouse gas accumulation may be tolerated, this article begins from the premise that some reduction is essential, to reduce the risk of catastrophic changes.

FEEDBACKS: CLIMATIC CHANGES THAT CHANGE THE CLIMATE

The warming of the earth will change physical, chemical and biotic processes locally and globally. The question is whether the net effect of such changes will be significant in speeding or slowing the warming. Any negative feedback will work to reduce the severity of the warming by reducing the concentration of carbon dioxide or other heat-trapping gases in the atmosphere. The possibility that there are negative feedbacks in the climatic system has been explored by the climatologists. Such feedbacks include increased cloud or snow cover that increase the reflectivity of the earth and reduce the energy reaching the surface. There is also the possibility that changes in the patterns of circulation of the oceans will change the rates of exchange of atmospheric carbon dioxide with the carbon held in various forms in oceanic water. These changes are less predictable and have not been incorporated into any estimates. Both include the possibility of effects well outside the current realm of systematic analysis.

There is a further set of potential negative feedbacks involving biotic interactions. The most important such effects has been assumed to be the stimulation of the fixation and storage of carbon by terrestrial plants by the increase in carbon dioxide in the atmosphere. Data in support of such a mechanism come largely from experiments with plants under controlled conditions. The data show that increasing concentrations of carbon dioxide often, but not always, increase the rate of growth of young plants. The effect declines with time and with increasing age of the plant. It appears to have little influence on forest trees. There is not direct evidence from natural communities confirming a general effects from the increase in the concentration of carbon dioxide in the atmosphere over the past century. Any such effect is assumed to be small; it has not compensated for the increase in carbon dioxide in the atmosphere to date in any measurable way. Positive feedbacks will speed the warming of the earth; the warming makes the warming worse. They include any effect that increases the release of carbon dioxide or other infra-red absorptive gases into the atmosphere or otherwise increases the retention of heat on the earth's surface. Changes in the circulation of the oceans have such a potential, but the effects are at the moment beyond prediction. If the circulation of the oceans were to change, it might occur suddenly and have profound effects on local and regional climates as well as on the composition of the atmosphere.

Biotic feedbacks are more predictable but still uncertain. The most important consideration is the ratio of gross production (total photosynthesis) to total respiration globally. Gross production is the total amount of photosynthesis carried on by plants without considering the other metabolic processes necessary to maintain the plants. These latter processes are usually added to the respiration of all animals and decay organisms and recognised as "total respiration". If gross production exceeds respiration, carbon compounds are accumulated in the large store of carbon held in plants and soils; if respiration dominates, the store of carbon is depleted. The relatioships among these sometimes confusing attributes of ecosystems.

The factors that affect these two processes have been discussed in earlier publications. In general gross production is easily affected by virtually any disturbance. The effect is commonly a reduction; it is rarely a stimulation. Specific, overt efforts are usually required to increase primary production and the effectiveness is usually limited. Such efforts include irrigation, fertilisation with nitrogen or potash, or removal of competition with other plants. How will the ratio of gross production to total respiration shift as the earth warms? As in most of science absolute proofs are elusive but insights are available from first principles of ecology, from the record of changes during the glacial period, and from current experience.

Two changes are occurring simultaneously and we wish to know which is more important. There is an increase in the concentration of carbon dioxide in the atmosphere. The increase is appreciable; it has been more than 25 per cent over the past century. At the same time there is a warming, apparently about 0.5 degree globally over the same period. Which of these changes has had a greater effect on the earth?

Strain and Cure have reviewed the evidence for the effects of increases in the concentration of carbon dioxide on rates of photosynthesis. There is little question that young, rapidly growing plants, especially herbaceous plants, usually benefit from increasing quantities of carbon dioxide in the air. The stimulation of growth is limited, however, to the early stages of growth, is small in perennials, and is reduced or disappears in later years. There is no evidence, presented by Strain and Cure or others, that there is a general stimulation of the growth of perennials due to the increase in carbon dioxde in the atmosphere. The suggestion that an apparent increase in growth of trees at high elevations in the Rocky Mountains of North America is due to increased carbon dioxide is unique and unsubstantiated. Differences in growth rates in such circumstances have normally been correlated with the availability of water.

Nonetheless, a warming accompanied by an increase in the amount of precipitation in the middle and higher latitudes can be expected to offer at least temporarily improved conditions for growth of certain trees and might result in stimulation of net primary production. Such an effect would be expected for a example in the central range of the boreal forest. A stimulation of net primary production under the most favourable circumstances of 20-25 per cent might occur. Net primary production, however, refers to the amount of organic matter remaining from total photosynthesis or gross production after the needs of the plant community for respiration have been met. An increase in net primary production does not assure that there is an increase in net ecosystem production, the amount of carbon added annually to a unit of landscape. Net primary production is the amount of organic matter available annually for all decay organisms and other consumers, including man. The storage of carbon in a forest (net ecosystem production) will hinge not only on the net primary production, but also on the total amount of respiration carried out by decay organisms and other consumers (respiration of the heterotrophs). In the normal course of plant succession leading to forest from in abandoned field or forest clearing total respiration of the ecosystem increases until it balances gross production. At that point the forest is no longer accumulating additional carbon and is considered to be at "climax", although further changes in the array of species may occur.

Respiration of especially sensitive to both temperature and the availability of water. If moisture is available, an increase in temperature may increase the rates of respiration by 10-35 per cent or more per degree C increase in temperature. This increase affects both the respiration of plants and that of the organic matter in soils. An increase in the rate of respiration of plants reduces net primary production and net ecosystem production; and increase in the decay of organic matter in soils reduces net ecosystem production. While data are few, there is enough carbon in soils of the boreal forest for such a stimulation of decay to equal or exceed any stimulation of net primary production. There is nothing inconsistent in a circumstances in which both net primary production and total respiration are stimulated by the combination of increasing amounts of carbon dioxide in the atmosphere, additional precipitation, and higher temperatures, with the net effect being an increase in the release of carbon from the ecosystem because the stimulation of respiration dominates.

I have discussed previously the transitions to be expected at the warmer and dryer limits of the distribution of forests as the climate warms. The rates of warming in these latitudes are expected to be high, possibly as high as 1 degree C/decade equivalent to moving 100-150 km (60-100 miles) in latitude. At the warmer and drier limit of the distribution of a forest the climate would change sufficiently to kill living trees and destroy the forest. There is no possibility that forests can be regenerated elsewhere at an equivalent rate. The net effect is a further release of carbon into the atmosphere as the warming progresses.

These considerations of forests and a parallel set of analyse of tundra suggest that the probability is high, that a warming will stimulate the respiration of terrestrial ecosystems, including the decay of organic matter in soils, sufficiently to exceed any stimulation of net primary production. The effect will be a further release of carbon as the earth warms and as the respiration of terrestrial ecosystems increases.

The analysis above seems to be supported by data from the Vostok Core, a sample of ice that reached down through the glacial-age ice of the Antarctic to sample state through the most recent 160,000 years. The core has provided data on temperature, carbon dioxide and methane concentrations throughout that period. The data show that as temperature rose, the concentrations of methane and carbon dioxide rose as well; as temperature dropped, carbon dioxide and methane concentrations dropped. The observations are consistent with the suggestion of a positive feedback; a warming cause an increase in the amount the of the heat-trapping gases in the atmosphere; a cooling results in a decline in their concentrations. But the data do not confirm the cause of the warming, the colling, or the causes of the shifts from warming trend to a cooling trend and back.

The warming during the 1980's has been abrupt and severe by any standards. It has been sufficient to be expected to generate a surge in the rate of accumulation of carbon as carbon dioxide and methane in the atmosphere. Data on carbon dioxide in the atmosphere suggest that such a surge is occurring.

Over the past 15 years there has been an annual accumulation of carbon dioxide in the atmosphere or about 1.5 ppm. This accumulation is equivalent to about 3.0 x 1015 gC annually, slightly more than 50 per cent of the release from fossil fuels alone, but probably considerably less than 50 per cent of the total release of carbon due to human activities. Recently the data from the monitoring station at Mauna Loa and at the South Pole have both recorded increases in the rate of accumulation of carbon in the atmosphere. The observation has extended over 18 months, according to Keeling, and suggests a new rate of accumulation of 2.4 ppm or about 5.0 x 1015 g C/yr, a 60 per cent increase above the rate observed in the past. Such a surge is not surprising in view of the speed of the warming observed over the last decade.

What is surprising is that the correlation has not been observed previously in response to earlier changes is regional and global temperatures. Whether the higher rate of accumulation will continue remains to be seen. A previous surge in 1973-4 proved transitory. The cause of the current surge will be subject of much speculation and research over the next years. At the moment the most probable explanation seems to be the biotic feedback outlined above, although the issue is far from resolved.

Which Impacts Should Worry Us Most?

There is by now a large literature on the impacts of climatic change on agriculture. The findings in the literature are not alarming, in large part because farmers are not dumb; they are accustomed to continually adapting to variations in markets and technology, as well as in environment and climate. For the most part, projected impacts of climate change on agriculture are small or moderate in comparison with impacts associated with fluctuations in government subsidies, international trade and other economic factors or with productivity gains that come with technological progress. Also, for a lengthening list of nations, agriculture is no longer the outdoor activity that is of most value. In nations such as the United Kingdom and Germany, agriculture is only about 3 per cent of gross national product, so impacts of climate change on agriculture pose little threat to national income. Travel and recreation may prove to be sources of jobs and revenue that are more vulnerable to climatic change than agriculture in several regions.

The consequences of sea-level rise are on a grand scale indeed, if one uses scenarios in which the sea level rises several meters. A quarter of Florida might be flooded if the ocean were five meters higher. The probable sea-level rise over the next 100 years—the rise that would be associated with a global temperature increase of 3 or 4 degrees Celsius—is about 70 centimetres. Of this rise, 33 centimeters are expected by 2050. Impacts from such rises are a concern to be addressed, but hardly merit images of the Great Flood. The impacts of sea-level may be scarcely apparent at all for the next 30 to 40 years. The fascination with sea-level rise probably originates in the fact that it could be disastrous, but it is mistaken to conflate coastal inundation with the likely impacts during the next generation or two.

Agriculture and sea level have captured much of the popular attention, but a more serious concern may be water resources. A 1989 comprehensive study of climate and water in the United States by the Panel on Climate and Water of the American Association for the Advancement of Science shows that the prospect of climatic change lights warning lamps of vulnerability in several regions. Vulnerability depends on the ratio of water supply to demand, and climate can affect both quantities. In a region such as the Missouri River basin or California, where the balance is already delicate, even small changes could bring substantial stress.

Another sector that looks worrisome is conservation of nature. Although agriculture and economic activity natural heritage—exemplified by national parks such as the Terai in India, Serengeti in East Africa, and Yosemite and the Everglades in the United States—cannot. Changes in variables, such as fire, that affect ecosystems, and changes in the delicate balance that maintains systems such as wetlands could have regrettable consequences.

It is interesting also to note that among the highest cost of *cooling* found in the CIAP study were loss of forest production and marine resources. Largest of all were the economic costs of impacts on health and of wages. The high estimates of costs of health impacts from cooling arose from findings that there is a covariation between several climate variables and expenditures for physicians' services. Mortality rates also were related to various measures of climate. The central hypothesis in the analysis of wage effects was that if the cost of living rises as one moves to different climatic regimes—as seems the usual case with a move to a cooler climate—then wage earners will have to be paid a differential to compensate for this difference.

In summary, the conventional wisdom seems to rely on preconceptions about which climatic impacts will be most serious. Many commentators and experts on global warming evidently ignore or disbelieve the results of the scholarly literature. The Global-warming issue may provide cultural anthropologists with an interesting case study of the ways we protect paradigms and reject information that does not fit. There is definitely a need for more systematic, through studies world-wide of impacts, together with adaptation, from which it would be possible to identify the greatest vulnerabilities.

The Importance of Extremes

Statements made about climatic change are often based on the notion that what matters more is not the average behaviour of the system, but its behaviour at its limits, which may be masked in consideration of averages. Important changes in extremes, it is noted, might come about in either of two ways. A shift in a climate mean (for instance, in the regional average spring temperature) could slide the complete intact distribution of climate events in such as way as to cross more frequently thresholds that are now rarely breached. Alternatively, the shape of the distribution of events might change in the direction of increased vulnerability.

So far, there are no more than a handful of papers in the scientific literature that directly examine changes in climatic extremes and their impacts. Linda Mearns, Richard Katz and Stephen Schneider in 1984 looked at the example of a change in the probability of extreme heat waves (five days or more in a row with a maximum temperature above a 95- or 100-degree Fahrenheit threshold). They examined what might happen in three US locations if the mean temperature were to increase by 3 degrees Fahrenheit, while there were not changes in the standard deviation of daily maximum temperatures nor in the autocorrelation of daily temperature variations. The results are worrisome; the odds of extreme heat waves would double on triple in each city. It would be useful to carry out more exercises of this type, emphasising thresholds that climatic variations may cross. In addition to heat-stress levels for crops, an obvious example of such a threshold is the melting of permafrost. It would also be useful to study whether there might be states of the climate system that would be characterised by more frequent and severe storms.

Will temperatures and precipitation become more variable with global warming? To the extent that there are research results touching on this question, they do not indicate an increase in variability. In fact, allowing for considerable noise in the results, they imply qualitatively that standard deviations of surface temperatures are more likely to decrease than to increase as a result of greenhouse warming. Precipitation variability, on the other hand, usually increases as climate warms. Of course, if the climate becomes drier on average, the chances of extremely dry weather are likely to increase, unless the rules of the system change.

In summary, we need to identify important thresholds and discontinuities, to study the likelihood that the altered natural system will reach these points, and to do more work about what consequences might, in fact, ensue. An "extreme," by itself, has little meaning. If summer temperatures in Siberia regularly breach 80 degrees Fahrenheit rather than 70 degrees, are there reasons to be concerned? Would, for example, large releases of biogenic gases be triggered from the soils?

The Greenhouse Effect and Climate Change: Two Scenarios

Any global warming resulting from a greenhouse effect will probably be uneven, with an increase of only 0.5^0—1^0C at the Equator, 2^0—3^0C at temperate latitudes and as much as 4^0—7^0C in the polar regions [17,18]. The result of such a distribution may be to shift climatic zones and rainfall patterns. Changes in the seasonal distribution of precipitation within regions could also be significant. Over the long run, the polar regions could experience the greatest changes, and there is the possibility that the polar ice caps may melt sufficiently to raise world sea levels. Thus, it is worth distinguishing two potential scenarios: the 'most likely' scenario in the short run of changes in regional precipitation and climatic zones, and a more 'dramatic' long-run scenario of a global sea-level rise.

It indicates that, as a result of global warming, the most extreme temperature changes would probably occur during winter in the high latitudes of the Northern Hemisphere. In contrast, temperature changes in the low latitudes will probably be less drastic and less seasonally variant. Although forecasting regional precipitation is full of uncertainties, the most likely scenario is for increased winter precipitation in the high latitudes, intensified rains in the low latitudes—except in semi-arid regions—and a decrease in summer rainfall in the mid-latitudes.

Thus for the mid-latitude regions of 30^0 to 60^0N, which include the major food-growing regions of North America, Europe and parts of Asia the rapid rate of change in temperature will probably have the most severe impact on forest ecosystems and on irrigated agriculture in the semi-arid areas of the mid-latitudes (eg the American Midwest), which will suffer from higher temperatures and increased drought in summer. In contrast, as climatic variability is already a major problem for the semi-arid tropical regions (latitudinal bands 5^0—35^0N and S), the predicted decrease in precipitation for one or more seasons will adversely affect agriculture and possibly the availability of fuelwood. This would be aggravated further by the decline in soil moisture from increased temperatures. On the other hand, increased rainfall intensity in the humid tropical regions would means more tropical storms and flooding, which would affect coastal and river regions and highly erodible soils in upland areas.

Such a scenario has been more or less confirmed by geological studies of past climate changes. For example, one early analysis suggested that there could be appreciable decreases in precipitation corresponding to increases in temperature and evaporation, particularly in latitudinal bands centred on 40^0N and 10^0S. On the other hand, precipitation may actually increase in regions between 10^0 and 20^0N and below 30^0S. More recent studies of past warm periods have refined these regional variations to reveal three interesting trends:

- In the US Midwest, the annual precipitation was 10—20 per cent less than it is today, with a climate pattern similar to that of the 1930s Dust Bowl period.
- In Africa and India, annual precipitation may have been as much as 300 mm (50—70 per cent) higher than it is today, with both the Rajasthan and Saharan Deserts covered by dry savanna and steppes.
- Winter temperatures over a substantial part of Europe were colder and more variable interannually; precipitation was decreased in spring and summer and increased in autumn and winter.

Perhaps the most dramatic consensus in recent years has been the general agreement in the scientific community that global warming should cause a rise in the global mean sea-level. The major causes would be the thermal expansion of oceans, and possibly the melting (or accumulation) of land ice. The sea-level has probably already risen 7—17 cm during the twentieth century, although it is not possible to attribute this solely to the greenhouse effect. On the basis of these observed changes, it is assumed that the predicted global warming of 1.5^0—3.1^0C would lead to a sea-level rise of 20—165 cm The 'best guess' for sea-level rise by the year 2030 is 20—40 cm; however, the rise could conceivably be as much as 1.5 m. The results would be erosion of beaches and coasts; land use changes, particularly in coastal and river regions; wetland loss; increased frequency and severity of flooding; damage to port facilities, coastal structures and water management systems. Low lying and island countries would be the worst affected by such changes.

In an extremely pessimistic scenario, a temperature rise of 2^0—4^0C due to an increase in atmospheric carbon dioxide to 600 ppmv or more would cause the polar ice caps to melt sufficiently to raise the sea-level of the major oceans by 5 m or more. However, disintegration of the West Antarctic Ice Sheet is not thought to be imminent, and if it were to happen, would take a century or more. On the other hand, some disintegration of the West Antarctic Ice Sheet has been observed; along a 1200 km ice front on the Sheet some 500 km^3 of ice are being deposited each year into the sea as huge icebergs. This should result in a rise in the sea-level of 1.5 mm per year, which has in fact been observed by independent studies of global tidal patterns. The above trend is inconclusive, and further research is required before a reliable assessment of the possibility and timing of this scenario can be made.

Short-run Effect: Agricultural and Ecological Disruption

The most likely impact of any global warming in the near future will be on world rainfall distribution. As rainfall patterns change, climatic zones will shift and the Earth's principal areas of agricultural cultivation and vegetation cover will be displaced.

Agriculture

The predicted changes in regional temperature and precipitation patterns summarised and discussed in the previous section suggest that several major producer regions are at risk from global warming. These include:

- the irrigated semi-arid areas of northern mid-latitudes, particularly in the US Midwest;
- the lowland areas and island countries of the humid tropics in Asia, the Pacific and Caribbean that are susceptible to excessive precipitation, violent storms and flooding;
- the arid and semi-arid tropics of Africa, South Asia and the Mediterranean climate of West Asia and North Africa that are already vulnerable to climate variability;
- rainfed upland and highland regions, particularly with poor soil conditions; and
- livestock raising in extreme northern latitudes.

The last four are especially singled out by Oram.

For example, Bryson estimated that the 0.1^0C rise in global mean temperatures between 1957 and 1970 due to increased atmospheric CO_2 should have led to a decrease of 86 mm of precipitation annually in the Sahel of West Africa. In fact, the actual decrease in precipitation in the Sahel over this period was 96 mm. Similarly, Glantz and Ausubel have argued that if atmospheric CO_2 accumulation increases the frequency, duration and severity of droughts in the Great Plains of the USA, more rapid depletion of the groundwater contained in the Ogallala aquifer, an underground geological formation of water-bearing porous rock, would occur. The aquifer is a source of groundwater reserves could have serious consequences for agriculture.

It summarises the expected flow changes in some of the world's major river systems as a result of global warming. All of the affected river systems form the basis of highly productive irrigated agricultural systems. Changes in the flows or storage capacities of these rivers would result in either less surface and underground flow or more frequent flooding which could be disastrous for irrigated agriculture. In many countries irrigated agriculture systems have been intensively developed in correspondence with the precise seasonal variations of delta flooding and runoff. Changes in seasonal flows would mean severe disruptions in cultivation. Moreover, many river systems do not have adequate water control and management facilities to cope with any variations in flooding or underground storage. For example, in the lowlands of the humid tropics, the result could be destructive flooding in the vast delta networks of Thailand, Laos, Cambodia, Vietnam, India, China and Bangladesh.

More detailed analysis of the impact of temperature rises on the river systems of the semi-arid western and mid-western states in the USA suggests that only modest changes in temperatures are necessary for severe disruptions in agriculture to occur. For example, a 2^0C warming combined with a 10 per cent decrease in precipitation would cause surface runoff into rivers to decline by between 40 per cent and 76 per cent. Especially for the Rio Grande, Colorado and Missouri rivers, the results could mean water requirements exceeding supplies by 20-270 per cent. In the region of the Colorado River, such a climate change would be particularly severe. Currently in this area, 85 per cent of the rainfall evaporates and only 15 per cent flows as runoff to feed the Colorado River. Even this flow has to be backed up by large volumes of reservoir water in order to meet the present agricultural demands in the region. As noted above, however, the region's main groundwater source, the Ogallala aquifer, cannot be expected to sustain any increase in demand over the long run.

However, with the possible exception of semi-arid irrigated agriculture in the western and mid-western USA, the overall impact of global warming on temperate agriculture in the mid-latitudes of the Northern Hemisphere is not clear. For example, for many crops, such as rice, wheat, alfalfa, maize, sugarcane and sorghum, productivity losses due to decreases in rainfall and increase in evaporation may be offset by the increasing photosynthesis and improving water efficiency in plants due to raised CO_2 levels. At higher northern latitudes, rising temperatures may allow lengthening of the growing season and thus possible expansion of rain-fed areas, whereas in the lower latitude zone of the 30^0 to 60^0N band agriculture might be adversely affected because of increased evapotranspiration.

Nevertheless, it is clear that farmers in the mid-latitudes of the Northern Hemisphere—in response to any greenhouse effect—will have to modify seed varieties and cultivation methods, change crops and move agricultural operations away from the worst affected areas. As Oram has pointed out, this in itself may cause significant declines in productivity first, because it is by no means certain that a rise in temperature would enable large new areas of land at higher latitudes to be brought into cultivation, because in higher latitudes land at the margin of cultivation is often of very low fertility, with low pH and anaerobic conditions due to low drainage; and second, as modern agriculture has become more sophisticated, there has been a trend away from trying to breed for wide adaptability toward tailoring varieties to favourable agro-climatic conditions.

Similarly, a study by the US National Academy of Sciences has concluded that, although the direct impact of CO_2 enrichment in the USA to 400 ppmv by 2000 may be to increase yields of well-tended crops by 5 per cent, the accompanying 1°C rise in temperature and decrease in precipitation by 5-10 per cent may actually produce a negative net effect on agricultural production. Other crop impact analyses also show that warmer average temperatures are detrimental to both wheat and maize yields in the mid-latitude regions of North America and Western Europe. Given current technology and crop varieties, a sudden warming of 2°C with no change in precipitation might reduce average yields by 3-17 per cent.

However, the greatest concern is the sensitivity to climatic change of the marginal agricultural lands in the developing world. The particular uncertainty for those lands in how global warming might interact with the increasing desertification that is currently affecting both rain-fed croplands and rangelands in the arid and semi-arid tropics. Currently, 1300 million hectares (ha) of rangelands (35 per cent of their dryland total) and 170 million ha of rain-fed croplands (30 per cent of their dryland total) are severely or very severely desertified, i.e. have lost more than 25 per cent of their productivity and require extensive reclamation.

The areas most vulnerable are the semi-arid and arid humid tropics—one of the regions at great risk from the greenhouse effect. By 2000, desertification is expected to have accelerated greatly in the rangelands of Andean South America and in Africa south of the Sudano—Sahel region, and to a lesser extent, in the Sudano—Sahel region and in parts of South Asia. For rain-fed croplands, desertification is already accelerating in tropical Africa, South Asia and South America.

In sum, any global impacts on agriculture and human welfare from a global warming will undoubtedly be unevenly spread. As Oram has pointed out, in analysing these impacts, 'the first step would be to look at global supply and demand for food with emphasis on cereals for food and feed, and on the major consuming and exporting countries'. The two key groups to look at would be the major Northern Hemisphere food-exporting countries and the low-income, food-importing developing countries.

Clearly, the ability to adapt agriculture to changing climatic and environmental conditions will also be unevenly distributed between those two groups. 'In a tight grain situation prices would rise and richer countries would have the first call; little might be left for concessional sales or aid to the developing countries. Internationally, the major food-exporting countries would probably benefit substantially; the food importers with weak bargaining power, and the poor in those countries, especially the landless laborers, would be the chief sufferers'.

For example, although the wealthier nations of the Northern Hemisphere could face drastic changes in areas of arable land, the distribution of water resources and the type of products cultivated, these countries tend to have a surplus of land available for production as well as accumulated surplus stocks of some produce, highly developed agricultural R & D infrastructure and techniques, efficient marketing, credit and information systems and extensive water management and control systems. It is more likely, therefore, that Northern Hemisphere nations will be able to meet most of their own agricultural and food needs despite any adverse climatic and environmental changes.

It is also likely, however, that the need for those countries to adjust their agricultural production to such changes may involve less production for export and thus less global supplies of many essential agricultural products. In addition, any adjustments to major changes in climate, precipitation and temperature will require extensive modifications in existing patterns of cultivation and water management. For Northern Hemisphere countries this may mean diverting resources away from other areas of economic activity and a corresponding rise in the costs of agricultural production.

The most serious repercussions, however, may be on the export and distribution of world food and feed supplies, particularly cereals. Cereals are the dominant crop in the global agricultural systems, and wheat is the major surplus food commodity both in world trade and aid. As shown the majority of global cereals, including wheat, are produced in the temperate zone of the Northern Hemisphere. Thus given the surplus production in places like the USA, Canada and Western Europe, it is not surprising that wheat has become a major feed crop in industrialised countries and the main fallback for much of the world in terms of reserve stocks and contingency aid. Wheat and barely are also extremely important staple crops in the developing countries of North Africa and the Middle East, as well as in Pakistan, Northern India and China. By contrast, in the semi-arid tropics—the region at high climatic risk from carbon dioxide induced climatic changes—maize, sorghum, millet, pulses and groundnuts are the major staples. But countries of that region are also heavily dependent on cereal (mainly wheat) imports and food aid.

Finally, although many countries of the lowland humid tropics—especially in Asia—are major producers of rice and other crops, for some of those countries self-sufficiency may be lost if their main food crops (e.g. rice, cassava, sweet potatoes, yams) are adversely affected by increased intensity of rainfall, tropical storms and flooding. This suggests that, during periods of global climatic instability, the failure or unwillingness of food-exporting Northern Hemisphere producers to supply, at reasonable prices, the rest of the world with cereal imports or aid could have important implications for global food security.

Thus the food security needs of the very poor in all developing countries, and the food-deficit low-income countries in particular, are vulnerable to the agricultural shocks and stresses posed by climate instability. Between 1980 and 2000, increased total food demand is projected to exceed the growth of food output in all developing regions except Asia, despite increases in per capita food production in Latin America, North Africa and the Middle East. Even though Latin America is expected to have the highest food production per capita growth rate, demand is projected to increase even more rapidly. With its rapid population growth, sub-Saharan Africa's food consumption is estimated to grow at 3.6 per cent year, substantially outpacing the projected growth in food output. Per capita food production in the region is expected to continue to decline.

Thus, even in the absence of disruptions to agriculture from climate change, many developing regions will continue to be dependent on food imports, and in some instances external assistance, to meet domestic consumption needs. The lack of food security—defined as the access by all people at all times to enough food for an active, healthy life—may therefore increasingly arise from a lack of purchasing power on the part of nations and households rather than from inadequate food supplies. The disturbing facts are that food security is already becoming worse in many developing countries, despite higher per capita food production; and, despite record levels of world food production and excess supplies, about 730 million people in developing countries do not obtain enough energy from their diet to allow them to have an active working life. About two-thirds of the undernourished live in South Asia a fifth in sub-Saharan Africa; four-fifths of the undernourished live in countries with very low average incomes.

Although all 65 low-income food-deficit countries identified by the Food and Agriculture Organisation of the United Nations (FAO) are potentially vulnerable to climate instability, those in that group that have failed to increase per capita food production are especially at risk. Increasing food production in those at-risk countries is essential to meeting their long-term food needs as their ability to purchase food imports is unlikely to improve in the near future. Moreover, data up to 1982 suggest that adequate growth in domestic cereal production and a healthy growth in export earnings generally go together, and both are important determinants of the ability of a country to ensure food security. Unfortunately, most of the at-risk countries have substantial agricultural lands in arid and semi-arid regions or at high altitudes—areas that face the most risk from a global warming.

To summarise, the economic impact of any greenhouse effect will most likely be in terms of the rising costs of agricultural displacement and adaption in the face of climatic instability, with potentially grave consequences for the pattern and distribution of global food production and for food security in many developing countries. As a result, sustaining even minimum nutritional standards for millions of people in the Third World will require a surplus in world agricultural production and the availability of that surplus to the Third World in the form of food or aid imports.

Hence any decline in production in Northern Hemisphere countries may either perpetuate a global food shortage or raise food prices beyond levels affordable by the low-income food-importing developing countries. In either situation, the ability of millions to survive in the Third World will depend on the political will of the Northern Hemisphere countries to ration existing global supplies at lower prices and to establish an international system to food security for the world's population. Northern Hemisphere countries will not only have to allocate more economic resources to adjust their own agricultural systems to the changing global climate, but will also have to be willing to expend additional resources in support of a global food security system and an expanded programme of agricultural investment in developing countries. The alternative is to allow developing countries to bear the brunt of the costs of global warming.

Forests

The impact of climatic change of the world's major ecosystems would also affect human welfare. This is particularly true for major forest systems, which are so crucial in stabilising regional climate. For example, the boreal forests found at higher northern latitudes may be particularly vulnerable. Although current evidence is inconclusive, some studies indicate that rising temperatures might cause the boreal forests to be replaced by either cool temperate forest or steppes, depending on the accompanying changes in precipitation. Warmer conditions at high latitudes could lead to large reductions in the areal extent of boreal forests and to a poleward shift in their boundaries. In contrast, a study based on Icelandic conditions estimates a net increase in growth of boreal forest systems as the climate warms. Nevertheless, any major change in the boreal forests, whether a net growth or decline, will most likely affect the overall ecological balance of the higher northern latitudes and lead to important feedback effects between local climatic zones and vegetation cover.

For the mid-latitude temperate forests in the Northern Hemisphere, a high rate of warming of 0.8^0—1^0C per decade will have major impacts—including large-scale forest dieback between 2000 and 2050. The result would be that more and more production forest would need replanting and increased management. Furthermore, there will also be impacts on recreational values if important national parks and wildlife habitats are decimated.

There could also be dramatic impacts on the major tropical rain forests of the world, such as the extensive Indonesian, Amazonian and Central African tropical forest systems. A typical rain forest system, such as Amazonia, is dependent on only small variations in average annual temperatures (e.g. 3^0C) and high annual rainfall (e.g. 1500-3000mm). Changes in these conditions would destabilise the unique climate and nutrient-cycling systems of tropical forests, and may also affect neighbouring ecosystems and climate zones. Because tropical deforestation is expected to increase over the coming decades, the effect of the additional stress induced by the greenhouse effect may be to cause a more rapid rate of tropical deforestation than might otherwise be the case. Thus tropical deforestation not only will be an important contributor to the greenhouse effect, tropical forests themselves may fall victim to any resulting global warming.

THE ROLE OF ENERGY

The relative significance of the various sectors and trace gases responsible for global warming, based on their respective contributions to the total warming commitment caused by human activity over the period 1980-2030, assuming current trends continue.

It is estimated that roughly 50 per cent of projected global warming over the coming 40 years will be caused directly by the energy sector. This proportion may increase, since action is already being taken on CFCs, one of the main greenhouse gases, through initiatives to combat the destruction of the ozone layer. Provided that substitutes found to replace CFCs also have a lower greenhouse impact, this will substantially reduce the industrial sector's direct contribution.

The origins and behaviour of two other main greenhouse gases, methane (CH_4) and nitrous oxide (N_2O) are still relatively unknown. Much of the increase in CH_4 levels is thought to be due to expanding agricultural activity in the Third World, and thus reducing these emissions would be extremely difficult. Current initiatives must, therefore, focus on CO_2.

Direct extraction of CO_2 from the atmosphere and from sea-water has been shown to be impractical, since the energy input required would effectively create more CO_2 than would be removed. The creation of biomass 'sinks' to absorb CO_2 through photosynthesis is a more promising extraction-based approach, but the land required and economic implications are formidable: particularly in the light of limited progress to date in reducing the rate of destruction of existing forests. Much attention has recently been focused on processes involving marine biota, but a consensus has yet to emerge on their precise role.

An increase in global biomass, as the only feasible way of increasing the rate of 'reprocessing' of atmospheric carbon, will be needed to ensure a stable carbon cycle in the long-term..In the immediate future, however, the priority is clearly to reduce the rate of CO_2 emissions, which are dominated by the energy sector. Climate change is, primarily, an energy issue. The remainder of this article focuses on the role of the energy sector in CO_2 abatement. It will be argued that this is a substantially more complex issue than other current problems of transboundary air pollution, such as acidification and ozone-depletion, and that this complexity provides both obstacles and opportunities.

The Montreal Protocol on Substances which Deplete the Ozone Later and the Helsinki Protocol on Sulphur Emissions in Europe are frequently cited as examples of successfully international initiatives on air pollution issues, providing a possible framework for a programme of action on CO_2. While these initiatives do provide valuable lessons, the qualitative differences between the problems may call for a somewhat different approach to be taken in reducing CO_2 emissions.

Energy Policy Scenarios

So far we have seen the relationships between the structure of Indian economy and CO_2 emissions as they are captured in the input-output matrix of 1983-84. We have seen which are the largest emitting sectors. After inverting the matrix, we also saw which final demand selectors are the most CO_2 emitting. How do these structural details help to formulate policy for CO_2 reduction? That is, given the structure of the system, how will it respond to external policies such as reducing CO_2 emission? We consider two types of policies for reducing CO_2 emissions:

- Reducing use of either coal or oil by conservation of the same amount of useful energy, so as to highlight the difference between these two conservation alternatives.
- Substituting oil for coal, a fuel with higher CO_2 emissions than oil, keeping the same energy use in the economy.

It is assumed that these policies can be effected without affecting the levels of production. Undoubtedly, it will take some time before the changes are made, and some think that, for this reason, a simulation of scenarios for, say, the year 2000 or beyond is necessary. However, this compounds two uncertainties—the accuracy with which the scenario for 2000 can be constructed and the accuracy with which the effects of policies can be captured. To avoid being diverted into another exercise of generating scenatios for the future, we consider the following problems in the base year itself, 1983-84, assuming these policies work instantly. What can we learn about the effects of CO_2 reduction policies? This question is addressed for two separate policies-energy conservation of coal and oil, and energy substitution of coal with oil.

Energy conservation has been a goal of India's energy policy for some time. Several organisations have been established. One of these is the Petroleum Conservation Research Association (PCRA) under the Indian Oil Corporation, which is the oil refinery and distribution agency. This represents the priority for energy conservation. Where conserving oil/products is seen as saving precious and scarce domestic oil resources which cater for only 50 per cent of consumption, and the foreign exchange required to purchase the remaining 50 per cent at international prices.

However, conserving coal in not one of the major goals of India's energy policy. Since coal generates more CO_2 per GJ than other energy resources, saving coal for the sake of global warming would require a shift in India's energy conservation policy. To assess the implications of this shift, we compare two scenarios, both of which save the same around of useful GJ (107 million GJ). To save this energy from coal, we would require a 10 per cent reduction of coal use (i.e. 10 per cent of 121 million tonnes of coal). To save this energy from oil, a 13.5 per cent reduction in oil use would be necessary. We compare saving 250 million GJ from coal with saving the same amount from oil. The comparison, along with base case results with no such policy. Highlights two scenatios in which 107 million useful GJ are saved from coal and oil respectively, assuming that the relative efficiency of oil to coal is 2.35. Since the cimomy uses much more coal than oil, despite the lower heat value, only 10% coal savings are required to save the same useful GJ as a 13.5% reduction in oil consumption. In actual amounts, this translates into 12.1 million tonnes of coal and 2.7 million tonnes of oil. It will be assumed that it is possible to do this without incurring capital costs, or that they have similar capital costs. Both results are compared with the base case (i.e. no energy savings). It can be seen that 6.6 million tonnes of carbon will saved if this energy is saved from coal, but only 2.18 million tonnes of carbon are saved if it is saved from oil. Both save the same amounts of GJ, and values of fuel used are much lower compared to the base case for both the cases. However, more valuable fuel in the oil-saving scenario is saved—Rs 7.7 billion compared to the base case. On the other hand, only Rs. 2.4 billion would be saved if this much energy were saved from coal. Clearly, it is in India's interest to save oil rather than coal. The marginal cost of the additional carbon savings from coal as opposed to oil is Rs. 1150 per tonne of carbon (US$I = Rs. 10 in 1983-84).

However, one should look beyond the amount of carbon savings and at details. For example, in the case of saving coal, it would be necessary to save carbon from the power sector (2.7 million tonnes of carbon), iron and steel (0.8 million tonnes), coal tar (0.7 million tonnes) and 0.2 to 0.3 million tonnes each from non-metallic minerals, railways and cement.

On the other hand, if one has to save this carbon from oil, a completely different sectoral strategy is required. Most attention will be required for the transport sector (0.9 million tonnes), iron and steel (0.7 million tonnes), and fertilisers (0.2 million tonnes). This diversion of attention to different sectors will also tranlate into diversion of efforts manpower, fuels, technology development etc. Thus the GHG reduction savings policy would be at variance with India's energy conservation policy, which has emphasised savings oil.

Fuel Substitution Scenarios

Currently India follows the policy of substituting coal for oil because of an abundance of coal resources and inadequate oil production, resulting in large imports. When oil cannot be substituted by coal directly, it is done through electricity production. For example, diesel pumps have been systematically replaced by electric pumps over the last 15 years, oil-fired biolers, and so on. Even diesel in transport is substituted by electricity through electrification of railways and by encouraging suburban electric railways when possible. The need for massive investment has limited the progress of these substitutions, but they are a part of strategic planning. These measures are not only for the limited purpose of fuel substitution, but also to speed up transport and reduce congestion. Therefore, we construct two fuel substitution scenarios as described above along with the assumption of relative efficiency of a factor of 2 between oil and coal. Unlike in the previous case, where energy was reduced (conservation), here only substitution takes place, i.e. energy in the economy in terms of GJ does not reduce. There are two equivalent scenarios: 10% of coal consumption substituted by oil and 13.5% of oil or gas consumption substituted by coal. The difference between the two can interpreted as the implications of changing the fuel substitution policy from its current policy.

It should be pointed out that currently much more gas is available to make it possible to carry out gas substitution. Indeed, gas-based power generation is already picking up rapidly. But gas is also priced as an oil-equivalent fuel. Therefore, the results discussed below are likely to be similar for gas substitution for coal. It can be seen that the current policy of substitution will lead to 12 million tonnes of additional coal and would cut nearly 3 million tonnes of oil, leading to 4.3 million tonnes and 8.7 million tonnes of additional carbon emissions compared to base case and oil substitution, respectively. However, it saves fuel worth Rs. 10.5 billion in fuel alone, but will save 8.7 million tonnes of carbon.

The magnitude of the trade-off is, in this framework, purely a function of the relative price (per unit of useful energy) of the two energy sources. This cost measure captures only the variable cost implications of fuel substitution, and ignores the differences in capital requirements of substitutions in different sectors of the economy.

In a wider sense, the costs of fuel substitution should also consider, (i) the opportunity costs of increased foreign exchange outlays on petroleum products and capital goods; (ii) the costs of diminished output of coal in terms of regional income and employment effects; and (iii) the costs (or benefits) of the change in the nature of local pollutants resulting from the substitution. It is difficult to assess a *priority* which of these various vompmnents will eventually dominate the cost of substitution. If the variable energy costs are the most significant element, then the trade-offs quantified here may be reasonable approximations of the true costs of carbon savings achieved through fuel substitution.

India's energy policy planners need to take not of the current debate on global warming. This concern has devalued India's coal resources. There will be pressures from international aid agencies to substitute gas and oil in place of coal. India needs to study its options carefully.

We have estimated CO_2 emissions for 60 sectors for India from the input-output matrix for 1983-84. The highest emissions are from the electricity sector (33%) followed by iron and steel (9.8%), road and air transport (8.6%), coal tar (8.1%), non-metallic minerals (4.5%), railways (4.3%), cement (4.0%), hotels and restaurants (4.0). While comparing direct and indirect emission coefficients from each sector, it can be seen that total emission coefficients, could be much larger than direct emission do coefficients for many sectors, such as electricity, iron and steel, cement, and so on. The Indira Gandhi Institute of Development Research (IGIDR) has carried out a detailed exercise concerning the possibilities of reducing carbon emissions by the power system in India. It may be possible to do that with modest investment. When one inverts the matrix and multiplies it with the final demand vector, an entirely different picture emerges. We get emissions by demand, where all indirect and embodied emissions are seen. The order of emission for final demand are: construction (22%), food crops (8.6%), road and air transport (6%), and so on. Construction by itself is not a very CO_2-intensive activity, but that picture changes when one considers the coal and oil embodied in construction materials such as iron and steel, metals and alloys, glass, cement and bricks. When multiplied by the final demand vector, it is the largest CO_2-emitting sector. Understandably, the construction needs of a growing developing country, which needs to build factories, offices, roads, schools, hospitals and houses, are considerable. On the other hand, food crops production, which is the first basic need, is not as energy-intensive because it is carried out with labour-intensive technologies. Electricity, which is more an intermediate good, moves to sixth from first place. This exercise brings out the important policy conclusion that increasing construction efficiency by optimising the use of construction materials, as well as by implementing construction technologies that use less energy-intensive materials, could be as important as improving energy efficiency for reducing carbon emissions. While the latter is well known, not enough attention is given to the former to reduce CO_2 emissions. Much of the construction is needed for infrastructure and production purposes.

Carbon is embedded in exports in transport and hotels, and it is embedded in imports, in iron and steel and electricity.

Policy scenarios of energy conservation for fossil fuels are also carried out. Two scenarios are constructed, both of which have 250 PJ (PJ = 10^6 GJ) of primary energy: one saves it from coal and the other saves it from oil. They show that while both coal and oil reduciton lead to reduced CO_2 emissions, to reduce energy from oil is more attractive for India due to savings in imported oil. India's preferred energy policy would be to save oil rather than coal. For example, saving 250 PJ from coal saves only Rs. 2706 million.

The same amount of useful energy savings (about 107 PJ) could be obtained from primary energy reduction from oil and would save Rs. 7710 million. This difference of Rs. 5634 million could inerpreted as incremental costs to India to follow coal-conserving policies. It is necessary to take such wide interpretations of indifferently costs where two development strategies are compared and not just to apply a cost-benefit analysis of coal conservation projects separate from other alternatives. Also at stake are the totally different sectoral approaches necessary to achieve this reduction.

For example, to reduce CO_2 from coal would require efforts for power (2.7 million tonnes), iron and steel (0.8 million tonnes), coal tar (0.7 million tonnes), and so on. On the other hand, reducing CO_2 from oil would require efforts in road and air transport (0.9 million tonnes), iron and steel (0.7 million tonnes), fertilisers (0.2 million tones), and so on. This diversion attention to different sectors will also translate into diversion of manpower, fuels, technology development, etc. Thus, a GHG reduction policy which would emphasise coal conservation would be at variance with India's energy conservation policy. Which has emphasised saving oil rather than coal. When coal substitution by oil and gas is promoted, the incremental costs of following such a strategy are even larger, because precious resources are substituted in place of cheap and abundant coal. Substitution of 10% coal in energy terms leads to additional costs of Rs. 10,416 billion. This would also be at variance with India's current substitution policy which promotes the substitution of coal for oil.

The main sources and uses of world energy. For developing countries biomass, used mainly by households, is the largest source of energy, and efficiency in its use will be important in controlling air pollution. Coal, oil, and gas are the next largest sources. Hydro-electric power provides 6 per cent of the energy needs of developing countries, while nuclear power provides less than 1 per cent.

As pointed out, fears that the world may be running out of fossil fuels are unfounded. The world's proven reserves of oil and gas in 1950 stood at 30 billion tonns of oil equivalent (btoe); today they exceed 250 btoe, notwithstanding a total world consumption of 100 btoe over the forty year period. Proven reserves of coal rose from 450 to 570 btoe in the same period. Reserves of natural gas have expanded more than fivefold since 1965 (despite a threefold increase in production during the period). They now amount to more than 100 btoe, almost as much as the world's given oil reserves; supplies in developing countries are strong and improving. Estimates of "ultimately recoverable" fossil fuel reserves worldwide are more than 600 times the present annual rate of extraction. All told, fossil fuel resources are probably sufficient to meet world energy demands for the next century, perhaps longer.

Policies to mitigate the effect on the environment of energy production and consumption take two complementary approaches. The first uses economic instruments and institutional reforms to encourage the more efficient use of energy.

The second is either to develop technologies that reduce the polluting effects of conventional fuels or to use less polluting substitutes. In discussing energy use for electric power generation and for transport—the two most rapidly growing categories—this chapter examines three scenarios: an "unchanged practices" scenario, with no environmental policies in place; a scenario that employs economic and institutional reforms to improve the efficiency with which fossil fuels are used; and a scenario that progressively adopts environmentally beneficial technologies. A combination of the second and third not only reduces local pollution appreciably but also improves economic efficiency. But using cleaner fossil fuels and technologies and improving efficiency will not by themselves solve the long-term problem of stabilizing carbon dioxide accumulations in the atmosphere. That will require a much greater use of nuclear or of renewable energy. This chapter (which concentrates mainly on local pollution) will show that solar energy, biomass, and other renewables are developing rapidly as environmentally and commercially viable energy sources.

Electric Power Generation from Fossil Fuels

More than half the world's consumption of coal and 30 per cent of fossil fuel consumption go to generate electricity. Fossil-fired power stations, in turn, account for two-thirds of the world's electric power-generating capacity, curently 2.6 million megawatts. In the 1980 electric power generation rose by 60 per cent in industrial countries and by more than 110 per cent in developing countries (where demand is expanding at 8 per cent a year and requires roughly 50,000 megawatts of added capacity each year). Under an "unchanged practices" scenario, in which pollution abatement technologies are not widely deployed, emissions of pollutants will increase more than fourfold in the next twenty years and tenfold in the next forty. Good policies would make this grimy prospect avoidable.

Economic and Institutional Reform

A second scenario considered in this section looks at the possible effects of price increases and institutional reforms. At present, under pricing of electricity is the rule, not the exception, in most developing countries. Prices, on average, are barely more than one-third of supply costs and are half those in industrial countries. Whereas average tariffs in the OECD countries rose by 1.4 per cent a year in real terms between 1979 and 1988, they fell by 3.5 per cent a year in developing countries.

Such low prices do not reflect improvements in the efficiency with which electric utilities supply their customers. On the contrary, losses during transmission and distribution, partly through theft, run at high levels: 31 per cent of generation in Bangladesh, 28 per cent in Pakistan, and 22 per cent in Thailand and the Philippines. (In the United States only 8 per cent of electricity is lost during transmission, in Japan, 7 per cent). These losses, the equivalent of about 75,000 megawatts of capacity and 300 terawatt hours 9300 billion kilowatt hours) a year, represent a loss to developing countries of approximately C30 billion a year through increase supply costs. Worse, by the end of the century, on present trends, aggregate losses would double. The reasons for persistent under pricing are largely institutional. The points made in Chapter 5 about the management of water utilities apply with equal force electric utilities. Governments frequently intervene in the day-to-day operations of utilities, and they worry the price increases will exacerbate inflation. Utility managers and their boards may have little say in pricing or investment decisions. Lack of accountability and transparency leads to poor management, either of the utilities themselves or of the state fuel companies that frequently supply them.

Subsidising the price of electricity has both economic costs and environmental effects. Low prices give rise to excessive demands and, by undermining the revenue base, reduce the ability of utilities to provide and maintain supplies; developing countries use about 20 per cent more electricity than they would if consumers paid the true marginal cost of supply. Under pricing electricity also discourages investment in new, cleaner technologies.

Because of the need to service the borrowing incurred to build new generating capacity, some developing countries are now starting to raise electricity tariffs. Some are considering or (in a few cases) implementing privatisation programmes, usually in the hope of tapping capital markets to build new capacity. Price increase may be easier in countries in which parts of the energy industry are privatised, and management is also likely to be improved.

Conservation: Sensible energy prices affect not just the generation of energy but also its use by industry and households. They create incentives for industry to use waste heat—for example, through co-generation, which combines power generation with the use of residual thermal energy for other purposes—and to improve efficiency in heating, motive power, refrigeration, and lighting.

A paradox of the energy market is that end users of electricity often appear to require much higher rates to return on the installation of more energy-efficient equipment than those that electric power producers require on new plant. This has led some countries to introduce subsidies for new types of energy-saving investments by electricity customers, paid either by the government or (as in several U.S. states) by the electric utilities themselves. Better information in the form of lebeling or advisory services to help customers make more informed decisions is also required. In developing countries industrial advisory services have sometimes identified ways of reducing energy consumption per unit of output and other costs. Such initiatives are important for improving energy efficiency, but their success, too, will depend greatly on prices that reflect the full economic and environmental costs of energy. These will, in themselves, help to make energy-efficient technologies financially more attractive to industry and individuals.

Technologies: By providing strong incentives to generate and use electricity in more efficiently ways, price and institutional reforms have the advantage of encouraging reductions in all polluting emissions (including carbon dioxide) per unit of output. However, low-polluting methods of generation are also required to reduc pollution significantly. The third scenario developed below therefore combines the efficiency reforms discussed above with the gradual adoption of environmentally improved technologies and practices. Technological advance has put developing countries in a better position to reduce all forms of pollution from electric power generation than the industrial countries were in as recently as twenty years ago. In industrial countries the capital stock takes about thirty years to turn over, and retrofitting is costly. Because developing countries are making new investments, they have the opportunity to install less-polluting plant right away.

There are, broadly, four technological options for reducing harmful emissions: (a) changing the fuel by switching to low-sulfur coals, oil, and gas; (b) cleaning the coal before combustion; (c) controlling emissions; and (d) using existing fuels more efficiency, low-emissions technologies. When coal is used, it is not unusual to find two or three approaches used in combination—for instance, to address the rather different problems posed by porticulate, sulfur dioxide, and nitrogen oxides.

Coal: Coal-fired stations are currently the main source of emissions from power stations because they make up more than half to total thermal generating capacity and because of the high sulfur content of coal in many regions. Combustion efficiencies are often poor, and modern emissions control technologies are not widely deployed; this gives rise to high emission rates of particulates and sulfur dioxide. The technological developments described that options are now available or emerging for reducing for all significant pollutants from coal (other than carbon dioxide) to low levels per unit of output. The costs of the options vary, but are not so great as to compromise the ability of developing countries to meet their growing demands if they pursue rational abatement policies.

GAS: Switching to natural gas, where it is economically available, carries many environmental advantages, Its use offers reductions in particulates and sulfur dioxide of more than 99.9 per cent in relation to conventional coal-fired boilers with poor or no emissions control technologies. The use of combined-cycle gas-fired boilers with poor or no emissions control technologies. The use of combined-cycle gas-fired stations also brings some reductions in emissions of nitrogen oxide per unit of energy produced. Current efficiencies (the proporation of energy converted into electricity from the fuel) of combined-cycle gas units are about 45 per cent and could rise to more than 50 per cent—almost twice those of conventional coal-fired stations thirty-five years ago. Construction times are also short (roughly four years). for many countries gas offers the prospect of both cheaper electric power generation and less local pollution.

These developments in the efficient use of natural gas for electric power generation have coincided with a remarkable increase in proven reserves over the past twenty-five years. In addition to proven reserves, there are several "unconventional" sources of methane that are thought to be vastly greater than conventional reserves in some countries-for example, coal-bed methane; "tight gas" formations in which the gas is held in rocks of low permeability and rock-fracturing techniques are required to bring about useful production; and some as yet uneconomical reserves such as shale gas deposits. The costs of exploiting natural gas reserves vary according to circumstances but have generally declined. Marginal costs in developing countries range from one-quarter to three-quarters of the cost of steam coal, the cheapest imported alternative energy source.

A major barrier to development has been the high fixed costs of exploration and production and of establishing a basic pipeline network. For those countries that have a narutal gas resource but have yet to exploit it, the development of gas-fired power station projects can provide the commercially justified starting point for the development of a widespread gas industry. Trade in natural gas will also be important from both commercial and environmental perspectives; there is enormous scope both for the shipment of liquefied natural gas to the main demand centers and for pipeline exports of gas from the former U.S.S.R., Europe, the Middle East, and North Africa.

FUEL OIL: Polluting emissions from the use of fuel oil for electric power generation can similarly be reduced to very low levels. Emissions of particulate matter are intrinsically much lower for oil than for coal and can be virtually eliminated by suing the technologies described. The use of low-sulfur fuel oil or flue gas desulfurisation can reduce emissions of sulfur dioxide by more than 90 per cent. Catalytic methods are also available for significantly reducing nitrogen oxides in exhaust gases. The costs of controlling emissions from oil-fired plant are lower than those for coal.

Regulation: To encourage electric utilities to employ pollution-reducing technologies, governments generally use regulation. This has been effective (although not always cost-effective) in industrial countries because there the pollution comes from a relatively small number of easily monitored point sources. In addition, the utilities are monopolies, are already regulated, and are perhaps more responsive to regulation than to taxes (on, for example, sulfur emissions) would help to encourage plants to adopt more cost-effective means of abatement.

Regulations has typically involved setting abatement standards. It is fortunate that the technology for addressing one of the most serious pollution problems of electric power production—emissions of particulate matter—is relatively simple and inexpensive. Increased use of gas-fired power stations will be important in this respect. Where coals is the preferred fuel, consturcting tall chimneys, siting power stations away from large population centers, and using the emissions control devices help to reduce disamenity and hazards to health from its combustion. They add less than 2 per cent to total supply costs and may be associated with reductions in costs. China, for example, has numerous small coal-fired power plants that emit three to eight times more particulates per kilowatt hour generated than do large plants yet have 30 per cent higher capitable costs, 60 per cent higher operating costs, and lower efficiencies than large plants. Given the costs to life and health of particulate matter emissions and the modest costs of reducing the emissions to low levels, the case for working toward high standards of abatement is unambiguous.

Unless natural gas is economically available, setting emissions standards for nitrogen oxides and sulfur dioxide will require a more critical examination of tradeoffs (as was done, for example, in Poland). Extensive studies in Europe and North America have found that the damage from these pollutants varies greatly with region. Much can be accomplished by using coal-cleaning technologies and low-sulfur fuels. Costs can also be reduced by a proper phasing of investments. In industrial countries and costs of flue gas desulfurisation are declining with experience; alternatively, it may be more cost-effective to postpone decisions until advance coal combustion technologies or new gas deposits are fully commercialised.

In the "unchanged practices" scenario, environmental policies are not in place, and rising demands for electric power are met at the cost of an exponential rise in pollution. In the second scenario, reforms to rectify the price inefficiencies and problems of a rectify the price inefficiencies and problems of accountability noted above are phased in gradually over twenty-five years. Pollution still rises (although more slowly), but there is less waste to capital, fuel, and operating resources in supply and less waste of energy in consumption—a clear case of good economic policies being good for the environment. The investment costs of expansion are lower and could even decline for a period as output expands, as a result of improved capacity utilisation and reductions in losses. The net benefits of electricity supply are also higher. Efficiency in energy production and use thus reduces pollution while raising incomes and welfare.

In the third scenario, in addition to energy efficiency, environmentally improved technologies and practices are gradually incorporated into the capital stock. Pollution rises initially on account of lags and difficulties in introducing new policies and practices but eventually declines as output expands. The savings in investment arising from improvements, in prices and institutional arrangements far outweigh any extra costs of pollution abatement.

Expanding Electricity Supply with Low CO_2 Emissions

Improving the efficiency of existing power plants and reducing electricity distribution and transmission losses could increase electricity delivered to end-users by a quarter. These gains would cost about one seventh as much as building new power plants and would not increase CO_2 emissions. Additional electricity generation can be provided by a range of technologies with low CO_2 emissions. These include efficient gas turbines, nuclear power and renewable energy sources. Renewable sources include geothermal, wind and solar energy-used to generate electricity directly or to power electrolysis of hydrogen used as a fuel. These renewable technologies produce virtually no CO_2 and are likely to play a rapidly growing role future power generation. This paper will briefly address only nuclear power and efficient gas turbines before turning to the potential of energy efficiency in developing nations.

Growing concern over global warming has prompted widespread interest in nuclear power because nuclear generated electricity is widely perceived to emit no CO_2 and is therefore viewed as an electricity source that does not contribute to global warming. Actually, the processing or uranium fuel—required to fuel nuclear reactors-does produce CO_2 at a level that one study estimates to be slightly more than one third that of electricity generated from efficient gas-fired electric plants. Large-scale nuclear programmes are being promoted as a cost-effective response to global warming. But in an era of limited funds, large-scale nuclear plant construction projects could divert funds from much more cost-effective solutions.

The Director General of the International Atomic Energy Agency (IAEA), has called for several thousand nuclear reactors to be operating by 2030 in order to make a substantial difference in slowing global warming. A large portion of these would have to be in developing countries. Yet experience in these countries shows that such a programme would have little chance of success.

After more than 30 years and tens of billions of dollars invested in nuclear power by developing countries, most of the nations have cancelled the programmes they began. In developing nations (excluding the newly industrialised countries of Asia) there are only 10 relatively small nuclear reactors providing a total of 3 130 MW of nuclear capacity, equal to about three large reactors. Third World nuclear programmes have been plagued with plant delivery and construction problems and delays and cost overruns that are even more chronic than those experienced in many industrialised nations. For example:

- Argentina's 698 MW Attucha II reactor is estimated to cost three billion dollars, or $4 297/kW, more than three times the average cost of constructing non-nuclear thermal plants in developing nations.
- Mexico's Laguna Verde plant, under construction for 16 years has seen an 11-fold cost increase.
- Philippine's 620 MW reactor cost $2.3 billion, or $3 710/kW to build. The Philippine government believes that the design was flawed, the construction poor and that consequently the plant will never be safe to operate. The Philippine government is suing the vendor.

A recent study, published in *Energy Policy,* compared nuclear power and energy efficiency as two CO_2 abatement strategies. The study found a nuclear cost of ¢13.5 kWh for nuclear power in the USA, compared with an efficiency cost of ¢2/kWh. In other words, investing in efficiency has the effect of reducing seven times more CO_2 than an equivalent in nuclear power, and it does so much more quickly.

Experience indicates that nuclear bower is not only more expensive but also is less reliable in developing nations than in OECD countries. The operating capacities of the 10 nuclear plants in developing nations are low and do not seem to be improving. The average operating capacity of these plants declined from 46 per cent in 187 to 42 per cent in 1988, or about one quarter below average operating levels of nuclear plants in industrialised nations.

The high-cost and poor performance of nuclear reactors in developing countries suggest that nuclear power is an even less cost effective strategy for reducing CO_2 emissions in developing nations than it is in industrialised ones. Because of severe capital constraints in developing nations, large-scale investment in nuclear power in these countries would actually increase CO_2 emissions to the extent it diverts funds from more effective CO2 reduction options, such as efficiency:

Advanced gas turbines

A new generation of efficient gas turbines promises to provide electricity with lowered CO_2 emissions. Efficient gas turbines promise to generate electricity with relatively low capital and operating costs and relatively low CO_2 emissions. A 100 MW inter-cooled, steam-injected gas turbine (ISTIG) is projected to have a capital cost of $400/KW, or roughly one-third the cost of US coal plants and one fifth the cost of US nuclear power plants. Switching from existing coal fired plants to gas-fired ISTIG plants would reduce CO_2 emissions by about 60 per cent due both to higher plant efficiency and to the lower CO_2 content of gas. The Worldwatch Institute estimates that in replacing coal-fired plants, steam injected gas turbines are three to five times most cost effective than nuclear power as a way to reduce CO_2 emissions. Commercial manufacture of gas turbines could be achieved in five years.

A promising potential application of efficient gas turbine technology is its operation fuelled with biomass, such as sugar cane or trees. If fuel were harvested on a fully sustainable basis, such power plants could generate electricity with no net release of CO_2.

Like a flat and soggy souffle that suddenly balloons into shape, "global warming" and the "greenhouse effect" swelled in the public and political consciousness during the summer of 1988. All it took was some heat.

Much of the United States was already suffering the hottest summer on record by June 23, when James Hansen, a NASA atmospheric scientist, appeared before the Senate's Committee on Energy and Natural Resources with an ominous message:

"Number one, the earth is warmer in 1988 than at any time in the history of instrument measurements. Number two, the global warming is now large enough that we can ascribe with a degree of confidence a cause and effect relationship to the greenhouse effect. And number three, our computer climate simulations indicate that the greenhouse effect is already large enough to begin to effect [sic] the probability of extreme events such as summer heat waves."

Hansen finally linked the 1988 heat wave to the "greenhouse effect," a century-old scientific theory, describing the natural process by which the earth's atmosphere holds in some of the sun's infrared radiation (heat) reflected off the planet's surface, thus making the earth habitable. He argued that mandate pollutants, especially carbon dioxide, are exacerbating the greenhouse effect by trapping too much solar heat. The result is global warming, the gradual rise in mean temperature that has been measured since accurate temperature records were first kept, about 1860.

Hansen's computer models and conclusion have since been challenged by other scientists, and the summer's heat and drought were also attributed to shifts in the Arctic jet stream and a Pacific ocean temperature current called "El Nino." But in

1988, to politicians and the press alike, global warming became the new reality, energy produced by fossil fuels the new issue.

Solutions were quickly offered, including energy conservation and alternatives such as solar. But nuclear power, and its strong constituency, soon elbowed other alternatives out of the congressional and media forums. Nuclear power had been touted as a "smokeless" antidote to air pollution as early as 1939, and in fanciful scientific speculation after World War II it was hailed as a potential energy source that would some day be so economical and convenient that it could replace all conventional fuels.

Although publicly descredited by the traumas at Three Mile Island in 1979 and at Chernobyl in 1986, nuclear power was proposed as a solution to the heat of the scorching 1988 summer because, unlike fossil fuels—coal, oil, and natural gas—nuclear power produces no carbon dioxide, the gas most responsible for the greenhouse effect.

Sen Wendell Ford, a Kentucky Democrat, was quick to make the connection. On the same day that Hansen testified, Ford asked another witness if natural gas or nuclear power produced more carbon dioxide. When told it was natural gas, Ford said, "so you would suggest that we go to nuclear then," and mused that many environmentalists may have swung "180 degrees" because of global warming, and may now condone nuclear power. Appearing that same month before a House committee, Edward Davis of the American Nuclear Energy Council called for a resurgence in atomic energy "to alleviate the greenhouse effect and to meet our growing demand for electricity".

On July 18, with temperatures across the country stuck in the nineties, *Newsweek* linked global warming and nuclear power in an article headlined "Not So Bad After All?" The article concluded that "nuclear energy can no longer be flatly dismissed by environmentalists." In August, *Fortune* reported: "Amid mounting evidence that the earth is warming because of the greenhouse effect, splitting atoms is getting new respect." A flurry of articles and editorials followed.

By that August, Colorado Sen, Timothy Wirth, a Democrat who chaired the June hearing, mustered 15 cosponsors for a comprehensive bill (S.2667) to combat the "greenhouse effect" by funding research on energy sources that produce no carbon dioxide, including "passively safe, modular and cost effective" nuclear power plants. Wirth challenged nuclear opponents to redirect their thinking. "Everyone must realise it will take a national and international commitment to develop nuclear and solar technologies," he said, "All energy sources must be looked at equally."

Renewable and Nuclear Energy

Fossil fuels will continue to be the predominant energy source for the next several decades, and the main task ahead will to use them in economically and environmentally satisfactory ways. But if the treat of greenhouse warming made it necessary to restrict the use of fossil fuels, could the world's demands for commercial energy still be met? More efficient use of fossil fuels and a switch from coal to fuels lower in carbon could substantially reduce emissions of carbon dioxide per unit of output. Beyond that, the options would be nuclear energy or renewable energy (primarily solar energy, biomass, geothermal energy, hydropower, and wind).

Nuclear power provides less than 1 per cent of the energy used in developing countries. That share seems unlikely to rise significantly. Quite apart from the abundance of fossil fuel reserves, which will act to depress demand for all alternatives, nuclear power has two handicaps: its costs and its environmental risks. Discoveries of fossil fuel reserves and progress in production and conversion technologies have helped to hold down the prices of fossil fuels. At the same time, the costs of nuclear stations have risen for a variety of reasons: long lead times and delays in seeking approval, meeting environmental safeguards, and constructing the plants; the costs and risks of disposing of radioactive wastes; and the prospective costs of decommissioning plants. Recent estimates (OECD 1989) show that fossil fuels still have lower costs than nuclear power, except perhaps at low discount rates.

While the costs of nuclear power have increased, developments in renewable energy in the 1970 and 1980—in solar, wind, and biomass energy, in particular—have led to remarkable cost reductions in these technologies. There is now a growing awareness that renewable energy is an abundant resource that can be harnessed.

Each year the earth's surface receives from the sun about ten times as much energy as is stored in the whole of the world's fossil fuel and uranium reserves. This energy—the equivalent of 15,000 times the world's primary energy demand—can be captured in solar-thermal systems, which produce heat for electric power generation and for domestic and commercial uses, or with photovoltaic systems, which produce electric power directly from sunlight. Both types of scheme have been considered for the production of hydrogen, which could be used as a transport, domestic, or industrial fuel. Solar energy can also be stored by growing plants and, in the from of biomass, may be used as a feedstock for the production of commercial fuels and electric power.

In the past there have always been two commercial drawbacks to solar schemes: the amount of land they require and their costs. Both are declining. In developing countries solar insulation is roughly 6,500 times the annual consumption of commercial energy. At current conversion efficiencies of 15 per cent, less than 0.1 per cent of these countries' land area would be required to meet, in theory, the whole of their primary energy requirements. In industrial countries the fraction of land area is 0.5 per cent. These areas are less than those currently occupied by hydro-electric reservoirs worldwide and are very small in relation to the area under crops. In fact, the land intensities of solar schemes average only one-twentieth those of hydro-electric schemes and sometimes considerably less—they are less than one-hundredth that of the Aswan High Dam, for example. Moreover, the ideal locations will often be sparsely populated arid areas, and the technology is modular and allows flexibility in the choice of sites. Thus, solar schemes suffer minimally or not at all from three problems that sometimes be set hydro-electric schemes—the inundation of arable or forested lands, ecological side effects, and the displacement of people.

The costs of all commercial forms of renewable energy have declined remarkably over the past two decades (as they did in the earlier part of this century for electric power generation from fossil fuels. The cost so of solar energy may well fall further. In high-solar-insolation areas the costs of electric power from solar energy seem likely to become competitive with those of nuclear power within the next ten years or so (even ignoring their advantages in reducing environmental costs) and probably with those of fossil fuels over the long-term.

The commercial development of renewables may thus be justified on non evironmental grounds. But if greenhouse warming makes it necessary to restrict the use of fossil fuels, several things could be done to promote the wider use of renewables. First, financial incentives could be put in place to encourage applications and market development. Environmental taxes (such as carbon taxes) on fossil fuels would favour renewables and encourage private researching development (R&D). Second, industrial countries, in particular, could allocate a greater share of their national energy R&D portfolios to renewables. Up to now, R&D has been heavily stewed in fever of nuclear power. The member countries of the International Energy Agency (IEA) allocate 60 per cent of their R&D budgets (which totalled $7.3 billion in 1989) to nuclear power, 15 per cent to coal, oil, and gas, and 19 per cent to electric power transmission and other areas but only 6 per cent to renewablws. International collaboration on research in renewable energy also merits support. Several developing countries—including Brazil, China, India, and Thailand—already have the nucleus of good research programmes. Third, applications in developing countries could be encouraged by expanding the global Environment Facility and concessional finance.

Other experts disagree that nuclear power is the answer. James G. Speth, president of the World Resources Institute, said that by the time nuclear power's many problems are solved, "we could have preferable alternatives available in the renewable energy area." Speth argued that "we are closer to inexpensive, durable photovoltaic [solar] systems than a new nuclear technology, and our research priorities should reflect this promise."

Timing is an essential element in the debate over nuclear power's potential for solving global warming. With the domestic nuclear industry moribund, and public concern still high over reactor accidents, nuclear proponents promise a "second generation" of nuclear plants of safer design. But these plants are barely onto blueprints, and it may be decades before they are build. Many of the environmentalists who endorse "inherently safe" nuclear plants know second-generation plants will not be available until the distant future, not in the next decade or so when global warming may be a measurable and continuing phenomenon.

"Only if all our other alternatives to climate disruption have failed," said senior staff scientist Jan Beyea of the National Audubon Society, "do I see public acceptance of a second nuclear era. Research into these reactors must be viewed as an insurance policy, not a component of our first line of defence."

In June 1988, Bill Keepin and Gregory Kats of the Rocky Mountain Institute concluded that investing in nuclear power would make only a "small contribution" to slowing global warming; it would be less expensive to invest in energy efficiency. To curb world-wide carbon dioxide emissions significantly, they warned, one new nuclear plant would need to be completed every two days for the next 38 years. "Such massive investments would be infeasible in the Third World, and would have severe economic impacts in developed countries."

Michael Oppenheimer of the Environmental Defense Fund argued that energy conservation and efficienty are more practical than nuclear power, and can be employed more promptly. And John Ahearne, the former chairman of the Nuclear Regulatory Commission, concluded in his 1989 study "Will Nuclear Power Recover in a Greenhouse?" That nuclear power could alleviate global warming only if fundamental changes occur in the US nuclear programme, including streamlined federal and state regulation; improved public acceptance; and standardised, safer, and more cost competitive designs.

Undaunted by their critics, nuclear power's advocates describe a very different energy future. As British environment secretary Nicholas Ridley concluded: "If we want to arrest the greenhouse effect we should concentrate on a massive increase in nuclear generating capacity."

The strongest link between global warming and unclear power was probably drawn by Hans Blix, director general of the International Atomic Energy Agency. Touring the United States during the 1988 heat wave, Blix praised the 400 or so nuclear power plants in 26 countries that produce 16 per cent of the world's electricity. "While 16 per cent does not appear to be a very large share," he said, "to replace it would require [the equivalent of the] entire coal production of either the United States or the Soviet Union."

In January 1989, *Nova* Broadcast a second programme on global warming, "Hot Enough for You?," highlighting nuclear power's possible role in slowing the greenhouse effect. Michael Oppenheimer, who had urged greater energy efficiency, conceded: "It seems likely that nuclear power will emerge as a possibility as citizens and policy makers assess their energy options in a greenhouse world." But William Moomaw, director of the World Resources Institute's Climate and Energy Programme, emphasised that nuclear power's most likely role will come with a "new generation" of reactor designs.

Fighting global warming with more nuclear power reamins a contentious issue for environmentalists. Delegates to a 1988 Toronto conference on global warming agreed that carbon dioxide emissions should be reduced by 20 per cent by the year 2005, but disagreed over which technologies might replace coal.

Last fall, global warming and a nuclear power revival were more conspicuously connected at a Montreal conference. "The cure for the greenhouse effect . . . is a huge shift to nuclear-generated power, speaker after speaker told the delegates at a World Energy Conference meeting," the *New York Times* reported on September 25. "But a few speakers predicted that more efficient electrical and electronic consumer products would make a big shift to nuclear power unnecessary, while other criticised the cost of reactor-generated electricity."

One study presented at Montreal predicted that nuclear power would generate up to eight per cent of the world's energy by 2020, down from an estimate of 12 per cent made six years earlier. Nuclear pioneer Alvin Weinberg, a former director of the Oak Ridge National Laboratory and a prophet of nuclear revival, said that global warming might be stabilised within a century—with stringent efficiency standards and energy from 5,000 nuclear plants. Safer reactor designs are essential, Weinberg said, because using today's designs would create an unacceptably high risk of a serious reactor accident.

Reporting in May 1989 on "Energy Technology R&D: What Could Make a Difference?," staf members at the US Energy Department's Oak Ridge National Laboratory concluded: "If a world-wide commitment is made to major reductions in carbon dioxide emissions in the atmosphere, the major technological options are improved energy efficiency and nuclear power." But for nuclear power to be deployed "on a massive scale"—several thousand gigawatts (billion watts) world-wide—research and development will need to demonstrate several techniques that do not exist today, among them: a commercial-scale nuclear fuel recyle; breeder reactors or other technology to extend existing fuels, uranium recovery from seawater, and perhaps a fusion/fission hybrid reactor.

The Energy Department itself is divided over the role that nuclear power might play in mitigating global warming. According to a May 12, 1989, *Wall Street Journal* report, some officials are privately skeptical about the greenhouse effect and have urged President Bush to slow steps to combat it. Yet publicly, Energy Secretary James D. Watkins is using global warming as an arguement for nuclear powers. At a recent public hearing to devise a National Energy strategy, Watkins said that the section on global warming "could become the energy strategy itself."

But last may, even the IAEA's Hans Blix narrowed the focus for nuclear power as a solution to global warming, pointing out that developing countries are unlikely to build nuclear plants soon. "If nuclear power is to be relied on to alleviate our burdening of the atmosphere with carbon dioxide, it is therefore to the industrialised countries that we must look," he told an Amsterdam symposium. "They are in the position to use these advanced technologies—and they are also the gratest emitters of carbon dioxide."

Both global warming and a nuclear power remedy have been studied and debated for many years, chiefly by scientists and policy analysts. The Conservation Foundation warned about the consequences of the greenhouse effect in a 1964 study, and the following year a President's Science Advisory Committee panel on environmental pollution argued that the greenhouse effect was a serious concern. In 1965, geophysicist Gordon J. MacDonald testified before the Senate Interior and Insular Affairs Committee that climate modification from carbon dioxide buildup posed a serious problem. This testimony prompted a lively response from New Mexico's Democratic Sen. Clinton P. Anderson, who also chaired the Joint Committee on Atomic Energy and saw nuclear power as an alternative to coal burning. And in 1969, when the Joint Committee held hearing on the environmental effects of electricity production, witnesses mentioned the greenhouse effect as a consequence of using fossil fuels and described nuclear power as an appealing alternative.

After the oil price shock of the 1970s, some viewed nuclear power as a reliable, domestic alternative to Mideast imports. But the inflation and recession that followed squelched US energy consumption. Flat demand and high interest rates caused a slump in the nation's ambitious nuclear power programme, and the slump deepened after the 1979 reactor accident. At Three Mile Island and the regulatory chaos the followed. In April 1979, shortly before the Three Mile Island accident, Congress's Office of Technology Assessment (OTA) mentioned the greenhouse effect and climate modification for the first time, in a report on coal use.

Throughout the 1980s, concern about the greenhouse effect grew slowly. It first gained attention in the State Department's 1980 *Global 2000* report. In 1983, the public television science series *Nova* included a programme on "The Climate Crisis," which mentioned nuclear power and solar enerlgy as two alternatives to fossil fuels. OTA made the connection between global warming and nuclear power in its 1984 report, *Nuclear Power in an Age of Uncertainty.* Occasionally, congressional hearings touched on the greenhouse effect in the 1980s, but more attention was focused on the problem of acid rain.

By 1986, the Atomic Industrial Forum (now the US Council for Energy Awareness) had seized on climate issues. In its report "Will There Be Energy After Chernobyl?" the pro-nuclear organisation predicted that atomic energy would overcome the legacy of the accident, citing worries about fossil-fuel pollution on climate and crops. "The fact that nuclear electricity does not contribute to either acid rain or the greenhouse effect has not been lost on the scientists. It still must make its way into the consciousness of the public." In 1986 more than 30 bills and resolutions were introduced in Congress to deal with global warming.

In 1987, the Global Climate Protection Act was passed. Hearings conducted by Senator Wirth highlighted the potential role that nuclear power might play in mitigating global warming. Frederick Koomanoff, director of the Energy Department's Carbon Dioxide Research Division, told Wirth's committee that "safe fission" was one of several energy sources that do not produce carbon dioxide. William A Nitze, the State Department's assistant secretary for environment, health, and natural resources, argued for "more aggressive" nuclear power development to diminish global warming. And Gordon J. Mac Donald, vice president and chief scientists of the Mitre Corporation, urged the United States to take a "leadership role" in developing nuclear power. "The problems of climate change underline the importance of restoring the nuclear energy option," he said.

Alan T. Crane, an energy policy specialist who headed the OTA's 1984 study on nuclear power, told the House Subcommittee on Energy and Power in May 1988 that the greenhouse effect was more worrisome than rising oil, natural gas, and coal prices.

"There is a significant, though not yet quantifiable, risk that the resulting climate changes will wreak devastating changes on agricultural production throughout the world, among other problems," Crane, said. "Such changes could lead to the death of far more people and cause far greater environmental damage than any nuclear reactor accident, and appear to be considerably more likely". Bearing this in mind, nuclear power "may prove an essential replacement for coal within the next fifty years."

3. Global Warming and Response of Developed Countries

A. USA

ENERGY USE and production account for by far the largest portion of emissions of greenhouse gases in the United States and the world. The US Environmental Protection Agency (EPA) has estimate that, worldwide, these activities were responsible for 57% of greenhouse warming in the 1980s. Other activities and their respective contributions include agriculture, 14%; land use and modification, 9%; chlorofluorocarbon (CFC) use, 17%; and other (nonenergy) industry, 3%.

Given this importance of energy activities; it is appropriate that efforts to forestall global warming have focused on these activities. Because the United States consumes the largest share of world energy and thus produces the largest share of greenhouse gas emissions, it is the target of many such efforts. A number of US government responses to global warming have been proposed or are now under way. For example, the 101st Congress has seen 11 bills dealing with global warming issues; a research programme on climate change has been promised $190 million for 1990; and the United States has signed the Montreal Protocol to control CFCs.

Despite these positive actions, it is important to note that none of the proposed bills has been signed into law; research on environmental issues and energy efficiency has been hindered for the last eight years by spending cuts; and the Montreal Protocol took almost a decade to achieve—similar international agreements for other greenhouse gas reductions are likely to be more complex and take longer to achieve. Yet the advantages of acting promptly on these and related issues are compelling. In addition to slowing the arrival of global warming, the benefits to be achieved by reducing energy use and using energy more efficiently are many increased energy security through lower oil imports, a more competitive economy, and a cleaner environment are just a few.

The paper that follows discusses US energy use and related emissions of greenhouse gases. Energy use in each sector is briefly characterised and several new technologies for energy use in that sector are described. Finally, national and state policies that offer potential to reduce energy use are discussed. The discussion is limited by space considerations to only a sampling of the many technologies under development and policy options that have been proposed.

Response from the United States Congress towards protection of tropical forests has been overwhelming. A number of bills have been introduced during the 100th and 101st Congresses concerning tropical forests, the greenhouse effect and climate change. The Tropical Forest Protection Act (H.R. 3010) was introduced by Congressman John Porter and congressman David Obey. Senator Robert Kasten and Senator Alfonse

D'Amato introduced the tropical Forest and Wetlands protection act (S. 1538). Both these bills have a primary focus on protection of tropical forests and wetlands study to determine areas unsuitable for agriculture and require determination by the US Treasury Secretary on institutional capacity to protect such forest resources in concert with such organisations as the Multi-Development Banks (MDB's) such as the World Bank. Continuing resolution H.R. 395 (public law 100-200) specifically direct department of treasury to promote environmentally sustainable economic growth through MDB's. Though World Bank and IMF loans cannot be converted to grants, are not saleable on the secondary market, and are not reschedulable, the Treasury Department recommends a number of initiatives including debt for nature pilot project.

Several US organisations and agencies such as the environmental Protection agency, the Forest Service of the US Department of Agriculture, Smithsonian Institution, National Academy of Sciences, National Science Foundation, a number of universities and several environmental organisations have started to include programmes on tropical forests as they relate to amelioration of climate change. In February 1989, EPA submitted a report to the US Congress on policy options for stabilising global climate including reducing demand on forest lands and increasing supply of forest products. The US Forest Service which initially started climate change research activity in its Fire Research Unit is now also looking towards initiating components on tropical forest conservation and use of reforestation to mitigate climate change in its International Forestry Programme.

Almost 25 per cent of the annual carbon emissions to the atmosphere is due to tropical deforestation. If future global warming is to be controlled, deforestation has to be curtailed. In order to reverse the greenhouse effect, existing forests need to be restocked and already cleared areas and unculturable waste lands afforested. To provide for the increasing human population, intensification of agriculture, development of wood reserves, and sustainable land-use systems are needed. Policies of each country must be reshaped to allow for protection of climate. Research and educational institutions need to be strengthened. Both policy-makers and the public must be motivated for concerted actions.

Reprioritisation of government programmes and increased funding for forestry development, climatic amelioration and environmental protection at the national, regional and international levels both by the public and private sectors would be necessary to protect this planet for ourselves and the future generations.

As developing economies grow, they will begin to catch up with the levels of energy consumption and industrial production of high-income countries. In today's industrial countries the main period of industrialisation saw rapidly increasing pollution. How far can the developing countries avoid repeating that experience and benefit from the ways in which the richer countries have learned to reduce pollution from energy use and industrial production even as output expands?

At present the omens are poor. The current levels of air pollution, water pollution, and hazardous assists in developing countries pose serious threats to human health, productivity, and welfare. These types of pollution arise mainly from the use of energy and from industrial production. If growth continues at present rates or higher—as it must if poverty is to diminish—then, on present trends, increased energy use and industrial production will add enormously to pollution.

The consumption of commercial energy in developing countries is rising rapidly and will soon dominate energy markets worldwide. Despite oil price shocks and financial crises, it tripled between 1970 and 1990 and now accounts for 27 per cent of the world total. Even if developing countries' demand for primary energy were to grow at a rate 1 to 2 per centage points lower than the trend growth rate, demand is likely to exceed 100 million barrels a day of oil equivalent (mbdoe) by 2010 and perhaps 200 mbdoe by 2030. Yet per capita consumption in these countries would remain much lower than in industrial countries.

The production and consumption of industrial goods have also increase rapidly in developing countries. In many, the historical and current peace of industrial growth has outstripped that of industrial counties and will continue to do so as per capita incomes rise. As incomes rise, the structure of consumption will also change. Manufactures have a high income elasticity of demand, and the structural shifts brought about by development are likely to put heavy pressure on the environment. The growth of manufacturing in developing countries averaged 8.0 per cent in 1965-80 and 6.0 per cent in 1980-90, compared with 3.1 and 3.3 per cent in the industrial countries for those periods. Manufacturing output will probably rise threefold in the generates most of these pollutants, as well as efficiently and wastes, which are becoming more numerous, toxic, and exotic as industrialisation proceeds. Like industrial countries, developing countries also need policies for dealing with "global pollutants" such as CFC and greenhouse gases.

In trying to skip the most polluting stages of industrialisation, developing countries have some special advantages. They are able to draw on advances in technology and management practices already made in industrial countries, under the pressure of increasingly strict pollution controls. And because they are expanding rapidly, they are generaly building new generating and industrial plants rather than refitting existing ones. They should therefore be able, with investment, to go straight to low-polluting practices. Developing countries are more likely to gain from such advantages if they encourage international trade and investment and if they adopt environmental taxes, laws, and regulations that make cleaner practices profitable and polluting ones unprofitable, thus creating a commercial interest in a clean environment.

A "Future Action Committee" of leaders from 9 different countries distilled from presentations and discussions a series of actions that might be taken to prevent or alleviate some of the effects of changes going on in the earth's atmosphere. Once finalised, this future actions report will be given wide publicity to the leaders of both developing and more developed countries. I shall make arrangements to have both UNDP and the World Bank provided with copies of the published report.

During the CHEMRAWN VII conference, a luncheon was held for Third World participants and for conference organisers. The enthusiasm of these participants was obvious as each was given an opportunity to speak. They felt they had gained much from their participation and indicated their interest in possible future nonitoring networks. CHEMRAWN officials were also impressed with these young people and have indicated their desire to encourage involvement of Third World scientists in similar future workshops and conferences.

All in all, these appeared to have been worthwhile events that will help alert Third World countries to global change problems and to the need for their involvement in monitoring and controlling these changes.

From a substantive point of view, the conference identified several areas of interest to persons concerned with international activities, examples being:

1. The atmosphere is a global entity whose characteristics are affected by all countries, rich and poor. In turn, the welfare of all countries is influenced for good or bad by these characteristics. Consequently, no one country or group of countries has the right to make changes that are detrirmental to the global community.

2. One of the constraints to better understanding the chemisty of the atmosphere and its effect on global changoe is the paucity of reliable monitoring data in many Third World areas. There is a dire need to set up monotoring networks that will include stations located in selected sites in developing countries. Training for Third World personnel and the maintenance of precise and quality records will be essential.

3. The nature of air pollution problems and steps necessary to solve them suggest the need to involve the private sector as solutions are sought and definitive action put into place to reduce air pollution.

4. The potential of atomic energy as a major source of energy is being re-examined. Its effect on greenhouse warming is nil compared to that of fossil fuel.

5. New knowledge of atmospheric pollution and of chemical reactions in the atmosphere result in an evolving concept of what is happening to influence global change. Examples of such knowledge coming from this conference are:

a. Chloro fluoro carbons (CFCs) have been found to have both a warming (greenhouse effect) and cooling (ozone destruction effect) influence, opening the possibility that these compounds have little net effect on global temperature. (But they still have implications for human health because of the increase in the ultraviolet light when they are destroyed).

b. Oceans absorb only about 25% of the carbon emitted into the atmosphere rather than the 50% considered here to for. This leaves the question of where the carbon goes.

c. Second and third order reactions of methane have effects opposite to first order reactions that have here to for been considered as being dominant. This finding, along with reports from India indicating methane loss from paddy rice lands is only about half previous estimates, may reduce methanes' importance as a greenhouse gas.

B. THE NETHERLANDS

The Netherlands is a highly industrialised country on the North Sea. Global warming and the resulting changes in climate and sea level are regarded as major causes for concern by the government and the public. The main reason is not so much the threat from sea level rise: the Dutch are used to coping with that and calculations show that the foreseeable rise in seal level would be manageable for them. Possibly more important is the feeling that mankind is threatening his very existence by overloading the environment with pollutants.

Following the recommendations of the World Commissionon Environment and Development, the Netherlands' government has decided to make "sustainable development" a leading them, for its overall policy. The recently formulated environmental policy plan has a broad basis in the government, as it is a co-production of four ministries (agriculture, economic affairs, environment, traffic and transport).

According to ongoing research, the contributions to the greenhouse effect in the Netherlands in terms of global warming potential seems to be in the order of 62% for CO_2 16% for CFC, 8% for CH_4 and 14% for other greenhouse gases (less than 1% of global greenhouse Gas (GHG) emissions. In its national policy, the Netherlands is already pursuing "no regret" measures. (Action that will benefit the environment even if there were no global warming). Furthermore, it has strengthened research on climate change and has set up programmes to increase public awareness.

In the international arena, the aim is to play an active, "catalytic" role, e.g. by hosting major international meetings (Den Hagg, Noordwijk 1989). The Netherlands hopes that the negotiations on the international Climate Convention will be concluded by mid-1992 during the United Nations Conference on Environment and Development (UNCED) in Brazil. This convention should include specific obligations with respect to greenhouse gas emissions, forestry and additional financial support to developing countries.

In 1989, CO_2 emissions were approximately at 182 million tonnes. The following targets have been set officially: Stabilisation of CO_2 emissions at the 1989-1990 level in 1994-1995 and subsequent reduction of 3-5% by the year 2000.

Achieving the stabilisation target implies a reduction of CO_2 emissions by 2% per year with respect to projected growth. This means that over the next four years 17 million tonnes of CO_2 emissions have to be avoided: 3 million tonnes through improving fuel mix, 10 million tonnes through energy savings, 2.5 million tonnes through traffic policy and 1.5 million tonnes through waste management. Some of the necessary measures are already in the implementation stage while others, especially in the transport and waste sectors, are being elaborated.

For all sectors, the policy of the European Community is of paramount importance. The EC has as a common goal for the stabilisation of overall CO_2 emissions in 2000 at present levels. However, concrete proposals for implementation are yet to come. In order to achieve the reductions in the energy sector, the national government works closely with the utilities sector; some cities have additional programmes. Measures include subsidies, pre-financing of energy saving investments, promotion of renewable sources, co-generation and district heating and major information campaigns. Relatively small levies on energy use are implemented as a source of financing these programmes. Meanwhile, the implementation of important "regulatory" charges is seriously being considered. Government expenditure amounts to U$$ 38 million annually with an eye to encouraging private investment on a much larger scale. The macro-economic effects of the CO_2 reduction policy have been calculated to be marginal.

In the traffic sector, emission reduction is to be achieved along three lines. The first line is to promote cars that emit less CO_2. Besides, the EC also aims at a reduction of CO_2 emitted by automobiles through tougher technical standards for new cars. The second line is to reduce car use through integrated "push and pull" measures. Pricing and other incentives will be used to raise variable motoring costs. Investments in new road infrastructure will be reduced while those in public transport will be increased. Furthermore, companies will be encouraged to draw up transport management plans to reduce commuter and business kilometres.

The third line is to take urban traffic measures, such as parking restrictions and stricter enforcement of speed limits.

In the waste sector, measures will be taken such as better use of energy from incineration and promoting the reduction of the waste volume through prevention, reuse or recycling. The latter type of measures will clearly lead to lower emissions during product manufacturing.

C. JAPAN

Japan holds 14 per cent of the world's GNP through its economic activities, occupies 0.27 per cent of land area and has 2.5 per cent of the world's population. This means that it has a deep dependency on foreign countries and the world ecosystem to supply its loving needs. This is especially true for agricultural products, where Japan can only provide 30 per cent of its food needs on original calorific basis. It imports almost all the wheat, bean and maize it uses from USA, Canada and Australia. half of the lumber it consumes is imported from Malaysia, USA, USSR and Canada. Japan's fish catch from the world's oceans amounts to 14 per cent of the world total. The other chief imported materials, consist mainly of primary products such as cotton, wool, natural gum, sugar etc., such primary industrial products come from developing countries, who are strongly dependent upon climatic conditions.

Japan is only self sufficient in 20 per cent of its energy demand. It has a small domestic resource of low quality coal (2.3 per cent of total energy), hydraulic power (4.1 per cent) and nuclear power (10 per cent). So this deficiency in domestic resources and dependency upon the world's ecosystem means that Japan has to maintain a population of 120 million and makes contribution to the world's economy through its industrial technology and services. Climatic change is certainly a challenge to all of the human race in this inter-dependent world, but it is taken more seriously by countries with scarce resources such as Japan, than ones with abundant resources which they can control.

Contribution to Global ecosystem

Japan 's estimated share to the world total of CO_2 emissions is 4-5 per cent. This is 4th largest in the world, behind USA, USSR and China. The per capita emission is 40 per cent of the USA and is relatively small amongest the other industrialised countries. Japan's lumber consumption accounts for 1.6 per cent of the world's, which is relatively small, when compared with its share of population (2.5 per cent).

Rice is the only agricultural product in which Japan is self sufficient. The rice fields occupy 232 km^2 and represents 1.5 per cent world's total rice field area, and is a potential source of methane.

Japan has set guidelines to phase out chrolofluorocarbons by the year 2000 as a response to the Montreal protocol and have also strictly regulated the emissions of CO NOx from industry and vehicles.

Impacts of Global Warming on Japan

Some preliminary research work has shown the possible impact of global warming on Japan. Agricultural production may increase a little, due to the warming of the northern regions. The impact on human settlements might be an acceptable level because Japan's location is along a wide span of climatic ones which range from subtropical to cold latitudes. Thus sheltering technologies have already been established which can adapt to any type of climatic condition.

Japan produces 90 per cent of its GNP from activities in its coastal area, and some part of this will be affected by a rise in sea level with potential high tides and floods. Local industries, such as tourism, sake fermentation, and primary products which are strongly dependent on particular climatic conditions, may well need to be relocated in the long tun.

Although the impact domestically might be considerable, the impact on other countries may dramatically affect Japanese industries and citizen's lives, because of Japan's strong dependency on foreign countries as described above. For example, the drought of 1973 in USSR accelerated its import of wheat from the US. This eventually caused a steep rise in price within the world market. Also, that year, there was a poor crop of counsellor in USSR, peanuts in India, and in addition there was a poor catch of anchovy in Peru. This induced "Tohu Panic" amongst the Japanese because of the tightness in their soyabean market.

Thus for these reasons Japan is keen to monitor the projected changes in climatical conditions as well as the effect on world trade mechanisms.

The Energy Position of Japan

Japan is self sufficiency in only 20 per cent of it's energy supply. This implies that it has a free hand about its choice of energy with the exception of some efforts to maintain domestic coal mines as well as the vulnerability within the international market. This situation has inevitably driven Japan since the Oil Crisis to (successfully) promote the development of energy conservation technologies.

The energy mixture shows: (1) the drastic decrease of energy consumed per GNP from 1973 to 1987; (2) the 12 per cent increase of total energy consumption in these 15 years and (3) the expanded use of nuclear and natural gas has helped to decrease the CO_2 emission-energy ratio to 0.88 in this period.

The tentative plan towards 2005 shows an additional decrease to 0.84 of the 1986 level in the emission of CO_2 per energy consumption. This would be achieved by promoting the use of nuclear and natural gas. However, the recent anti-nuclear movement may be an obstacle in achieving this figure.

Technological Progress within the Environmental and Energy Fields

The rapid industrial development within small settlements caused serious pollution problems in the 1960—70's. Technology was the only way in which a compromise between economic growth and environmental protection could be reached. The regulative standards on SO_2 and NOx emissions from both stationary and mobile sources were set at a slightly higher level than was attainable by the technologies of that time. Thus this lead and spurred on R & D activities and competition among the various industries concerned. For example, exhaust gas treatment facilities that have passed exceedingly strict standards are now exported world-wide as a result of the concern over acid rain and other related environmental issues.

Emission control techniques such as these also helped to achieve energy conservation at the same time. After the Oil Crisis, the export of Japanese automobiles greatly expanded because their high energy efficiency was welcomed by the customers. More than 90 per cent of Japanese automobiles have small engines under 2000cc, and a 10-40 per cent greater fuel economy in comparison with other vehicles imported into Japan. The energy consumption of Japan's steel industry is 25 per cent lower than the world average and has decreased by 10 per cent in the last 10 years. The total energy conservation of 2.7 per cent per year is the best within OECD countries, and shows Japan's continuing effort, which is encouraged because of its economic and geographical conditions.

The Japanese experience is a good guiding example of how to respond to global warming: the goals when appropriately sets give incentives for technological progress, as well as showing that the environment and energy consumption needs be not conflicting concepts.

Principle for Evaluating Global Environmental Technology

Technological developments can enlarge the range of political options available. Evaluations based on new and different concepts from conventional ones should be applied. Technologies which prevent and adapt to global warming in this new environmental era should be selected.

The key characteristics of global environmental issues are "irreversibility" and "closed". Besides these, conventional evaluation criteria such as "feasibility", "cost", "effectiveness" are also essential. International equity and consensus amongst countries are also other important criteria which should be considered. "Irreversibility" of the environment implies that the technology being applied should be one which is technically well established, has a secure safety operation and has also been assessed for any adverse effects. Thus we should be prudent about the introduction of large and complicated technologies into areas that do not have a sufficient technological infrastructure. This also applies for the application of climate control technologies which may have the potential for causing new disasters.

Another characteristic of "closed" suggests that the technology should be self sustainable in terms of generation, country and energy. For example, it is not "closed" generationwise if we accumulate radioactive waste for which the final treatment has not established. This then means that we transfer to the next generation the risk of radio activity and the responsibility for developing that technology.

The shipping and dumping of toxic waste to foreign countries is not regionally closed. The use of imported synthesised liquid fuel from coal does not mean that there will be a decrease in CO_2 emissions from the total energy cycle. The method of energy analysis should be fully applied in the development of new technologies if it is to ensure that self sustainability in energy and a sunstantial gain in energy are attained. The introduction of such new criteria will, in due course, promote "local" technology. This should be taken into deliberate consideration when transfering technologies from the industrialised countries to the developing ones. Developing countries should establish their own self development programmes based on their local characteristics. Exporting countries should assist and promote this self-supporting effort of importing through organising cooperative R & D projects. Also the transfer of technologies that have developed in a different environment should not be forced. For example, large and centralized technologies may not be suited to their specific needs which may need instead rather small and diversified technologies. So it is not necessarily the newest and speculative ones that may be needed but rather mature and fixed ones that can be used prospectively under global environmental limitations.

The Japanese Response to Global Warming

As a follow up to "The World Commission on Environment and Development" recommendations (1987), the environment Agency of Japan set up an "ad hoc advisory panel on the Global Environment". The final report published in June 1988 gave a comprehensive response from the Japanese Government on global environmental issues including that of climatic change (JEA, 1988a).

Parallel to the debate on the Ozone layer problem, public interest on global warming started to increase in 1988. The "Panel on Global Warming" which was established in June 1988 summarised the State-of-the Art of scientific dimensions in this matter in November 1988, and indicated what medium and long term guidelines the Japanese Government should take. Other Ministries followed this. For example, the Ministry of International Trade and Industry set up 3 advisory committees on energy, pollution control and technological related aspects in July 1988. The Meteorological Agency's advisory group recognised the global warming trend in its report issued in February 1989.

In response to the international cooperative activities by UNEP/WMO, the Japanese Government expressed a positive attitude, and participates in the Intergovernmental Panel for climate Change (IPCC) as a vice chairman of the 2nd Working Group and as co-chairmen of sub groups in the 2nd and 3rd Working Groups.

The Japanese Government is hoisting on September 11th-13th of this year a world conference "On Global Environment and Human response towards Sustainable Development" where "Global Atmospheric Change" and "Development and the Environment" are the main issues due to be discussed.

The report of the Committee entitled "Policy Recommendation Concerning Çlimate Change" suggested the following guidelines, and was used as a basic concept for the Japanese Government's response to the work of the IPCC.

(1) Indications about warming are extremely difficult to detect since it will vary according to the locality, and the mechanisms of warming are still not understood. In sum, the extent and timing of the warming, the effect on the global environment, as well as its social and economic impacts are not at all clear at this stage. Furthermore, little investigation has so far been done regarding the technological feasibility of counter-measures which may be taken against the effects of warming.

(2) However, global warming will seriously affect the whole world and it is a gradual, complex process which will have an irreversible impact on the environment. Hence appropriate provisions must be made in advance.

(3) The panel recommended that the Japanese Government should support basic scientific research, the development of technological countermeasures, and further investigations on the social and economic impacts of global warming together with the execution of any feasible countermeasures.

(4) Recent discussions on the problem of global warming have focused primarily on carbon dioxide and reducing the emissions of CO_2 by the setting up appropriate criteria. However, in order to achieve international agreement on a programme aimed at reducing CO_2 emissions, further discussion by experts is required because of the current deficiency in data correlating the rate of reduction to the control of global warming, and also because each nation's available energy sources and ability to achieve energy reduction will differ. Provisions for reducing CO_2 emissions are recommended to be as flexible as possible in order to gain a wide spread acceptance from all nations.

(5) Medium/long term guidelines for technological development

(i) Provisions for energy utilisation which contribute to the emission control of carbon dioxide emissions. These would include energy conservation, fuel con-version, energy substitution, improvement of heat-electricity conversion efficiency, etc.

(ii) Provisions for the development of technologies which contribute to the emission control of carbon dioxide including the recovery of exhaust gases, development of global biomass, conservation and recycling of resources such as wood.

(iii) To minimise the effects of the trace gases (methane, ozone, nitrous oxide, chlorofluorocarbons, etc.), provisions to limit CFCs, to reduce tropospheric ozone concentrations by regulating photochemical air pollutants, and to reduce methane concentrations by controlling carbon monoxide and hydrocarbon emissions.

(6) Necessity of having a standing international organisation.

(7) Support of developing countries. Although industrialised countries are mainly responsible for the discharge of greenhouse gases, developing countries will take some of this responsibility in the future. Moreover, developing countries may well suffer from the most serious damage caused by warming. Due to technological and financial difficulties and the lack of expertise in developing countries, there will be a great need for support from other countries. To assist developing countries in taking appropriate and timely action, it is the responsibility of developed nations to provide adequate information regarding the global warming problem; The developed nations must also assist the developing nations in taking feasible action to provide technological and financial assistance. It is recommended that this support should be carried out extensively.

Some policy options are now under discussion in response to the worked of the IPCC's third working group.

(1) Economic measures: charging for emissions of CO_2, Transfer of Pollution rights (internationally & domestically), subsidisation of R & D, financial aid towards the conservation of energy.

(2) Financial measures: Establish funding systems which promote the transfer of technology to the developing countries.

(3) Educational measures: strengthening of educational activities at the citizen level.

(4) Regulational measures: strengthen the regulations on exhaust gas emissions from stationary and mobile sources.

(5) Technical assistance through Overseas Development Aid.

(6) International Law-frame work convention and detailed agreements.

A high energy price certainly promotes energy conservation. The price of Japanese gasoline is three to four items more expensive than in the USA, including a tax of about 40 per cent. One reason why the Japanese have attained savings in energy is due to this price mechanism.

To allow membership countries to choose options suitable and manageable for each of their backgrounds, a loose international agreement is preferable. The concept of international Transfer of Pollution Rights (TPR) would be a rather acceptable option from this point of view.

Research Project in Japan

Organised research work was started in 1979 by groups funded by the Ministry of Education, culture and Science, and from 1982 the research programme on climatic variation was organised under the World Climate Programme. However, so far these have been rather small scale studies.

In 1984 the Science and Technology Agency (STA) surveyed world-wide research trends and recommended the promotion of research works on CO_2 issues. In 1988 the National Institute for Environmental Studies (NIES) started research on the atmospheric chemical reaction mechanism, and STA initiated an international research programme on atmospheric-ocean-climatic changes for the Pacific Ocean region.

The rapid rising of this issue in 1987-88 spurred governmental organisations to initiate research projects in each of their responsible fields as shown. Recently there has been a slight tendency of increasing the budget for joint international research work, including research carried out under the ODA budget.

It is necessary to have data and a common method which may be applied if the international cooperative research efforts are to be effective in this global warming issue.

Already IGBP and IPCC are establishing cooperative research networks, and Japan has positively participated in these programmes.

It is important and beneficial to discuss the possibility of establishing a research network for the pacific Asian basin, where there is now a deepening in the mutual dependency amongst the countries of this region. The rationale and implementation of this cooperative work might then be:

(a) Application of common methodologies to individual regions (communication by holding conference)

(b) Transfer of methodologies and technologies (conference, communication, education.

(c) Regionwise integrated research (research works in individual regions and establishment of a head quarter to summarise these).

(d) Establishing a common data base (individual survey and a function as a centre).

(e) Survey of local conditions for introducing technologies into a specified region (joint survey between concerned countries).

(f) Making a common international cooperative policy (conference).

(g) Solving conflicts among countries (joint study, discussion at conference).

The research items in the global warming field that need international cooperation are illustrated.

Research Works Being carried out in the National Institute for Environmental Studies (NIES)

NIES has a five years research plan which started in 1988. The research themes now carried on are as follows:

(1) identification of the origin and measurement of greenhouse gas emissions-measurement of the stable carbon isotope ratio, methane from rice fields, nitrous oxide from field crops, hydrocarbons from vegetation;

(2) measurement of trace gas behaviour:

- development of a measuring method for tropospheric trace gases;
- feasibility study for designing a monitoring system in the East Asia region;

(3) analysis of the photo-chemical reaction in relation with the greenhouse effect:

- analysis of the reaction mechanism of low hydrocarbons in a photo chemical smog chamber;
- estimation of the greenhouse contribution to radiation using a one-dimensional radiation model;

(4) analysis of the transfer mechanism between the biosphere, ocean and atmosphere:

- experimental analysis of plant growth under high CO_2 concentrations using a plant environmental simulator;
- experimental analysis of the role of aqua-organisms using a microcosm;

(5) policy option analysis and establishing a data base system:

- development of energy and environmental model to verify policy options;
- development of a gaming simulator using a knowledge based simulation system (Kainuma 1989);
- structuring of a data base on the greenhouse effect to support research planning and model building (this research is carried out

with close connection with JEA's policy making group under IPCC 3WG work);

(6) development of methodologies to assess environmental and socio-economic impact:

— development of methodologies to estimate the impact on hydrology water resources vegetation, human health and industry (This impact study corresponds to the work of the IPCC 2WG. Planned field is east Asia Region);

— development of impact monitoring method by using data from remote sensing activities;

Some international cooperation is proceeding (including that in planning stage) as follow:

(1) identification of the sources of greenhouse gases:

— survey of emission sources like methane from rice fields in the east Asia region (planning joint research with east Asian countries and the USA);

(2) ecosystem monitoring by remote sensing technique:

— monitoring of deforestation using Landsat data (a joint project being carried out with Thailand);

(3) impact study:

— preparing report for IPCC 2WG (Section V);

— estimating the impacts mainly in the East-Asia region (feasibility study for cooperative work in the region);

— workshop on climatic impact (planning: January 1991, Tsukuba with UNEP, WCIP);

(4) development of methodology for estimating socio-economic impact:

— joint study with research fellow from the UK (STA research fellowship);

(5) information exchange of impact study and policy option:

— planning under Japan/US environmental cooperation agreement;

(6) cooperative work to establish research centres for environmental studies:

— advise on the planning of environmental information and research centres in china through the ODA and in Thailand.

Japan has so far responded to the environmental and energy problems in a way induced by its economic growth within a limited land area. This has shown that Japan greatly depends on the world ecosystem and is responsible for preserving it. Also, at the same time it can potentially contribute via its technological and managerial capacity on confronting this global environmental crisis.

Some restrictions can be a powerful driving force in promoting technology, goals once appropriately set can give incentives to the innovators, a saving in energy is equal to saving the environment. Early preventive action results in a better balance of accounts in the long run, sometimes bold decision making under uncertainty is necessary. These are a few of the lessons Japan has learnt in the process of struggling for a compromise between the environment and the economy. There were some tragic episodes caused by pollution in its industrialising stage, although it still has many environmental problems to solve.

The importance of scientific analysis is obvious, and we should look patiently at what scientists are revealing and carefully try to understand the implications of it.

Meanwhile, the irreversible characteristics of such global environmental phenomena and the predicted shortness of time left to respond to the changes that will be imposed upon us, we must to prepare for the future, by developing technology, by proposing policy options to secure the rational usage of technology, and by making a consensus amongst countries for a cooperative action plan which responds to it. Also, we need to make an early warning network system, establish a new concept by global environmental technology may be evaluated, to set appropriate goals which will challenge and continue R & D efforts with international cooperation. And these are the subjects which must now be fully discussed.

In addition to these, we should advance practically our existing knowledge and technologies to ease and delay the forthcoming change. The transfer of technology with consideration of the locality in which it is to be used is one thing we can do immediately. Also, Japan's experiences suggests that we can watch carefully and prudently, deliberately prepare for the future and sometimes boldly and practically proceed.

D. NEW ZEALAND

New Zealand has actively participated at Ministerial and official levels in the Hague Declaration and the work of the WMO/UNEP Intergovernmetnal Panel on Climate Change and cochairs with the Netherlands the Coastal Zone Subgroup of the Response Strategies Working Group. New Zeland supports a global convention on climate and institutional means of control through the International Court of Justice. New Zealand supports the existing UN agencies as the means of implementation and a strong Pacific input to the policy development work.

The New Zealand Government initiated a three-phase Climate Change Programme in June 1988. The Facts Working Group has prepared climate change scenarios for New Zealand and the Impacts Working Group is currently receiving expert responses to the scenarios in terms of the biophysical environment.

In July the social and economic impacts will be assessed and an integrated impacts assessment developed by March 1990. The Policy Working Group is currently identifying key issues on which the impacts work can focus. CO_2 and other greenhouse gas emissions are being studied for New Zealand which it appears is a net consumer of CO_2. Energy management issues will dominate in this context. The policy Group report by June 1990.

The hallmark of the NZ Climate Change Programme is a consultative process with expert and community representatives.

New Zealand is very pleased to be able to take part in this important conference in the Pacific context with which we strongly identify. Thank you to the organisers for your invitation and support to attend.

New Zealand identifies culturally and politically with the Pacific Basin. Its policy response to the climate change issue has two main foci.

The international community by representing the Pacific context, and domestically by focusing on the peculiaritis of an isolated island nation which is economically developed, but dependent on its primary production-agriculture, forestry and fishing-sectors for two-thirds of its export earnings.

Small nations like New Zealand in association with other small nations can play a very effective role internationally by keeping global issues on the international agenda and by raising the peculiar vulnerabilities of island nations in such fora as UN, UNEP/WMO, IPCC, SPREP, OECD, IUCN.

On 9 May in Geneva the Secretary for the Environment, Dr. Roger Balkeley, represented New Zealand as cochair of the Coastal Zone Management Subgroup of the Intergovernmental Panel on Climate Change (IPCC), Response Strategies Working Group. He sought their leave for me to report to this meeting about the work of the Coastal zone Management Subgroup. I will also outline New Zealand's domestic response to Global Climate Change.

New Zealand's International Response

New Zealand is taking a very active role in the international activities on global climate change at Ministerial and official representative levels.

New Zealand was an active participant and signatory to the Hague Declaration seeking global institutional means which provide legally binding solutions to the global climate issue subject to the control of the International Court of Justice. New Zealand supports further work on a global climate Convention through existing institutions in the UN system such as UNEP/WMO and the IPCC. It is important that the work of Pacific based agencies link into the international work of existing institutions.

New Zealand supports the SPREP work on climate change and ways in which the Pacific input to this work can be strengthened. The SPREP Intergovernmental meeting in July will further address this issues. The WMO/UNEP Intergovernmental Panel on Climate Change (IPCC) will provide an integrated assessment of climate change scientific findings, impacts and response strategies by mid 1990. Work will follow on a global climate convention and protocols.

IPCC Response Strategies Working Group Coastal Zone Management Subgroup

The Coastal Zone Management Subgroup goal is:

"To provide information and recommendation to national and international policy centres, enabling decision making on:

(a) coastal zone management strategies for the next 10-20 years;

(b) long-term policies dealing with limitation and/or adaptation with respect to climate change/sea-level rise."

The Subgroup will take into account technical, economic, social, cultural, environmental, legal, and institutional aspects.

For the purpose of the Coastal zone Management Subgroup, the "coastal zone" has been defined as:

"The area of sea-land interface extending seaward as far as land-based anthropogenic activity directly affects the sea, and extending landward as far as the marine environment affects the land in a physical, environmental, social cultural or economic sense."

A report is being prepard for June 1990 addressing:

- the extent and timing of sea level rise and associated climate change effects including regional changes in climate, changes in variability and extremes, transient predictioins and uncertainties.
- the impacts of sea level rise and associated climate change effects of land use, ecosystems, water resources, and human activities.
- the context within which the problems arise, for example, the relationship with existing problems, timing and national (site specific), regional and global aspects such as acceptable levels of risk, costs and benefits associated with preventative responses and the relationship with other natural disasters.

- available adaptive options and costs for coastal areas and activities at risk. Land loss and flood risk have been identified as critical factors. National and cultural differences were recognised as important when identifying appropriate response strategies.
- social, economic, legal, institutional, environmental and cultural implications of adaptive options. The NZ delegation considered that the following issues needed to be addressed by countries when responding to sea level rise and coastal zone problems associated with climate change.
- a consensus needs to be developed in the expected range of rates of sea-level rise.
- institutional means need to be established to make this information widely available.
- persuasive means need to be developed, as well as legislation and regulations, including fiscal measures.
- areas prone to inundatioin through a rising sea level need to be identified and mapped.
- contingency measures need to be generated to cope with extreme events.
- planning needs to be developed and/or improved to cope with social implications of sea-level rise.
- future development of the coastal zone needs to be controlled, including land-uses.
- technical options available need to be inventoried .
- financial, social, cultural, environmental, and legal implications need to be identified.
- geographical considerations need to be taken into account (tropical or temperate coasts, continental coasts or island systems).
- international implications (forced migration) need to be considered.
- possible funding for adaptive options, in particular mechanisms to assist developing countries to implement adaptive options.

The Netherlands as cochair identified the following criteria to be considered when identifying problems:

- vulnerability
- people at risk
- economic values
- environment
- per cent of Gross National Product

It was suggested that scenarios need to be developed covering the periods between the present and 1990 (existing problems), the present and the year 2050 and the present and the year 2100.

In addition, the following categories of responses were suggested for the analysis:

- technical measures leading to a lessening of risk for lives, economy, environment
- non-technical measures, leading to abandonment of regions at risk (resettlement, land use planning).

Long-term costs in terms of lives, economy and environment need to be weighed, in particular the financial implications in coastal zone management for developing nations where neither the technological nor the financial ability to cope with the threat of sea level rise and associated coastal effects exists.

The work plan is ambitious with a mid-April 1990 draft reporting date.

To facilitate a comprehensive input to

the programme by as many different geographical situations and nations, two workshops are planned:

- Workshop 1 in November 1989 to be held in USA (sponsored by USA) on Adaptive Options and Policy Implications for Coastal States in the Temperate Zone.
- Workshop 2 in late 1989 or early 1990, to be held in Australia (sponsored by Australia, New Zealand and the Netherlands) on adaptive Options and Policy Implications for Coastal States including island nations in the tropical zone.
- The focus in Workshop 2 will be on possible severe problems for Pacific and Indian Ocean countries. Liaison with SPREP will ensure maximum country participation.

The Response Strategies Working Group Steering Committee met from 10-12 May in Geneva and identified the following key issues to be addressed:

- policy options to limit global warming
- legal initiatives towards a global convention on climate
- ways of increasing involvement of developing countries in the work of IPCC.

Several key conclusions were drawn from the scientific evaluations:

- revolutionary changes will be required in energy consumption patterns and technological development if CO_2 levels in the atmosphere are to be stabilised by the end of the century at twice pre-industrial levels or lower.
- changes required would be radical improvement in energy conservation and energy efficiency in domestic, manufacturing and transport sectors, savings in energy conversion, a shift from fossil fuels to renewable energy resources such as solar and wind power and photovoltaics, use of less carbonated fuels, e.g. natural gas and biomass from plantations, a rapid phaseout CFCs and a halt to tropical rainforest destruction and increased afforestation.
- the steering group decided to initiate work on a global climate convention in accordance with the UN General Assembly resolutions to so do. The elements of a convention will be reported by June 1990 as part of the IPCC report to UNEP/WMO and the 45th UN General Assembly. It is anticipated that UNEP would take up the negotiation of a global convention and the necessary protocols.

Preparatory work will begin shortly on

- legal measures and processes
- technology transfer and development issues
- financial resources
- public education and implementation measures.
- economic (market measures).
- directions were given to all subgroups of RSWG to make special efforts to encourage participation by developing countries including funding assistance to attend workshops and meetings.

NZ Climate Change Programme

The Government agreed in June 1988 that a coordinated climate change programme was required by New Zealand to respond effectively to the climate change issue which has potential widespread implications for New Zealand society. Three working groups on climate change have been set up under a Steering Group chaired by the Secretary for the Environment, Dr. Roger Blakeley.

The establishment of the New Zeland programmes followed the Ministry for the Environment seminar "Climate Change—The New Zealand Response": held on 29-30 March 1988 which addressed what the New Zealand response should be, including:

- the need for further refinement of the climate change predictions;
- the need for planning to take into account the predictions of climate and sea level changes;
- the development of reviewable planning guidelines for local authorities to work from;
- the need to inform and educate the public on the issues;
- the need to initiate a programme to reduce greenhouse gas emission and to recognise the links with energy policy.

The following working groups were set up in June 1988 to:

- produce a report on the scientific basis for the predictions of climate change. This was completed in December 1988 (Facts Working Group convened by the Royal Society of New Zealand).
- develop an interim data base of all New Zealand climate change information and assess the impacts of climate change using the Facts Working Group predictions in order to produce an interim impact assessment report to be completed by March 1990 (Impacts Working Group convened by Ministry for the Environment).
- develop policy on how New Zealand can respond to climate change (including mitigation and adaptation strategies) for presentation to Government by mid-1990 (Policy Working Group convened by Ministry for the Environment).

The Ministry for the Environment provides overall coordination for this programme and there is significant involvement by many other departments. These include the New Zealand Meteorological Service, Department of Scientific and Industrial Research, Department of Conservation, Ministry of Agriculture and Fisheries, Ministry of Forestry, Department of Maori Affairs and Ministry of Womens Affairs.

In addition it was agreed that the working groups should consult with a wide range of interests outside government from other institutions and agencies and from the community including tangata whenua.

The Facts Working Group has published an abridged version of its report. The following conclusions were reached:

- The most likely prospect is an accelerated warming trend, so that global temperatures average 0.5-1.0^{0}C higher than present by 2025, and perhaps 1.0-2.0^{0}C higher than present by 2050. In New Zealand latitudes the annual average warming should be in the same range as the global average values. There is some possibility that winter temperatures will rise faster than summer temperatures in more southerly parts of the country.
- No meaningful predictions can, at present, be made about wind changes although on average the prevailing westerly winds could decrease, particularly in the winder period.
- The snowline is expected to retreat by about 100-150 metres for every 1^{0}C rise in temperature. However, whilst New Zealand temperatures have increased by 0.5^{0}C since the 1940s no retreat of snowline or advance or treeline has yet been measures.
- An urgent need exists to improve regional computer models to help determine possible future variations to the dominant westerly wind over New

Zealand, and to further explore atmospheric circulation patterns during past analogous climates.

- The most likely mean sea level change in the New Zealand region, consistent with the expected temperature changes, is estimated to be a rise of 10-20 cm by 2025 and a rise to 20-40cm above present levels by 2050. This does not take account of any vertical movements of the land due to tectonic forces. No significant mean sea level change is expected to result from changes in large-scale atmospheric circulation patterns.

New Zealand scientists at the Meteorological Service and Department of Scientific and Industrial Research especially are taking an active role in the climate modelling work being done in the Southern Hemisphere in collaboration especially with Australia and also in the measurement of trace gases at clean air sites.

The agricultural, forestry and conservation agencies are actively involved in impacts research with respect to their sectors. The Steering Group of the New Zealand Climate Change Programme and the Royal Society of New Zealand have a coordinating role over climate change research to ensure continued support and its focusing on the key political issues.

The Impacts Working Group includes a comprehensive network of New Zealand's leading experts on the broad range of physical, biological, economic and social issues likely to be affected by climate change. Currently, first order impact scenarios are being prepared for workshop in July at which time the social and economic impacts will be considered. The Impacts Working Group are reporting to a core group that has been charged with preparing impact scenarios from the information the experts provide. The core group will also consult widely with special interest and community groups throughout New Zealand.

The Policy Working Group is currently scoping the key policy issues to enable the impacts work to be focused on most likely effects requiring policy responses. A wide range of natural and physical resource based departments and service sectors are addressing the issues along with sectors based on cultural, social and external relations and trade interests. Work to date is of a preliminary nature as the policy group awaits the impact scenarios.

Several preliminary themes have already emerged:

- institutional/legal issues
- research/data/monitoring issues as they relate to limitation and adaptive responses
- operational policy issues, eg health, energy, agriculture, transport
- information/education issues
- transboundary/migration/Pacific issues
- cultural issues (tangata whenua)
- economic issues
- social issues.

The key focus that the New Zealand policy programme will have prior to the impacts scenarios completion, will be to address the range of limitation responses as they relate to CO_2 and other greenhouse gas emissions. With international initiatives developing on a global climate convention, New Zealand needs to be in position to address how a convention and protocols might affect its emissions policy and in particular its energy management.

In the meantime the preliminary indications from the impacts group are:

- largely negative impacts could be experienced by freshwater fisheries due to water temperature over current changes.

- positive impacts would be experienced by the agricultural sector with opportunities to respond to a change being presented.
- increased vigilance at New Zealand's borders could be necessary to avert possible importation of unwanted pests.
- changes in soils could be subtle, increased erosion and soil fertility changes could be associated with climate extremes being amplified.
- coastal, high altitude, "island" ecosystems, could be at gratest risk.

International fora have emphasised the need for national and regional scale assessments of the effects of global warming and atmospheric deterioration.

New Zealand and other Pacific island nations have unique problems and national programmes such as New Zealand's will help shape the international responses and focus attention on the urgency for action considering a range of effects and responses appropriate to geographic location and cultural circumstances.

E. AUSTRALIA

The prospect of global warming as a consequence of increasing concentrations of greenhouse gases in the atmosphere in a matter of the highest concern for Australia. It shares its concerns with all other countries of the world but is potentially placed in a situation of particular vulnerability because of the combination of:

- its economic dependence of agriculture and fossil fuel exports;
- the expectation that large areas of its agriculturally productive land will be subject to hotter and drier conditions;
- the extent of its coastline (over 36,000 km in length) and the concentration of its major population centres along it;
- the likely extension of cyclones and other extreme climatic events into previously unaffected highly populated and unprepared areas of the country;
- the extension of the area of the country where tropical diseases—human and animal—are likely to occur; and
- the potential shrinking of habitat for unique flora and fauna.

Australia is therefore affording the highest priority to responding to the threat of global warming through a three-phase strategy of:

- further research, particularly to provide regional assessments of likely climate changes and their impacts;
- development of policies to reduce the build-up of greenhouse gases in the atmosphere so as to reduce the greenhouse effect; and
- development of policies to deal with the effects of climate change due to the greenhouse effect, by means of minimising the impact of such changes or adapting to those impacts.

This paper discusses the contemporary state of knowledge of the likely impacts of global warming on Australia and recent Government initiatives to address the greenhouse threat. In discussing potential impacts, it draws heavily on papers presented at a national conference, *Greenhouse 87: Planning for Climate Change.* The conference papers are based on a working scenario of climate change in Australia to the year 2030, which was developed for the conference, and are published in a collected volume (Pearman 1988). There is still considerable uncertainty associated with this scenario and the predictions based upon it. Nevertheless, the fact that they are even possibilities underscores the need for concerted action on the part of governments.

Climate

Australia is the lowest, flattest, and, apart from Antarctica, the driest of the continents. Approximately one-third of its land mass lies within the tropics. The climate of Australia is predominantly continental but the insular nature of the land mass is significant in producing some modification of the continental pattern.

Extreme minimum temperatures are not as low as those recorded in other continents but extreme maxima are comparatively high, reaching 50^{0}C over the inland. Average annual temperatures range from 28^{0}C to 4^{0}C with average monthly maxima ranging from in excess of 40^{0}C in the north-west to around 15^{0}C on the west coast of Tasmania. Based on the best currently available scenario (Pittock 1988), by 2030 temperature increases are expected to be about 2^{0}C in inland areas of northern Australia and up to 3 or 4^{0}C further south. Near the coast these increases could be moderated by the lagging behind of sea-surface temperatures, and it is probable that winter and overnight minimum temperatures will warm more than summer and daily maximum temperatures. The continent is relatively dry, with 80 per cent of the area having a median rainfall of less than 600 millimetres per year and 50 per cent less than 300 millimetres. The rainfall pattern is strongly seasonal, with a winter rainfall regime in the south and a summer regime in the north. A likely greenhouse effect is that the summer rainfall regime will intensify and push further south. Reduced rainfall is expected in the southwest of Australia but the implications for rainfall in southeastern Australia are not clear at present (Pittock 1988).

Climatic discomfort, particularly heat discomfort, is significant over most of Australia. During summer, prolonged high temperatures and humidity around the northern coasts and high temperatures over the inalnd cause physical discomfort. In winter, low temperatures and strong cold winds over the interior and southern areas can be severe for relatively short periods. Westerlies may strengthen in the southern areas due to the lag of ocean temperatures behind that of the continent (Pittock 1988).

Tropical cyclones develop over the seas around northern Australia between November and April. On average, about three directly affect the Queensland coast and a similar number the north and north-west coasts. Pittock (1988) predicts that by 2030 tropical cyclones might form 200-400 km further south, based on increases in sea-surface temperatures. This would mean that major populations centres such as Brisbane might be affected.

Coastal Impacts

In common with other countries, the coastal zone of Australia is likely to be significantly affected by global warming. Coastal areas, particularly in northern Australia, characterised by low gradient tidal flats and beach ridge plains would be most affected, with a rise in sea level of 1 m producing shore line retreats in excess of 1 km (Short 1988). However, in a preliminary investigation of the impact in arise in sea levels of 0.2-1.4m over the next 40 years, Cocks et al. (1988) concluded that, with the exception of increased flooding in coastal settlements, the impact on settled areas of coastal Australia will be small.

Australia's island external territories would be markedly affected by sea level rises. The situation is particularly bleak for the Cocos and Keeling Islands, a group of 27 small Indian Ocean coral islands. The maximum elevation of these island is less than six metres and they would suffer severe land loss, in addition to threats to their freshwater lens. Adverse secondary effects would include impacts on wildlife, especially the large seabird nesting population on North Keeling Island, and the need to resettle the

450 inhabitants, all of whom live on Home Island. This would almost completely disappear if the sea level rose 1.5 metres.

The availability of water, or the lack of it, is the most important single factor determining land use and rural production in Australia. Permanent rivers and streams flow in only a small part of the continent, with about 80 per cent of Australia being significantly dependent on groundwater supplies. About 80 per cent of all water used in Australia is surface water. Even in areas of high rainfall, large variability in stream flow means that, for local regional development, most streams must be regulated by surface storage. However, in many areas evaporation is so great that storage costs are high in terms of yield. Extreme floods also add greatly to the cost of water storage, because of the need for adequate spillway capacity.

The climate changes outlined above may impact adversely on Australia's water resources, particularly in the southern part of the continent, which contains the bulk of its population and agriculture. Pittock (1988) has marked regional and seasonal changes in the water balance can be expected.

This is supported by modelling studies indicating a considerable amplification of the effect of climatic change on catchment hydrology and reservoir performance. The summer rainfall regime regions are likely to experience problems associated with high runoff volumes, and those regions in the south of Australia subject to winter dominated rainfalls will suffer as a result of too little runoff (Nathan *et al.* 1988). Sadler *et al.* (1988) estimate that the yield of the sources for the water supply of the major southwest city of Perth will decline by 35-45 per cent over a 50 year period as a result of projected hydroclimatic changes. Agriculture remains a vitally important sector of the Australian economy. Although its relative importance as an export earner has been declinging since the late 1950s, it still accounts for significant percentage of export earnings.

76 million hectares of the Australian continent, or about one tenth of its total area, are suitable for arable agriculture, having regard to climatic, soil and topographic constraints (Nix 1978).

Climate change will impact on agriculture through changes in temperature, rainfall and water availability. Increases in rainfall in subtropical areas could result in the extension of cultivation of summer growing crops such as sorghum (Russell 1988). However, Nathan *et al.* (1988) suggest that reductions of up to 100 mm of soil moisture during the growing season in winter wheat areas may threaten the viability of production in marginal areas.

The innermost 70 per cent of the Australian continent is arid. The major use of this land is the grazing of native pastures by sheep (in the southern arid areas) and cattle (in the central and northern areas) on an extensive scale and at very low densities. Graetz *et al.* (1988). conclude that expected climatic changes in the south of the continent will have deleterious effects on the wool industry because of a decline in pasture nutrition values and other related factors.

This picture is complicated by other factors such as storm and cyclone damage, salinity resulting from groundwater changes, soil erosion, and the directly beneficial effects on growth of increased atmospheric carbon dioxide levels. Russell has also observed that the ENSO phenomenon has been shown to have substantial effects on the productivity of Australian agro eco systems, and that with a southward extension of summer rainfall this effect is likely to increase.

It is unclear at this stage whether the above factors will result in a net increase or decrease in agricultural productivity. However, it is clear that the significant climate variability (on both a year-to-year and long-term basis) which has characterised Australian agriculture since European settlement will intensify and that adaptability will have to increase accordingly. In this regard, Australia is fortunate, as it has a better basis for adaptability than those countries with relatively stable climatic patterns and government supported rigidities in their production and trade systems.

Natural Environment

Australia is one of the seven mega-biodiversity countries of the world, which between them account for the great majority of the world's biodiversity.

Predicted climatic changes are expected to have significant effects on natural flora and fauna as a direct result of those changes and secondary effects such as sea level rise. Main (1988) has noted that the resilience shown in the past by Australian ecosystems to disturbances depended in part on the lack of barriers to movement of the flora and fauna. However, this is no longer likely to be the case, particularly for smaller conservation reserves, since many are surrounded by settled or agricultural areas unsuited to the passage of many animals and native plants, and ecosystem distruption is to be expected. Main has further concluded that some species will be restricted to environments at the limits of their physiological or ecological tolerances and hence at risk of extinction.

Busby (1988) has undertaken detailed studies of one endangered species, Potorous Longipes, the long-footed potoroo, using the species-climate model BIOCLIM. He concludes that the continued survival of this species is doubtful because, after climate change, the only area predicted to have a suitable climate for it is one over 200 km away from its present highly restricted distribution.

Many low lying islands on the Australian continental shelf retain species no longer or rare on their mainland range and are designated as nature reserves. These may be affected by sea level rises and changes in storm intensity. The above discussion outlines possible physical impacts of climate change due to the greenhouse effect. It is not comprehensive, and does not attempt to address the significant downstream effects for energy, public health and tertiary industries (eg, insurance and tourism).

Nevertheless, the scope of affected areas make it apparent that this is an issue in which governments cannot operate effectively in isolation. There needs to be constant dialogue between policy-makers, planners, researchers and the wider community. Non-government organisations, polling and research organisations and the media are valuable sources of information on community interests and closely monitor the activities of government agencies; academic work contributes to policy and programme development; while the participation of industry and the general public is essential for acceptance and the effective operation of government initiatives.

Australia is a federation of six states. It also has two self-governing internal territories and a number of external territories. Legislative powers are divided between the Australian Commonwealth Parliament and the eight state and territory legislatures. Responsibility for those matters relevant to the greenhouse effect are shared between Commonwealth, state and local governments. This makes effective national coordination difficult but fortunately there is recognition of this and efforts are being made to ensure harmonised national action.

An overarching principle in responding to global climate change is that of sustainable development, as articulated in the report of the World Commission on Environment and Development, 'Our Common Future'(WCED 1987). The Australian Government has endorsed the general thrust of the WCED Report. It is apparent that the ultimate solution to global warming is beyond any one country and cannot be treated in isolation from those areas which are the focus of 'Our Common Future', including population, food security and energy.

Australia does not have the population pressures of other countries and it is a net exporter of energy and food. However, it cannot rest on this. To do so would ignore the impact on it of events elsewhere, and its responsibilities as a member of the world community. It is accepted in Australia that this involves particular regard for the countries and the circumstances of the region (Indian and Pacific Oceans and Asia) and Antarctica.

National Greenhouse Programme

As noted earlier, at a national level in Australia, there are three broad areas to be pursued:

- further research to provide regional assessments of likely climate changes and impacts of these changes, and to reduce scientific uncertainty;
- development of policies to deal with the effects of climate change due to the greenhouse effect, by means of minimising the impact of such changes or adapting to those impacts; and
- development of policies to reduce the build-up of greenhouse gases in the atmosphere so as to retard the greenhouse effect.

An important part of these goals is development of the flexibility of prepare for an uncertain future, likely to contain unforeseen developments and surprise.

In April the Government has recently announced funding for an accelerated programme of research to enable the development of regional climate models to allow better prediction of regional climate changes. This programme will support the Commonwealth Scientific and Industrial Research Organisation (CSIRO) and the Bureau of Meteorology to undertake core research in:

- general circulation modelling of regional climate change for Australia and the region, detailed estimates of regional climate change for selected areas of Australia, and modelling oceanic processes relevant to climatic change;
- measurement of atmospheric greenhouse gases relevant to validation of climate change models;
- interaction of the atmosphere and the oceans, including the role of the southern hemisphere oceans in climate change; and
- interactions of the atmosphere and the terrestrial biosphere, including biospheric inputs to models, satellite modelling of Australian region climate change and implications for ground water recharge, rainfall and temperature.

Additional funds will be provided to the Permanent Committee on Tides and Mean Sea Level to establish 7-8 state-of-the-art 'super' monitoring stations at appropriate sites around the country to serve as baseline stations for the existing monitoring network and provide high accuracy sea level and geodetic reference sites. This enhanced effort will complement the substantial amount of research already underway in the areas of atmospheric/climate, oceanic, primary production, natural resources and geological/paleosedimentary studies. This study made possible the Greenhouse 87 Conference. However, an accelerated programme of research is vital to ensuring a better understanding of the local and regional patterns and impacts of global warming and thus allowing appropriate responses.

The Antarctic Division of the Commonwealth environment department (DASETT) is currently undertaking a number of greenhouse related projects. Its Glaciology Deep Drilling Programme will drill over the next few years through various thickness of ice, with the object of extracting atmospheric gases representing the atmosphere of earlier times. The aims are to identify the content in the atmosphere of greenhouse gases on the following time scales:

- pre-industrial Revolution to the present (last 200-300 years);
- during the last deglaciation (30,000 years to the present); and
- over the last 500,000 years, recording changes that took place over approximately four glaciation-deglaciation cycles.

The Australian Antarctic stations undertake regular atmospheric monitoring for analysis by CSIRO in Australia and lake level measurements to determine changed precipitation regimes.

Additional research will be undertaken within the context of Australia's participation in the International Geosphere-Biosphere Project (IGBP). Key areas in which Australia is looking to contribute include the Joint Global Ocean Flux Study, the World Ocean climate Experiment and sea level rise.

A National Greenhouse Advisory Committee is currently being established. This committee of eminent scientists will provide expert advice to the Government on greenhouse issues and advise on priority areas for further research. It will also set objectives for dedicated research grants scheme to commence operation in 1990, and promote public dissemination of greenhouse information. Its work will include ascertaining more precisely the full extent of relevant research in Australia, its strengths and weaknesses, and significant gaps. The Government is also looking at greater investment in data bass, which would improve our situation considerably.

In common with other countries, Australia is still assessing preferred approaches to limited emissions of greenhouse gases. This assessment work is considering all greenhouses gases. It is likely that the preferred approach will involve many small gains across the spectrum of possible limitations, rather than reliance on major, and highly disruptive, gains in a single area.

Reduction of emissions through wider adoption of existing technology in relation to energy conservation and energy efficiency is a favoured short-term strategy since:

- it is estimated that it could generate 10-20 per cent savings of CO_2 emissions, from the expected year 2000 emission levels;
- it would reduce energy input into manufacturing; and
- it would improve Australia's domestic energy situation as the production of fossil fuels declines.

Australia is accelerating on-going programmes of research into energy conservation and renewable energy forms, principally through its National Energy Research, Development and Demonstration Programme. Although they are minor contributors to total energy use, Australia has considerable experience with some forms of wind and solar power.

In the area of reduction of methane emissions, Australia is looking at diverse possibilities such as reducing methane emissions from livestock (Australian research into changing the efficiency of animal stomachs is considered to have potential benefits in this area), encouragement of 'dry' rice production and reduction in leakages from gas circulation systems.

Changes in the motor vehicle fleet offer considerable potential, given the additive effect of reductions in CO_2 N_2O and CO emissions achievable from reduced tail-pipe emissions. The latter is significant in view of the part played by motor vehicle emitted CO in upsetting the traditional hydroxyl radical mediated balance between methane and CO.

Reductions in N_2O emissions may also be achieved by reduced use of nitrogenous fertilisers.

Australia is highly aware of the severe threat which the greenhouse effect places on the Pacific Island countries. Consistent with a an announcement of its Prime Minister at the 1988 South Pacific Forum, Australia has undertaken a feasibility study into the establishment of a network of stations to monitor climate change in the South Pacific. Following on from this study, the Government is likely to fund a programme of sea level and climate monitoring in the South Pacific.

In other initiatives this year. Australia has provided funding to the UNEP Climate Impact Studies Programme specifically to stimulate impact assessment work in the Australian/Asian/South Pacific region. It also agreed in December 1987 to a cooperative research programme with New Zealand. However, the detail of this programme is still to be developed.

The focus of international activity in the area of impacts of and responses to global warming is the Intergovernmental Panel on Climate Change (IPCC), which was established jointly by the United Nations Environment Programme (UNEP) and the World Meteorological Organisation (WMO) in 1988. The IPCC has three working groups focusing on science, impacts and responses (limitation and adaptation). Australia is a vice-chair of Working Group II (Impacts) and is actively contributing to the other two working groups. The IPCC is to report in July-August 1990 and will provide the comprehensive international statement to date, although the time available to is too short for any substantial new work to be undertaken.

To contribute to the work of the IPCC, Australia will be sponsoring in cooperation with New Zealand and the Netherlands a workshop on sea level rise and tropical coastal zone response strategies in early 1990. The workshop will focus in particular on the Indian Ocean and South Pacific regions. Substantial assistance will be provided to allow for attendance by representatives of developing countries.

Australia is also committing additional resources to participation in the WMO World Climate Programme. This programme will be reviewed by the second World Climate Conference, to be held in 1990. It is anticipated that this major conference will allow for review of greenhouse activities that have occurred not only as part of the World Climate Programme but in other fora, including the IPCC, as well.

Australia considers climate change to be an important part of the OECD agenda. The OECD is a significant forum for energy matters in particular, given the large contribution to total emissions of its member countries, and the Environment Committee and the International Energy Agency are examining options for reduction of emissions of carbon dioxide.

The build-up of greenhouse gases in the atmosphere is a global problem. Unilateral action by most countries, ahead of action by the major industrialised countries, is likely to have little effect. Australia is a signatory to the Declaration arising from the Hague Summit on Protection of the Environment held in July 1989. It is intended as a catalyst for international activity, with the signatories undertaking to broaden international support for its principles, to achieve a strengthened institutional basis for international action and acceptance of the need for support and assistance for developing countries.

A comprehensive international treaty is being pursued as the best way to get the necessary concerted international action. The recent treaty and convention to limit the use of ozone depleting substances (the Montreal Protocol), while indirectly of considerable importance to the greenhouse problem, given the significance of chlorofluorocarbons as greenhouse gases, is also an example of what can be achieved.

It is likely that the best way ahead is to secure the agreement of all countries to the need for concerted international action as a first step, and then to follow this up with protocols for detailed responses. Securing even broad international agreement is likely to dwarf the difficulties associated with negotiating the convention for protection of the ozone layer. A major issue will be the transfer of technology to developing countries and the question of compensation and aid for countries which are particularly disadvantaged by the impacts of climate change and internationally agreed responses. This is inextricably linked with already existing development assistance issues, as many climate change impact issues parallel existing problems related to population pressures, health, food security, land degradation and natural disasters.

Limitations on emissions of carbon dioxide is also necessarily a major issue, given that this gas contributes around 50 per cent of the greenhouse effect. However, action in this area is synonymous with changes in the way we consume energy, and therefore with fundamental aspects of industries, lifestyles and national economies. This has enormous significance for the highly industrialised developed countries and for the rapidly developing countries of Asia. The resultant changes will flow on to countries which at the moment are neither major producers nor consumers of energy. It is worth noting in this regard that emissions limitations are an area of considerable significance to Australia because of the dependence of its economy on fossil fuel exports. It is the world's leading exporter of coal, accounting for around 25 per cent of the world coal trade, and is becoming a major exporter of LNG. Energy industries (including uranium) represent 25 per cent of its total exports.

- Climate change due to the greenhouse effect is one of the major issues facing policy-makers in Australia. The predicted climate changes are likely to have a considerable impact on a wide variety of sectors and areas including coastal communities, agriculture, manufacturing and construction industries, tertiary industries (eg. tourism, insurance, transport and finance) and public works and services.
- Priority needs to be given to research to prepare soundly based regional assessments of the likely climate changes to assist policy-makers in the development of strategies to deal with the impacts of those changes. Adequate monitoring of those changes which are actually occurring is also essential.
- Interaction between researchers, planners and policy-makers is central to the development of effective strategies.
- Opportunities for collaborative regional and international research programmes should be explored.
- As an initial step, policy-makers and planners should examine the implications, and time frames, of the possible scenarios for climate change and assess the ability of existing mechanisms to cope with and plan for such impacts.
- Although the magnitude of climate change remains uncertain, it is already apparent that steps must be taken to reduce emissions of greenhouse gases. However, given the possible economic consequences of such limitations, it is essential that limitations be devised so as to meet their end-purpose with the minimum of cost, having regard to the position of all affected countries.
- International agreement is essential to reduce the build-up of greenhouse gases in the atmosphere. It is essential that the developing nations by fully involved in the development of any such agreement.

4. Global Warming and Response of Developing Countries

GLOBAL WARMING is the greatest crisis ever faced collectively by humankind; unlike other earlier crisis, it is global in nature, threatens the very survival of civilisation, and promises to throw up only losers over the entire international socioeconomic fabric. The reason for such a potential apocalyptic scenario is simple: climatic changes of geological proportions are occurring over timespans as short as a single human lifetime. The World Bank, regional development banks, and other development assistance agencies will need to reappraise their policies in light of the impending global warming. In particular, developing countries will need assistance in the transition phase from traditional fossil fuels to more appropriate energy forms and in promoting the preservation of forests and reforestation.

Certain atmospheric emissions resulting from various human activities are unambigously responsible for this crisis. They include carbon dioxide-the necessary product of combusion of fossil fuels, CFCs (chlorofluorocarbons) from refrigeration and halons from firefighting systems, methane from the anaerobic digestion of organic matter, and nitrous oxide from increased use of chemical fertilisers. The problem is compounded by deforestation which contributes to increased carbon dioxide and other greenhouse gases emissions due to wood combustion, as well as to decreased carbon fixation in biomass due to reduced tree cover. These gases absorb and partially trap the heat radiated by the earth, reradiate some of it back to the surface, and this leads to a warming of the lower atmosphere. The current global heat balance is thus upset and leads to a warming of the earth's surface. Scientific meetings in 1985-87at Villach, Austria and Bellagio, Italy, were critical in developing an updated scientific consensus on global warming.

The World Commission on Environment and Development, in its report, called upon governments "to initiate discussions leading to a convention. . . ." on measures to limit global warming and sea level rise. The Commission added that "if a convention on containment policies cannot be implemented rapidly, governments should adopt contigency strategies and plans for adoption to climate change".

The UN General Assembly's first debate on "Our Common Future" in October 1987 was marked by references to global warming dramatised by a plea from the President of the Republic of the Maldives for international action to prevent the disappearance of his nation beneath rising sea levels. Presently, the contribution of carbon dioxide to global warming is roughly fifty per cent: the other half is due to the other gases (methane, CFCs, nitrous oxide, etc.). These latter gases are one-thousand to ten-thousand times more effective than carbon dioxide and consequently are dangerous even at their present trace concentration levels.

In addition, positive feedback effects are very important in global warming. A consideration of the interactions of warming with the large biotically-controlled pools of carbon on land suggests that there will be further release of carbon dioxide and methane from these sources with increase in temperature. The cause of this additional release is the stimulation of the respiration and the decay of organic matter. Over the past fifteen years, the rate of accumulation of carbon dioxide in the atmosphere has been about 1.5 ppm per year.

However, measurements conducted over the past 18 months indicate that the rate has increase to 2.4 ppm. This is consistent with the positive feedback hypothesis. The atmosphere is already committed to a warming of 0.7 to 2 degrees C due to the emission of greenhouse gases upto the early eighties. At current rates of emissions, committed global warming will increase by 0.2 to 0.5 degree C every decade, and by the end of this century, the cumulative surface warming will be large enough to rise above the background climate fluctuations. Much of the committed warming due to current emissions will be stored in the oceasn and will show up in the atmosphere several decades later. However, this can lead to an upsetting of the temperature gradients in the world. Consequently, wind and ocean currents, and precipitation patterns will be affected. The higher temperatures will also lead to a rise in the sea level. Sea level rise of the order of 5 to 24 cms per decade can be expected as against the background increase of the order of 12 cms per century. This implies that coastal areas presently less than 1 to 3 metres above the sea level will be threatened by the middle of the next century. The most-affected would be the river deltas which are also the most populated areas in the world.

Thus human activities have opened an era of rapid climate changes that, if unchecked, promise an extraordinary reduction in the potential of the earth to support a reasonable quality of life for all.

This topic has been reviewed recently in several international conferences. The conclusions have universally pointed to the need for early action to deflect or stop these climatic changes. The WMO/UNEP scientific conference at Villach, Austria in 1985 and the Villach-Bellagio meetings of 1987 both expressed the extent of the consensus that exists among scientists that the warming will proceed rapidly and presents a serious threat to the human race. The Canadian Conference on the Changing Atmosphere held in Toronto in June 1988 produced a clear and detailed statement that drew on the earlier conclusions to call for early action to reduce emissions of carbon by reducing use of fossil fuels and by improving management of forests. The purpose was to reduce rates of change of climate in the next decades to rates similar to those experienced over recent centuries. All of these discussions of both effects and potential corrective action have recognised the special interests of the developing nations.

This conferences is the first arranged to address the particular concerns of the developing nations, which are struggling to improve the standards of life of their people. Nearly 4 billion of the present human population of 5 billion live in developing countries. They need accelerated economic growth but on an ecologically sustainable basis. It is against this background that the participants of the International Conference on "Global Warming and Climate Change: Perspectives from Developing Countries", who met at New Delhi, India, from February 21 to 23, 1989, present the following analysis and recommendations.

There are three distinctly different but strongly inter-related parts of the problem: (i) Global chemical pollution; (ii) The greenhouse effect of these pollutants; (iii) The global climate change, resulting from the greenhouse effect induced by the pollutants. Significant scientific progress has been made in understanding and observationally documenting many of these changes. This progress has led to an international consensus among scientists on the significance and the seriousness of the potential global scale warming and the accompanying rise in sea level. The predicted warming rates for the next several decades are unprecendented in terms of climate changes of the last several thousand years.

Global Chemical Pollution

Instrumented observations of the air have demonstrated that:

(1) The concentrations of several gases, such as Carbon-dioxide; Methane' Chlorofluorocarbons amongst several others, have increased significantly during the last century and are continuing to increase substantially.

(2) The increases in the pollutants are caused by a variety of human activities including:

- Fossil-fuel combustion
- Other industrial activities
- Deforestation, biomass burning, and the accelerated decay of organic matter in the soil

(3) The pollution is global in extent and spreads through the strata of the atmosphere.

The Greenhouse Effect of the Pollutants:

(1) The gases trap the heat radiation from the earth, and, hence, the observed rise in the gas concentrations has increased the heat trapped in the planet. This so-called greenhouse effect is a well understood phenomenon and is based on sound physical principles.

(2) Until the 1960's, carbon-dioxide increase was the major source of heating. This picture has changed dramatically in the recent during which several non-carbon-dioxide gases contributed as much as carbon-dioxide to the increase in the planetary heating.

(3) Thus as time goes by, the problem is getting not only larger in magnitude but more complex in character.

Global Warming and Climate Changes:

(1) The most direct effect of the increased trapping of heat radiation is a global warming.

(2) The warming will not be globally uniform but will differ significantly between geographical regions; in addition, the warming may vary during different seasons. As a result, the altered temperature gradients will alter the pattern of winds and precipitation distribution regionally. The details of these localised changes are not clearly understood.

(3) The observed global temperature records, that include ocean and land temperatures, reveal a warming trend during this century; the magnitude of the warming trend is within the range predicted by models. Furthermore, the later half of the decade of the 1980's registered the warmest temperatures on record.

(4) The warming of the oceans, as well as the melting of ice sheets and glaciers resulting from the warming of the land will lead to a rise in the sea-level.

Major issues that need to resolved:

(1) *Biosphere-climate interactions:*

The biosphere controls the atmospheric concentration of greenhouse gases like CO_2 and CH_4 and biotic processes like respiration are regulated by temperatures. Interactions between the biosphere and climate can play a significant role in determining the future concentrations of greenhouse gases like CO_2 and methane.

(2) *Cloud-climate interactions:*

One of the largest sources of uncertainty in predictions of regional and global climate changes is the response of clouds to the warming. The tropical monsoon cloud-systems are one of the biggest factors that regulate the global heat budget and these clouds respond significantly to small changes in ocean temperatures. There is an important need for a focused analysis of this problem.

(3) *Future atmospheric concentrations of greenhouse gases:*

Another large source of uncertainty is the rates of future increase in the concentrations of the greenhouse gases. This will depend on the scenarios adopted for future energy demands and supplies and other human activities.

(4) *Deforestation, biomass burning and emission of particles:*

We have to assess how these localised changes, which have profound influence on the regional climate, interact with the global scale warming effect of the gaseous pollutants.

(5) *The role of the oceans:*

The oceans play a dominant role in governing the timing and the rate of the warming, because of the enormous heat capacity of oceans compared to land. We are at the very early stages of understanding the interactions between the greenhouse warming and the dynamics of the oceans. Significant improvement in understanding this problem is needed to improve the predictions of regional climate changes.

Potential Impact

Global Warming is occurring at a time when many of the world's life-support systems are already stressed by the growth of population, industrial development and need for agricultural land and the unseasonable exploitation of natural resources. These stress are caused both by careless and short-sighted actions and as a consequence of poverty and under-development. The include increasing air and water pollution, deforestation, soil erosion and salination, among others.

A disregard for long-term consequences of industrial development and population expansion have resulted in air and water pollution, deforestation, and soil erosion among others.

On all of these changes, global warming and associated climate change will bring additional consequences, such as:

- higher temperatures
- changes in precipitation and storm activity
- widespread run off
- reduction in fresh water availability
- global rising of mean sea level

The consequences of these climate changes will affect every aspect of society and the material environment. Their impacts will cause a strain particularly in developing countries, where already, in many cases, existing conditions allow for only marginal existence for both people and ecosystems.

Areas of particular concern include:

- agriculture
- water availability
- human health
- human habitation
- natural ecosystems, including biological diversity

Global food supplies may be maintained through shifts in productive regions and technological advances. However, changes in local production and food distribution may aggravate circumstances which are already unacceptable in some areas. Particularly in low latitude regions where precipitation is already highly episodic. Sea-level may rise above one metre within the next century, which is about eight times as fast as occurred over the last century. Since already 70 per cent of the world's coastlines are eroding, this problem will aggravate an already difficult situation in many areas where half the global population resides.

Effects of seal-level rise include:

- loss of land and human habitation
- penetration of salt into drinking and agricultural water supplies
- beach erosion
- loss of wetlands and wildlife habitat, including air fauna
- damage of infrastructure including harbours, cooling water facilities, coastal defence systems, roads and other infrastructure

Lower latitude coastlines, frequently found in developing countries, are particularly vulnerable to these effects due to the particular morphology of these coastlines.

Island states would be particularly vulnerable to sea level rise and are in grave danger of facing serious climate aberrations long before the sea level rises to a point of total submergence.

Tropical storms may occur with greater intensity and will certainly penetrate further inland due to sea-level rise, resulting in greater loss of life.

Ecosystem

Sea-level rise will devastate coastal ecosystems such as mangrove seaways, which no longer migrate inland because of human habitation near the coast.

Terrestrial ecosystems will need to move poleward in response to the warming. However, the rate of warming may exceed the ability of ecosystems of migrate, (or corridors for migration may not be available); so loss of species or recution in numbers can be expected.

In particular, canopy forests may suffer substantial declines. Loss of carbon from forests and soils due to increased respiration and reduced biomass would add substantially to the build up of CO_2 and thus to the rate of warming.

Direct and indirect consequences of climate change such as increased air and water pollution, spread of tropical disease vectors and decreased fuelwood availability have important consequences for human health. Since climate change and rising-sea will occur at rates which far exceed historical values, adjustment to these changes will be difficult and costly for human societies in the less developed countries.

The potential in developing countries for large percentage increases in emissions of greenhouse gases (GHG) over the next decades has resulted in suggestions that control over such emissions should be considered as part of near-term development planning. Sometimes the hope is expressed that such restrictions can be part of a win-win strategy, i.e. that concerns over global warming only lead us to take on programs such as reforestation, energy efficiency, and fuel switching, for which there are many other good reasons to undertake anyway. While undeniably true in many circumstances, this begs a crucial question: does the threat of climate change through GHG emissions justify efforts to alter the development pathway now being followed in poor countries?

Many observers apparently believe that the implications of global warming are so grave that, in some fashion, restrictions on

GHG emissions should be considered along with the other numerous and already burdensome restrictions on development, e.g. lack of capital, lack of skilled personnel, and lack of infrastructure. This paper, although accepting that the threat is important enough to warrant changes in present development pathways, draws the opposite conclusion. Rather than impose new constraints, it takes the position that the potential for global warming is actually one of the strongest arguments for *reducing constraints and accelerating development*. From the standpoint of the poorest two-thirds of the population in developing countries, concerns about climate change probably justify a policy response in which not only do GHG restrictions have no place, but GHG emissions rates may actually be stimulated locally.

I follow two lines of argument to reach this conclusion. Firstly, I argue that the best way to protect the poorest groups is to bring them as soon as possible to a minimum level of development at which they can be expected to adequately handle environmental stress. Secondly, I argue that the extra GHG picture and thus will not be critical in determining climate change.

To a first approximation, the impact of climate change will be like having more and, perhaps, more severe natural disasters, e.g., floods, droughts, tropical cyclones, and storm surges. Much evidence indicates that there is a marked difference in the impact of such events depending on the level of economic development. Cross-sectional evidence, for example shows that the same size natural disaster will have vastly different consequences in different parts of the world, depending on the level of development. For illustration, a tropical cyclone of the size that may kill 10 people in Florida might kill 10,000 people in Bangladesh. A similar ratio will be seen in the number left homeless. Only in the dollar amount of property damage do disasters appear larger in developed regions.

Longitudinal studies provide similar results. The same size hurricane that killed many people in the USA or Hong Kong in the early part of the century, for example, kills few in either place today. This is true in spite of the great increase in population density that has occured over the period.

The reasons for this effects are fairly obvious: better warning systems; more capable and organised fire, police, and medical protection; more and better roads; an educated and literate public; more substantial housing; a reduction in the number of people living at the margin with little savings or other means of security during periods of households or community disruption; and so on.

It is important here to distinguish the *stress* (analogus to the dose in toxicology) from the *consequence* (the response). Although the dose of natural disasters can be similar, the vulnerability of the population and thus the response is markedly higher in developing countries. In short, the poorest countries suffer greater loss of life, and the richer countries show more property loss (as measured by the usual indicators).

Although the analogy with natural disasters is not perfect, there is little reason to expect that the consequences of climate change will not have a similar distribution. All nations will experience stress, but few people will die of it in developed countries where the economic damage will be highest. Many, however, will die or become homeless and ill in the developing world. Climate change is an economic threat in developed areas but, in addition, a threat to life itself in poor areas. This argues that the best approach for a developing-country planner worried about climate change may be to accelerate those aspects of development that will assist their poorest and most vulnerable populations in attaining the levels of health, flexibility, education and resilience that will allow them to cope with environmental stress of all sorts, including climate change. This may offer a much more effective least-risk development pathway than would one attempting to rank development projects on the basis of greenhouse-gas emissions.

The marginal increase in resilience (reduction in vulnerability) to environmental stress, however, is not uniform across the development spectrum. At high levels of development the additional increase of resilience per unit development is much less than at low levels. The implication is that it may be possible to define some sort of minimum resilience as a basic need. Development planning could then be designed to bring poor populations as rapidly as possible to this level, whatever the cost. Above this level, however, a range of other considerations such as GHG emissions might come into play.

The establishment of such a level would obviously require some careful investigation and will be geographically and perhaps culturely dependent. To illustrate the concept, however, it is instructive to examine the two most commonly available indicators, life expectancy and infant mortality. In a rough way, a cross-sectional analysis shows a slowing of the marginal increase in resilience when per capita energy use reaches 1.0-1.5 tons oil equivalent per year. In addition, such a minimum level of resilience is a necessary, if not sufficient, condition for achieving low birth rates. Parents who see many neighbour children fail to survive young childhood are not likely to wish to limit their own fertility.

For illustration, let us take 1.0 ton/year per capita carbon emissions (mainly as CO_2) as the amount required to reach a minimum level of resilience. This level is somewhat less than the present global mean and lies in the gap between the rich and poor countries within which lie only a few countries such as South Korea. We can then examine the global GHG implications of policies designed to achieve this level in developing countries as fast as possible. For this purpose, consider a simple two-nation world consisting only of India and the USA. (These are actually the second largest developing nation and developed nation respectively, each with about 22 per cent of the population of its group.)

The important GHG parameters over time are analogous to those of taking an automobile trip: distance travelled, speed, and acceleration. For GHG emissions, these are: cumulative amount added to the atmosphere, annual rate of emissions, and annual rate of increase in emission rate. Table 1 lists these parameters at the present (1990) for India and the USA as well as projections at four points in the future given continuation of current conditions.

The most critical parameter for determining the extent of global warming is the cumulative amount of GHG in the atmosphere. The amount placed into the atmosphere since 1900 as part of the US industrialisation process is equivalent to about 260 tons per person in 1990, more than 40 times that of the average Indian. The present rate of emissions as well as the rate of increase in emission rate (acceleration) are also substantially higher in the US at present. Indeed, as shown, it will not be until 2013 or so until the acceleration becomes equal at about 30 (Kg/year)/year. In addition, even if the present rapid rate of economic and energy growth continues in India, it will not be until nearly 2070 before the rate of emissions reaches that of the US, then at about 7.7 t/year.

Under present conditions India would reach the emission rate of 1.0 t/capita-a year, which is our arbitrary indicator of minimum resilience, in about 2024. At that point, however, India would still have been responsible for only a minor part of the total atmospheric accumulation, less than 5 per cent of that of the USA.

A question arouses, therefore: is it really worthwhile to spend much effort to try to reduce the small addition that India will make in the next generation, which, even under these conditions of rapid growth in India and extremely small growth in the USA, will be less than 10 per cent of the US additions during the same period (18 versus 190 tons, 1990 to 2024)? Rather, is it not likely that the huge benefit to be gained by bringing resilience up to a minimum standard far outweights any GHG savings to be squeesed out of India's energy system during the same period?

Furthermore, if it were decided that the global community should limit GHG total emissions to their present level, it would still be possible to bring up the developing world to this minimum level without undue hardships in the presently developed world. All the countries in the world now below 1.0 t/capita-year, for example, could be brought up to that level and the total rate kept constant merely by reducing the average emissions of the developed countries to about 2.2 t/capita-year, the present level of Japan. This does not appear to be impossible, or indeed, even difficult from a technical standpoint.

I follow the example of the climate modelers and emphasise that this (two-nation, one-pollutant) model is greatly oversimplified. It considers, for example, none of the other greenhouse gases, no sources other than fossil fuels, no distributional effects, and rigid growth rates. In addition, of course, energy use, income, health, and resilience are actually related in complicated ways. Efforts toward improving our understanding and ability to influence these relationships should have the highest priority.

Nevertheless, at least one conclusion seems fairly robust: large ratios of cumulative emissions between developed and developing countries will remain until for into the next century. Changing the scenarios would not likely significantly change this basic nature. It is these cumulative emissions that are the problem and it thus seems appropriate that the burden of slowing and reducing emissions should be borne according to the amount emitted by each economy. In analogy to the national debt, this might be termed the *natural debt*. Over the years, just as economies take on a financial debt in order to grow faster, they take on a natural debt for the same reason. In this case, economies have borrowed the assimilative capacity of the environment by releasing waste gases faster than they can be removed naturally. Just as there are economic disruptions if the national debt becomes too large, so too with the natural debt.

Developed countries are the holders of the largest natural debts (260 tonnes per capita for the US, for example compared to 6 tonnes for India, in 1990) and are best able to devote some of their Wealth to paying them off. They should not think of this as a sudden extra cost being imposed, but as the inevitable need to take responsibility to repay the loans that helped them achieve their present state of wealth.

The Role of Developing Countries

An equitable response to the special needs of developing countries is crucial to removing greenhouse threats to the global climate. On the one hand, developing countries have caused little of the problem and industrialised countries must bear the bulk of the blame. On the other hand, as economic development accelerates, Third World countries may account for the preponderance of greenhouse gas emissions by the middle of the next century. An international solution that provides incentives for the participation of developing countries while fairly distributing the responsibility for implementing solutions is essential to a successful global strategy for combating greenhouse warming.

The consequences of the greenhouse effect strongly suggest that it is in the self-interest of Third World countries to re-examine expeditiously their energy priorities. Developing countries, with fewer resources to adapt to environmental disturbances, stand to suffer disproportionately from a rapid climate change. For example, the productivity of common rice varieties falls off dramatically at temperatures just a few degrees higher than those currently prevailing in many rice-growing areas.

Tapping the tremendous potential for conservation and improved end-use efficiency in the developing world would contribute to a solution for greenhouse warming while meeting much of the Third World's growing energy needs. This strategy also avoids other serious environmental and social problems, such as land degradation, local air pollution, and population displacement, that accompany the building of fossil-fuel-fired power installations. By the year 2020 it may be possible to achieve a universal standard of living far beyond that necessary to satisfy basic needs with little or no increase in global energy consumption from today's levels. However, many developing countries use energy in a highly inefficient manner. Macroeconomic policies in many developing countries, such as electricity price subsidies, discourage conservation and efficiency improvements. Firms in Brazil, where electricity prices are subsidised, manufacture energy-efficient air conditioners for export but cheap, inefficient models for domestic consumption.

Investments in efficiency gains are extremely attractive from many points of view. They require less capital and less foreign exchange than do comparable amounts of new power supply, contributing to overall economic productivity. Through efficiency and conservation, developing countries could avoid at least $1.4 trillion in power-supply expansion costs between now and the year 2008.

Efficiency investments represent a major opportunity for donors like the United States and the World Bank to assist developing countries in making energy choices that both avoid mistakes made earlier in the developed world and reduce risks to the entire planet from greenhouse warming. Nonetheless, development assistance in the environmentally sensitive energy sector often exacerbates the threat of greenhouse warming by emphasising conventional sources of energy, such as massive fossil-fuel-fired power plants. The World Bank, which controls an annual energy lending portfolio averaging $3.5 billion, is one of the principal donors supporting power generation projects in the Third World. Through measures such as pricing reforms and improvements in the operation of existing power plants and distribution systems, the Bank has already made a commitment to encourage conservation and the efficient use of energy. There is, however, considerably more that the Bank can do.

The Bank requires preparation of a "least cost plan" to precede investments in the energy sector. Current methodologies for these studies primarily address strategies for increasing energy supply. Support for demand reduction measures, such as end-use efficiency improvements, which are often economically as well as environmentally superior to investments in supply, have not consistently been considered as alternatives to conventional power generation projects in Bank energy-sector strategies. Expanding the universe of alternatives to include demand reduction options would simultaneously help developing countries reduce the rate of growth in their power-generating capacity and reduce greenhouse-gas emissions without sacrificing the energy needed for economic development. Additional staff trained in strategies for encouraging end-use efficiency improvements would significantly increase the Bank's capabilities in this crucial area.

Forest policy is another area where development assistance can provide benefits to Third World countries while simultaneously cutting emissions of greenhouse gases. While there has been great concern in North America and Western Europe about destruction of tropical forests, donor countries historically have devoted little capital to conservation of this crucial resource and have earmarked even less for the creation of new forest areas. Case studies have documented that projects financed with little regard for the integrity of natural resources by donors such as the World Bank have seriously exacerbated forest loss in key countries such as Brazil and Indonesia. Industrialised countries can also help to reverse tropical deforestation and encourage reforestation through changes in domestic policies. Developed countries provide the primary market for tropical hardwoods, virtually all of which are unsustainably harvested, and firms based in industrialised countries often reap the profits of this trade. Governments of industrialised countries can take a serious look at controlling trade in tropical woods and compensate exporting countries for lost revenues through alternative investments.

The Third World debt crisis present major opportunities for encouraging better forest management in developing countries. As the market value of such debt has fallen, a number of private banks have sold debt owed to them by Third World governments to private conservation organisations, which have then forgiven the debt in return for specific promises by the governments concerned, such as a commitment to conserve a particular area and to support its maintenance with a stream of payments in local currency. Such "debt for nature" swaps are already in place in Bolivia, Costa Rica, and Ecuador, and more are under negotiation. Governments can adopt policies, such as tax incentives, that encourage creditor banks to sell debt for swaps. Creditor governments can reduce interest or principal on sovereign debt in return for promises of policy reform in this critical sector.

Coordinating policies on the international level to fight greenhouse warming will maximise environmental and foreign-policy benefits. Unilateral reductions in releases of greenhouse gases by large emitters such as the United States and the Soviet Union will go a long way toward arresting global climate disruption. However, a multilateral consensus strategy will further the crucial goals of creating incentives for universal participation and establishing an equitable balancing of responsibility for solving the problem.

Existing international mechanisms are an important part of such a strategy. A reassessment of the Montreal Protocol, a process that is provided for by the document itself, is the most expeditious way to eliminate the contributions CFCs and halons make to the global warming problem. The World Bank's institutional structure also includes mechanisms for member countries to redirect priorities in the critical energy and forest sectors.

The remainder of the greenhouse problem could be handled most effectively through a multilateral treaty, with standards binding under international law that would require each country to take prescribed actions to reduce and halt greenhouse warming. Considerable precedent is now in place for multinational environmental agreements containing strict regulatory standards. In addition to the Montreal Protocol, which is an ancillary agreement to the 1985 Vienna Convention for the Protection of the Ozone Layer, several other international agreements establish requirements for controlling emissions of specific air pollutants. The Protocol on the Reduction of Sulphur Emissions or Their Transboundary Fluxes by at Least 30 per cent and the Protocol Concerning the Control of

Emissions of Nitrogen Oxides of Their Transboundary Fluxes—both auxiliary agreements to the 1979 Convention on Long-Range Transboundary Air Pollution—set out precise regulatory limitations on releases of specified chemicals. Principles established in the case law of international tribunals and in the 1972 Stockholm Declaration adopted by the United Nations Conference on the Human Environment also discouraged countries from acting in ways that could harm the environment in another's territory.

A multilateral treaty designed to arrest global climate change should satisfy several basic requirements. First, it must require reductions in releases of greenhouse gases of a magnitude and speed sufficient to stabilise the earth's climate. The most important gas to control is CO_2 for which global reductions of at least 50 per cent are necessary. Participating countries should accomplish these reductions by means of environmentally and economically sound technologies that do not present unacceptable risks to public health or global security. The creation of new forested areas might be encouraged by allowing credits against reductions of CO_2 emissions that would otherwise be required and by provisions establishing or promoting forestry programmes. Because the agreement could be expected to cover a large number of emissions sources, it should mandate strict mechanisms for enforcement through reporting of emissions, on-site audits, and internationally controlled remote sensing.

Second, the responsibility for making reductions must be distributed equitably. Among the criteria that could be applied is relative national wealth as measured by per capita GNP. Another test could be per capita emissions of CO_2 with the highest reductions required of those countries with the highest emissions per unit of population. Another possibility would be to require the imposition of a fee for carbon emissions, either as a primary mechanism for achieving reductions or as a supplementary measure. Any of these formulas would require proportionally greater cutbacks by the wealthiest countries and leave the poorest countries with the fewest constraints on CO_2 emissions. All countries would be encouraged to use existing energy supplies more efficiently.

A treaty should also require a commitment from the wealthier countries for increased research in to nonCO_2 energy supply technologies and development assistance to help poorer countries meet the requirements imposed on them by the agreement. One mechanism for generating the necessary capital is to require countries to contribute to a fund in proportion to their CO_2 emissions. Restricting access to this fund to those countries that accepted the obligations of the treaty would create incentives for broad participation.

Considering the importance of the resources at risk, it would be nothing short of reckless to continue with business as usual. A failure to respond to the threat of greenhouse warming would amount to an affirmative decision to wager the health and well-being of current and future generations against over-whelming odds.

At the 1992 Earth summit in Brazil, which was attended by more heads of state than any other meeting in human history, the UN Framework convention on climate Change (included as the Appendix to this volume) was opened for signature. By mid-October 1992, 158 nations had signed it. To become law, it must be ratified by national legislatures of at least fifty countries, a process that may take two years.

Unfortunately, the convention did not contain any specific provisions for funding its implementation. This lack is a major obstacle to its realisation. The questions of how to decide who should pay and how much it might cost are the central topics of this book. The participation of developing countries in a Climate change Convention will determine whether the world responds prudently to the greenhouse effect. Even if the wealthy states radically reduce their greenhouse gas emissions, the poorer states will replace and eventually surpass them as major contributors to the greehouse effect. Action by members of the Organisation for Economic Co-operation of Development and other industrialised countries can significantly slow the rate and reduce the magnitude of global warming. But unless the developing countries also act, the threat remains to everyone. Based on current trends, big poor countries like China, Indonesia, India, and Brazil will become major carbon dioxide contributors. They are already big methane gas emitters even though their per capita output is small.

As is argued below, the Climate Change Convention itself is still mostly symbolic. Unresolved issues include the practical implementation of the Convention in protocols to the Conventional technology and resource transfer; obtaining commitments from parties to limit carbon emissions; and the design and implementation of abatement strategies. All this and much more remains to be settled in protocols to be negotiated now that the Convention itself has been signed.

In this chapter, we do not tackle all these important issues. Instead, we postulate that the major determinant of developing country participation will be the terms offered by the developed world. The need for the rich and poor nations to work together to respond to the greenhouse effect could create a new political-economic interdependence between them. Alternatively, as Norwegian analyst Anne Kristin sydnes warns, it could pretend' another twenty years of fruitless North-South bargaining. The authors of this chapter examine the grounds for, the scale of, and possible conditions on possible resource transfer agreements from rich to poor states that will be central to any successful greenhouse management regime.

In this chapter, we undertake four tasks. First, we review the basic scientific understanding of the greenhouse gas effect that gave rise to the Climate Change Convention. Second, we describe the content of the Convention and note its limitations. Third, we review the novel negotiating difficulties that will arise in the course of developing effective protocols under the Convention. Fourth, we summarise the key issues for the ongoing negotiations under the rubric of the Convention as presented in this book. In the latter section, we also provide a synopsis of each chapter of the book.

The Greenhouse Effect

Planet Earth's capacities far dispersing. dilutiny, and degrading most human-generated pollutants are large, but limited. As pollution rates increase, the natural processes that absorb and assimilate pollutants are eventually over-whelmed, leading to rising concentrations of pollutants in the environment. Depending on the pollutants, this overloading can create local disruptions in human health the ecosystem sustainability or, eventually, even global effects, such as climate change.

The authors of this chapter focus on the largest and oldest of all human pollutant releases, carbon into the atmosphere. Since their mastery of fire, human beings have disrupted the global carbon cycle by burning wood and other biomass at greater rates than occur naturally. Being about half carbon, biomass upon combustion releases carbon dioxide, methane, carbon monoxide, and other carbon-containing pollutants, which must be transported and broken down by natural processes.

Some of this release has been the direct result of using wood and other biomass forms such as crop resides as fuel. Another part is due to the clearing of biomass so that the land could be used for farming or other human purposes. Through out most of human history, however, it is thought that human biomass combustion did not create large disturbances in the atmosphere. That is, natural process such as re-growth replaced sufficient portions of the burned biomass to prevent significant build-up of these carbon-containing pollutants in the atmosphere.

The industrial revolution in what are now the economically developed countries increased the combustion of fossil fuels, which are mostly carbon. Today, fossil fuel combustion is the major source of atmospheric carbon releases although biomass burning has also probably increased. The rates of release became such that in the second half of the twentieth century, it became clear that atmospheric levels of important carbon-containing gases, particularly carbon dioxide and methane, were steadily increasing over their natural levels.

The atmospheric concentrations of these gases are far from those thought to be toxic or otherwise of must acute concern. Their impact is more subtle, for they act to blanket Earth, keeping in more of the sun's warmth than otherwise would be the case. This effect is indisputable because humans have observed the warming due to natural levels and variations of these same gases. Indeed, without the existence of these natural amounts of 'greenhouse gases, 'Earth would be too cold for life.

The higher-than-natural rates of greenhouse gas releases resulting from human fossil fuel and biomass combustion are boosting carbon dioxide and methane levels at grater rates than has occurred in recent Earth history. Earth's natural systems may not be able to cope with the extra heat being absorbed except by an overall increase in temperature, that is, global warming. This warming may in turn be associated with significant disruptions in local weather, such as patterns of precipitation and cloudiness. It may also have global impacts through thermal expansion of the oceans and melting of glaciers to cause sea-level rise. It may even distribute ocean current patterns and marine and terrestrial ecosystems.

The nature and magnitude of global warming and associate climate change and sea-level rise resulting from greenhouse gases released by human activities are not known with certainty. The global atmospheric/ocean/climate systems are extremely complex, so much so that even the largest computers can only model a small portion of them at one time. Some natural processes seem to reduce the effect. Extra heat, for example, leads to more evaporation, which leads to more clouds, which leads to more sunlight being reflected, which leads to less heat. In contrast, other processes may enhance the effect. Extra heat, for example, leads no more snow melting, which leads to less sunlight reflection, which leads to more heat. Thus, contemporary estimates of the global effects are imprecise and so uncertain as to be not usable for predicting effects at any one place and time.

Nevertheless, a growing number of scientists believe that there is a significant chance that damaging levels of global warming will occur sometime before the middle of next century if existing trends of greenhouse gas emissions are continued. The most authoritative source for this view is the scientific report of the Intergovernmental Panel on Climate Change (IPCC).

The IPCC consists of scientists form many countries working together under the aegis of the World Meteorological Society and United Nations Environment Programme and who have reviewed and summarised the available knowledge and uncertainties about greenhouse gases and climate change. As part of its conclusion, the IPCC predicts that unless emissions patterns change, there will be:

> a rate if increase is global means temperature during the next century of 0.3^0 C per decade (with an uncertainty

range of 0.2-0.5° C per decade); this is greater than that seen over the past 10,000 years. This will result in a likely increase in global mean temperature of about 1°C above the present value by 2025 and 3°C before the end of the next century.

Its collective judgement is that:

> Rapid change in climate will change the composition of ecosystems; some species will benefit while others will be unable to migrate or adapt fast enough . . . The effect of warming on biological processes . . . may increase the atmospheric concetrations of natural greenhouse gasses.

IPCC impact report concluded that this warming cloud cause major and mostly negative local impacts on agriculture, forestry, water resources, natural ecosystems, air quality, and coastal zones among other sectors important to humanity?

Recognising the there are still significant uncertainties, particularly about the timing of impost, and that some countries, notably the United States, have timing of impacts, and that some countries, notably the United States, have not yet officials acknowledged a need to take immediate steps, this book nevertheless starts with the premise that the world community decided when it signed the Climate Change Convention that a serious effort must be made to reduce the probability and magnitude of adverse impacts from global warming. According to the IPCC, a successful response strategy is likely to be one which recognises that:

- Climate change is a global issue, effective responses . . . require a global effort;
- Effective responses may have considerable impact on human kind and individual societies;
- Industrialised countries and developing countries have a common responsibility indealing with problems arising from climate change.

Although the developed countries have emitted far more greenhouse gases to date, based on the current relationships between economic development and greenhouse-gas emissions, and on the sorely needed increases in economic welfare that are required, large poor countries such as China, India, Indonesia, Nigeria, and Brazil will also turn into major greenhouse-gas contributors. Whether developing countries are engaged adequately in the implementation of a Climate Change Convention will be a major determinant of how quickly and completely the world responds to the dangers presented by human-enhanced greenhouse warming. Action in the Organisation for Economic Co-operation and Development (OECD) and other developed countries alone can slow the rate and magnitude of global warming in the short term, but unless action is also started in developing countries, significant threat remains. The fact—often discussed in this book—that there are attractive greenhouse amelioration projects in developing countries, however, does not necessarily mean that developing countries should pay for them.

The challenge facing humanity, therefore. is to find ways that the many benefits accompanying economic oddment can be attained by the world's poor without simultaneously emitting the amounts of greenhouse gases that have accompanied such economic development in the past, and to reduce dramatically the emissions from rich countries at the same time.

What was decide at Rio?

The Climate Change Convention signed at Rio sought:

1. to stabilise "greenhouse gas concentrations in the atmosphere at a level that would prevent dangerous

anthropogenic interference with the climate system",

2. to do so quickly enough "to allow ecosystems to adapt naturally to climate change",
3. "to ensure that food production is not threatened",
4. "to enable economic development to proceed in sustainable manner".

To paraphrase, the world's leaders committed themselves to work together in first slowing and then stopping the growth of greenhouse gases in the atmosphere at concentrations that would not be so greatly above natural levels as to significantly threaten human welfare and natural ecosystem balance. In this process, other important goals must not be sacrificed, including provision of adequate food supplies and eradication of poverty.

Abatement Commitments

Developed countries committed themselves to a vaguely worded'aim of returning individually or jointly to their 1990 levels (of) these anthropogenic emissions of carbon dioxide and other greenhouse gases not controlled by the Montreal Protocol.' Developing countries which signed the treaty undertook to provide an inventory of greenhouse gas emissions and sinks and a national climate change response strategy. They undertook no specific commitments to reduce their greenhouse gas emissions.

Indeed, developing countries asserted their right to increase their greenhouse gas emissions. As the preamble to the treaty states:

- All countries, especially developing countries, need access to resources required to a achieve sustainable social and economic development.... In order for developing countries to progress towards that goal, their energy consumption will need to grow taking into account the possibilities for achieving greater energy efficiently and for controlling greenhouse gas emissions in general, including through the applications of new technologies to terms which make such an applications economically and socially beneficial.
- In effect, parties from developing countries made any future possible abatement commitment on their part contingent upon the developed world demonstrating its intent to make available atmospheric space for future growth in emissions from the poor countries to fulfil more stringent commitments in the future. This refusal to commit to reductions enabled the developed world to leave vague the exact scope and scale of their commitments to provide funds and technology.

The climate Change Convention addressed four key issues that determine who will pay for the costs of incremental abatement in developing countries. These are the provision of financial resource, technology transfer, an interim financial mechanism, and the definition of incremental abatement cost. The following sections briefly outlines the agreement reached on each of these concerns.

For nearly two decades since the landmark 1972 United Nations Conference on the Human Environment in Stockholm, Sweden, the principal environmental worries facing most nations have concerned air and water quality, soil conservation, and adequate food supplies. Although some environmental progress has been made in recent years, notably in improving air and water quality in industrialised nations, enormous problems remain to be solved. In particular, environmental deterioration in developing countries continues at an alarming rate.

Equally alarming, new large-scale environmental problems have appeared that threaten to aggravate international relations, behaviour, and security. These problems include acid precipitation, Arctic haze, the depletion of stratospheric ozone, species extinction, and global warming. Unlike previous concerns about air and water pollution, these threats have implications that are truly global in scope—they know no political distinctions and they obey no international boundaries. As a result, finding satisfactory solutions will be especially difficult.

The issue of global climatic change, the so-called greenhouse effect, epitomises these concerns. With the exception of the consequences of nuclear war, no other environmental problem has the scope or the potential for such widespread societal impacts. Recently, analysts and policy-makers have begun to express concern about the international implications of climatic changes. In 1987, the World Commission on Environment and Development (WCED) stated:

"Environmental threats to security are now beginning to emerge on a global scale. The most worrisome of these stem from the possible consequences of global warming caused by the atmospheric build-up of carbon dioxide and other gases... Slowing, or adapting to, global warming is becoming an essential task to reduce the risks of conflict."

Similarly, the Toronto Conference on the Changing Atmosphere concluded in June 1988:

"The Earth's atmosphere is being changed at an unprecedented rate by pollutants resulting from human activities, inefficient and wasteful fossil-fuel use and the effects of rapid population growth in many regions. These changes represent a major threat to international security... The best predictions available indicate potentially severe economic and social dislocation for present and future generations, which will worsen international tensions and increase risk of conflict between and within nations."

Other signs that the issue of greenhouse warming has reached the international political agenda can be seen through statements by General Security Gorbachev, Prime Minister Brundtland, Prime Minister Mulroney, President Reagan, and President Bush and members of his cabinet.

Climate change will affect more than just traditional East-West concerns. Indeed, while the principal responsibility for the production of greenhouse gases lies with the industrialised countries, some of the worst societal impacts will fall on developing countries. This discrepancy has the potential to worsen existing tensions and to raise new disputes between industrialised and developing nations and between poorer nations themselves.

This article identifies severe risks facing both developed and developing nations from rapid climatic change and proposes actions to prevent or mitigate the worst international fractions that may arise. If nations act today to slow the rate of climatic change, it may be possible to increase the time available to understand and prevent the worst impacts, and to reduce the possibility of international disputes and conflicts. Environmental stresses are both a cause and a consequence of political tensions and international disputed. Many factors affect the connections between environmental stress, poverty, and security, including inappropriate development policies, resource constraints, population pressures, and economic inequities.

Although the links are complex and poorly understood, it is widely acknowledged that resource constraints or environmental degradation can lead to conflict when other pressures and tensions exist between states. In this way, resources act as *triggers* (e.g. Middle East oil), as *tools* (e.g. weapons links to energy, financing, access to materials), as *consequences* (e.g. weapons links to energy, financing, access to materials), as *consequences* (e.g. targeting of dams, nuclear reactors, or oil fields), and as *roots* (e.g.

economic competition, pressures, and tensions) of conflict.

Among all the major environmental threats, global climatic change appears to be the most likely to affect international politics because of its wide scope and magnitude. Climatic change may cause either a direct shortage of a resource such as water or food, or they may lead to the degradation of resources held either in common or exclusively (9-12). In the first case, disputes and conflicts may arise through competition for a limited resource, by the perception that competition may arise for that resource, or by threats of action. In the second case, resource degradation may provoke a reaction by the holder of that resource or may lead to political or economic exploitation. These effects will have consequences for international relationships, behaviour, and policy (3, 6, 13).

A 1.5 to 4.5^{0}C rise would almost certainly alter most characteristics of the global climate. Rainfall patterns are expected to shift, perhaps causing Dust Bowl-like conditions in the US grain belt. Ocean currents are also projected to shift, dramatically altering the climates of many areas. A diversion of the Gulf Stream, for example, would transform western Europe's climate, making it far cooler than it is today. The timing and patterns of monsoons would likely change. The extent of sea-level rise is somewhat more uncertain. Most estimates project a rise of about one meter over 100 years. Such a rise would inundate many coastal regions, erode beaches, damage water supplies through saltwater intrusion, and destroy ecologically vital swamps, marshes, and estuaries. The fishing industry would be affected. Growing seasons would change, and this, together with altered rainfall, would force the relocation of some agricultural areas. Even the local effects of warmer temperatures would be considerable. Washington D.C., for example, is projected to experience twelve days of 100^{0}F or hotter temperatures, instead of one, the current average. And instead of thirty-six days of 90^{0}F or hotter, the capital would swelter through eighty-seven. Patterns of energy demand, construction, transportation, water management, and coastal enginnering would all be affected. Over a longer period, perhaps one to two centuries, a large global warming could melt the land-based ice over Greenland and Antartica. Estimates of the resulting sea-level rise are very uncertain, ranging up to 24 feet. After just a 15-foot rise, however, major parts of San Francisco, New Orleans, Boston, Miami, New York, and most of the rest of the US coastline would be underwater.

Positive consequences are also expected. Some plants would grow more quickly, stimulated by the additional carbon dioxide. Rainfall might increase in what are now arid but potentially fertile regions, especially sub-Saharan Africa. Conditions for agriculture would also improve in northern areas with appropriate soils and adequate rainfall or irrigation. As profound as these effects are, consequences we cannot now predict—either positive or negative—may well have the greatest impact.

Some climatologists believe that another characteristic of a rapidly warming climate could be larger and more frequent weather extremes. Hurricanes and other violent disturbances, droughts, and extremely wet periods might all occur more frequently. This past year's drought in the Southeast, which caused an estimated $1.75 billion in damage in just 6 of the 11 affected states, is a sobering reminder of the extremely high costs of such weather anomalies. The Great Lakes region, meanwhile, has experienced above-average fainfall in 17 of the last 18 years, perhaps due to a change in the path of the jet stream. The rising water level in the lakes is projected to cause $100 million to $150 million in damage this year and far greater consternation and political maneuvering as lakeshore governments struggle to anticipate the effects of future higher levels. A climate in disequilibrium is in itself a costly and socially difficult situation.

There are many areas of great uncertainty about greenhouse mechanisms. In particular, the behaviour of clouds, the mechanisms of ocean circulation, and various oceanic and atmospheric feedback loops may change some of these projections. However, scientists are reasonably confident of the probable effects over large areas. For example, mid-continental regions of the United States are likely to become drier. But predictions of more localised effects—for example, the likely climate in Nebraska—are highly uncertain. But at least in their broad outlines, the theory, causes, and impacts of the greenhouse warming described here appear reliable.

The greenhouse warming is part of a bigger picture; it is closely linked to stratospheric ozone depletion and acid deposition. These three phenomena are tied together by many of the same chemical species, by the same human activities that generate them, and by the effects of one phenomenon on the timing and severity of the others. The leading parts in each are played by the same actors, though sometimes indifferent roles: carbondioxide, nitrous oxide, methane, carbon monoxide, ozone, nitrogen oxides, sulphur oxides, the hydroxyl radical, and CFCs. These interconnections have important implications for control strategies.

For example, as noted earlier, carbon dioxide is the largest contributor to global warming. At the top layer of the atmosphere, however, it cools the stratosphere and slows the rates of certain reactions that determine the rate of ozone depletion. Methane's role is more complicated. In the stratosphere it acts as a chemical "sink" for chlorine atoms, thereby reducing ozone depletion. It also reacts with other compounds in the stratosphere to form carbon dioxide and water, thereby amplifying the greenhouse effect because water vapour is also a greenhouse gas. In the troposphere, methane leads to higher ozone levels, though by different chemical pathways depending on the concentration of nitrous oxides. Sufphur oxides are converted by ozone, hydroxyl, and other chemicals to sulfates, which cause acid deposition. Sulphur oxides also affect the fate of many other of the leading actors. Ozone in the troposphere acts as both a greenhouse gas and as a pollutant, affecting plant growth and human health. In the stratosphere, however, it absorbs incoming ultraviolet radiation, playing a vital role in protecting human, plant, and animal life.

Moreover, many of the same human activities are involved in these phenomena. For example, a refrigerator eventually releases CFCs, which contribute to both stratospheric ozone depletion and global warming. And if the refrigerator runs on electricity produced by coal burning, its use also

- produces carbon dioxide (which contributes to warming) and nitrous oxide (which contributes to both warming and to ozone depletion):
- produces sulphur and nitrogen oxides, the precursors of acid deposition (nitrogen oxides also contribute to warming by increasing the concentrations of tropospheric ozone);
- produces carbon monoxide, which enhances warming by reacting with hydroxyl radicals, which re the principal troposphric sink for methane, a greenhouse gas.

As these examples of the mind-numbing chemical complexity of the atmosphere illustrate, the use of names to designate individual phenomena creates an illusion that these are distinct chages. They are not. Acid deposition is just one of the results of a far more complex mix of air pollutants that interact in many ways to affect plant, animal, aquatic, and human life. Greenhouse warming and ozone depletion are so closely tied together that they should be treated as a single phenomenon. In fact, the atmosphere might most usefully be visualised as a volume of fluid in which all the chemicals mix. Above all, as the costs and benefits of controlling any one of these problems are weighed in the political arena, the effects on the other two must also be considered. Luckily, most emission reduction work in the same direction, helping to solve all three problems.

Beyond this intimacy with ozone depletion and acid deposition, the key factor in considering a policy response to global warming is the question of timing. Is the greenhouse effect a problem for our children? Given the scientific uncertainties, is too soon to act, or might we wake up one day and find that it is too late? What are the consequences of acting later rather than sooner? Can we expect to have significantly reduced the uncertainties in time to significantly affect the outcome? Paraphrasing Senator Howard Baker's immortal question: What will we know, and when will we know it?

The 1980 have seen a dramatic change in the nature of the environmental issues in which decision-makers are interested. Almost exactly a decade ago, we (EAPI/EWC) orginised a meeting to help identify priorities for a new programme dealing with "The environmental Dimensions of energy Policies." A group of distinguished persons from 11 countries in Asia, the Pacific, and North America suggested that we concentrate our initial efforts on developing environmental guidelines for offshore oil and gas exploration and production, and for the mining, transportation, and combustion of coal (Saddiqi, 1979). There was some discussion of the longer-term climatic effects of the increased use of fossil furls, but the view expressed by the group was that it was too early to try to get policy-makers interested in the subject. As recently as three years ago, when I submitted a paper on a possible "law of the Atomosphere" to a well-known law jorunal, the paper elicited no response for several months, and was then returned with more or less of a yawn!

In this closing year of the decade, the situations changed so drastically that there is a conference or a workshop almost every week dealing with some aspect of climate change. If one might be permitted to borrow a phrase from the heyday of the energy Crisis of the 1970, "The sun never sets on Climate Change conferences." Even though there is a great deal of redundancy in having so many meetings on the same topic, there are significant benefits as well. One of them is that awareness of the "Greenhouse Effect," and of its implications for the earth's climate, has become widespread amongst the public and decision-makers in a large number of countries.

It is no longer necessary to describe the nature of the "Greenhouse Effect," and the role played by human activities in contributing to it. With your permission, then, I shall simply summarize here the relative importance of different gases in terms of contributing to global warming, and then focus on efforts to reduce the one most directly related to energy use, i.e. carbon dioxide (CO_2). According to current estimates (Unquote Commission, 1988), carbon dioxide is responsible for about half of the "Greenhouse Warming" at present.

In subsequent sections of this paper, I shall discuss present energy use in Asia, and possible changes in the relative importance of different sources of energy on that continent in response to climate change concerns.

In addition to using energy more efficiently, their contribution could come in two other important ways:

(1) By switching, where possible, from, oil and coal *to energy sources that emit less or*

no carbon dioxide, such as natural gas, hydronuclear-, and geothermal-power, and renewable sources such as solar-and wind-energy.

(2) By stopping the deforestation that is currently taking place in many of the countries, and pushing a strategy of rapid afforestation.

Each of the two approaches mentioned above has significant difficulties associated with it, which should not be under-estimated. Even though natural gas emits 30-40 per cent less CO_2 than oil or coal for the same amount of heat generated, its main constituent is methane (CH_4), a gas that is an order of magnitude worse than carbon dioxide in terms of the "Greenhouse Effect." Unless strict measures are taken to ensure against leaks in the system, the benefits of switching to this source could be negated.

The technical and pereceptional problems associated with the use of nuclear power, such as the long-term storage of nuclear wastes, reactor accident and nuclear weapons proliferation (see, for example, Albright and Ferveson, 1988), are too well known to need a detailed discussion here.

Although solar- and wind-energy are already in use in a number of locations, their use on a large scale is still not competitive with fossil fuels. The expected large scale is still not competitive with fossil fuels. The expected large costs to essentially all countries that could result from a global climate change might provide an adequate rationale for investing heavily in the development of new (such as nuclear fusion) and renewable energy sources. Many in the scientific and technological communities had lamented the cut-backs in such research and development that occurred during the early years of this decade, when oil prices declined. Those charged with decision-making respons, unfortunately, to short-term needs, Shereas the development of new energy technologies requires long-term commitments. Several factors are contri-buting to the deforestation taking place in Asia and much of the developing world. Firewood is often the only source of energy available for cooking or heating in many rural areas of South-and South-East Asia. The need for agricultural land as populations expand has been an important source of foreign currency in many countries, and the practice continues, although on a smaller scale, even when it is officially banned.

In the limited space available here, we shall examine the extent to which countries in Asia might be able to switch from coal or oil to energy sources that emit less carbon dioxide.

Least-Cost Emissions Reduction Planning—the Ideal Approach

With respect to its energy-related investments, the primary function of the Global Environmental Facility (GEF) is to develop cost-effective approaches towards addressing the global warming problem. Investments should be directed towards that mix of technologies which can achieve the maximum reduction in emissions for a given investment. The identification of such a mix requires information on the unit cost of emissions reduction for the various technologies and the magnitude of the emissions reduction achievable with these technologies.

If this information were available, one could adopt a least-cost emissions-reduction strategy based on cost-emissions-reduction-technology (CERT) curves. These curves are constructed by choosing the technology with the lowest unit cost of emissions reduction as the first element of the mix, exploiting its emissions reduction potential, choosing the next most expensive technology as the second element of the mix, and so on until the desired emissions reduction target is achieved. Thus, one can either estimate the total investment that is required for the mix of technologies to achieve a given magnitude of emissions reduction or one can identify which mix of technologies is likely to achieve the maximum emissions reduction for a given total investment.

Least-cost-emissions planning is a worthwhile approach to move towards, for several important reasons:

- it takes into account both the cost-effectiveness of a technology in reducing greenhouse gas (GHG) emissions as well the potential impact of that technology with regard to emissions-reduction;
- it treats the supply-side and demand-side options for reducing GHG emissions on equal terms and does not discriminate against either of them;
- it ensures that different technologies are compared and prioritised on the basis of their cost-effectiveness (unit costs and potential emissions-reduction);
- it provids some idea of how much reduction in emissions is achievable (say, in percentage terms) and what cost;
- it constitutes a powerful heuristic for developing an investment strategy and portfolio for reducing GHG emissions.

Unfortunately, least-cost emissions-reduction planning is easier recommended than implemented because of the many conceptual and methodological problems in computing the costs of interventions and in estimating the benefits or effectiveness. In the first place, there has to be an agreed methodology of computing the costs of an emissions-reduction technology, and in particular the incremental costs over and above the conventional technology. Also, there can significant reduction of net CO_2 emissions from energy production and use without there being a stabilisation of carbon accumulations and a diminution of the threat of global warming. Thus, the benefits of a technology cannot be separated from objectives, for instance, whether the objective is reduction of net CO_2 emissions or stabilisation of carbon accumulations.

In particular, there are a number of specific questions:

(1) how to allow for the prospective costs of GHG emissions of costs, bearing in mind inter-generational concerns on the one hand and the importance of cost-effectiveness on the other;

(2) what discount rate is appropriate to use in comparisons of costs, bearing in mind inter-generational concerns on the one hand and the importance of cost-effectiveness on the other;

(3) how to estimate the benefits of innovation, in particular the contribution of an investment to reductions in the costs of future investments when cost-curves (as a function of investment) are steep;

(4) how to reduce the large transactions costs in some types of GEF projects having good economic potential.

Costs of Stabilising Carbon Accumulations

Consider the present situation in which carbon accumulations are rising over time due to the dependence on conventional fossil-fuel energy technologies. If a safe limit has to be set on the level of accumulations, then it would be necessary to switch eventually to non-fossil alternatives. Suppose, for heuristic purposes, the switch is assumed to take place at the time T in a step-functin fashion, then the marginal cost of energy consumption would change from the fossil-fuel value of "f" to that of the non-fossil fuel alternatives, "n". Then, the present value of the extra marginal cost at the time t = 0 is

$$C_0 = (n-f)\ (1+r)^{-T}$$

Thus, the actual marginal cost of fossil fuel consumption is f_0+C_0 and C_0 is the *shadow price* or carbon tax necessary to bring about investment in the non-carbon alternatives.

The shadow price increases with time such that

$$C_t = C_0 (1+r)^t$$

but this increase can continue only until it equals the difference (n-f) between the marginal costs of the non-fossil fuel alternatives and the fossil fuels. Thereafter, the non-fossil fuel alternatives become the cheaper and chosen option.

Consider any activity (such as energy efficiency or use of a low-carbon emitting energy resource such as natural gas) that would delay the time at which the carbon accumulations constraint is reached so that the transition to non-fossil fuel alternatives is shifted from T to T + Δ T. As a result, the present value of the "delayed" marginal cost at the time t = 0 is

$$C_0 = (n\text{-}f) (1+r)^{-(T+\Delta T)}$$

which is related to the "undelayed" extra marginal cost C_0 thus:

$$[C_0/C_0] = (1+r)^{-\Delta T)}$$

From this expression, it is seen that C_0 is less than or equal to C_0 depending upon whether Π D is greater than or equal to zero. Thus, the term C_0 is the *shadow price* or marginal benefit to be attributed to "buying time" through measures such as energy efficiency, use of a low-carbon emitting energy resource such as natural gas, other emissions reduction measures and sequestration.

Incidentally, this elementary treatment shows that *the best strategy of addressing the problem of global warming is a two-pronged strategy of (1) "buying time" with technologies that delay global warming, and (2) preventing global warming with non-fossil fuel technologies.* Since the first prong of the strategy (viz., energy efficiency, use of a low-carbon emitting energy source, other emissions reduction measures and sequestration) is effective only to a limited, albeit significant, extent, it is essential to back them with the second prong of non-fossil fuel alternatives which may therefore be called *backstop* technologies.

The step-function switch to non-fossil fuel alterntives is, however, an over-simplification because

- the time taken to switch to the non-fossil fuel alternatives in order to comply with the carbon emissions would be quite long.
- the costs of the renewable energy alternatives are declining while in certain markets (particularly in the non-electric markets) the costs of fossil fuels may rise slowly over the long term.

As a result, the programme of introduction of non-fossil fuel alternatives would have to be brought forward to address the marginal increase in CO_2 emissions. In other words, the investments on these alternatives will have to be distributed over time. But how these investments are distributed over the future is a matter of estimating the scope for substituting renewable energy for fossil fuels.

Perhaps the best assumption is that the investments could be scaled up in proportion to the estimated output of renewables, N_t, in each period t of the duration T over which the programme of renewable energy investments is phased in (roughly 50-70 years):

$$d_t = N_t/\text{Sum}(N_t)$$

For instance, if the d_ts sum up to about one-third in the first half of the programme and two-thirds in the second, more investments would be devoted to the later phases when the renewable technologies are fully developed without starving the earlier phases of the programme when the technologies have to be developed. An illustrative time profile of investment in the non-fossil fuel or backstop technologies during the phases in which they are being developed and substituted for fossil fuels.

This time profile is essential for estimating the shadow prices to be placed on reducing carbon emissions, but in addition, it is necessary to know the prospective costs of the backstop technologies relative to those of fossil fuel.

A satisfactory portfolio cannot be determined without an analysis of how relative costs are changing. The most promising types of investments in renewable energy are still small scale, and costs are declining with investment and technical progress. The transactions costs of demonstrating and developing new approaches are also initially high. What is relevant for the GEF therefore is not only (1) their current cost, but (2) the prospects for reductions in costs of the technologies in question, and (3) the contribution that the GEF can make to cost reductions.

The costs of the backstop technologies relative to fossil fuels vary greatly with market and application. They also vary over time. For some applications, renewable energy is already the least-cost option, e.g., the use of biomass for cogeneration, of wind energy in favourable locations, and of photovoltaics for rural electrification and the provision of power in electricity distribution networks. But substituting renewable energy for fossil fuels on a large scale would likely raise costs. Table 2 presents an illustrative assessment of the long-term costs of using renewables for electric power on a large scale.

There is growing consensus that, for electricity generation, the backstop technologies may eventually become competitive with fossil fuels, at least in the high insolation regions of the world. It is the provision of substitutes for solid, liquid and gaseous fuels (or non-electric energy) where the costs of the backstop technologies are highest. And, non-electric energy currently comprises 60 per cent of the primary energy markets in the industrialised countries and over 65 per cent in the developing countries. The main backstop technologies are biomass-drived fuels (ethanol and methanol), hydrogen (via electrolysis), or further electrification of the energy markets which will depend crucially on developments on storage technologies.

The effective costs to consumers of the electrification and hydrogen alternatives may eventually be lower than for two reasons. The first is that the efficiency of electric motors powered by batteries and fuel cells is much greater than that of the internal combustion engine; it is potentially around 60 per cent as compared with 20 per cent in the latter. Second, they have other environmental advantages because there are no emissions of harmful gases (assuming the electricity supplies are eventually based on renewables). Biomass-derived energy is also land intensive and its production on a large scale would place great pressure on land resurces; this would eventually raise costs and make electrification or the use of hydrogen more attractive, if only as a complement to biomass fuels. Nevertheless, it seems that (a) biomass-derived liquid fuels would be the main backstop technology for some time, and (b) that, whatever backstop technology is used eventually, and even allowing for the efficiency factor just mentioned, its costs would likely be higher than those of fossil fuels.

Since the shadow price, C_t, to be attached to carbon emissions is greater for the *marginal* backstop technologies than for the most promising non-marginal options, the question arises as to whether the higher or the lower shadow prices should be used. Here, the view is taken that the shadow prices should best based on the costs in the non-electric markets where the substitutes for fossil fuels are likely to be more expensive rather than in the electricity markets where the renewable energy options have good prospects of becoming competitive with fossil (and nuclear) fuels in the long-term.

The logic underlying this view is that the more promising of the backstop technologies whose costs would be less than $f_0+C_{0.m}$ would then be given added weight when the present value of the costs are compared. At the same time, some marginal technologies with costs close to $f_0+C_{0.m}$ would not be excluded; only the outliers would be left out, pending further developments. Further, those applications of the backstop technologies that can be identified as having costs lower than f_0 in the so-called "niche" markets, would have the highest returns. Using the cost estimates, the estimates of shadow prices to be placed on carbon emissions.

Until more reliable estimates of costs are available, a reasonable basis for cost-effectiveness studies might be to take a rounded value of C_0 = \$35/tonne of carbon rising at 10 per cent per year up to a limit of about \$600 per tonne. Alternatively, if all comparisons of costs are made in present value terms at t = 0, a constant undiscounted figure of \$35/tonne might be used.

The Cost of Capital and the Discount Rate Problem

A general principle of environmental policy-making is that policies that address an environmental problem *directly* are both less costly and more effective than those that address it only indirectly. For example, taxes or regulations on pollution, such as on the sulphur or lead content of vehicle fuels, or to take another example, on the treatment and disposal of spent nuclear fuels, are far more effective in reducing pollution than, say, a general tax or a restriction on energy use. Indirect measures may penalise 'clean' and 'dirty' fuels alike, and have only small effects on pollution while raising costs appreciably; direct measures in contrast may reduce pollution to low levels at a comparatively low cost.

The same principle is applicable to proposals to lower the test discount rate when deciding upon investments intended to address the global warming problem. Such a proposal has been shown to have the following defects:

- it is a blunt instrument of policy;
- it may also lead to decisions that contradict the aims of the policy;
- it favours capital-intensive solutions over labour-intensive ones, and
- it may sometimes work against investments with more immediate promise (such as renewables) by giving added weight to those with remote prospects (such as nuclear fusion).

In general, therefore, *lowering the discount rate is no substitute for direct measures to address a pollution problem, and will not guarantee the required results.* While blunt instruments may sometimes serve a useful purpose, the preferred approach is

(1) to reflect environmental concerns directly in policy with proper regard for the welfare of all parties, including future generations, and

(2) to use a discount rate equal to the opportunity cost of capital.

This is the approach suggested for the GEF.

The Reflection of International and Inter-generational Concerns

Two rules might usefully be followed to best reflect in GEF policies the international and inter-generational concerns about global warming.

The first is to *base the GEF portfolio on the types of investments that would most likely be needed in a "worst-case" scenario of global warming.* "Middle-of-the-road" scenarios are not relevant for GEF for two reasons. First, they ignore the *asymmetry* of the risks and the point that what eventually might happen, in the absence of appropriate policies, could be much worse than projected. The idea of a precautionary policy, like insurance policies, is to prepare for "worst-case" scenarios. Second, given the ambiguities in the evidence in the possible extent and consequences of climate change, and that some relationships (e.g., ocean-climate interactions and ocean currents) have not been modelled and estimated in a way that has been universally acepted, it is necessary to pay as much attention to the *variance* as to the *expected value* of the independent variable. This too argues for "worst-case" scenario.

The GEF, therefore, must be part of a precautionary policy in which the aim is to leave the international community better placed to respond to the "worst-case" global warming problem should the need arise. If its investments are based on a seemingly 'risk-neutral' position or a 'most likely' outcome, this would be to work against the purpose of the GEF, which is to prepare for contingencies. It follows that GEF's own investments should spport the development of the various technologies and practices that would have to be turned to a large scale in event of global warming requiring fundamental changes in policy.

The second is to *base GEF decisions on a carbon accumulations constraint* of the form derived in Section 8 to provide preliminary estimates of costs (the shadow price to be placed on carbon emissions). Introducing a carbon accumulations constraint into the analysis of energy investments amounts to sing the *resource depletion rule* often used in the past for fossil fuels. The rule has become discredited in recent years, because each time a limit to reserves was thought to be approaching, the limit was extended by new discoveries (there is a moral to this). In the context of global warming, the resource in question is not the *availability* of fossil fuels but the *safety margin*. This safety margin (defined by the limit to the use of fossil fuels set by concerns about climate change) can be deemed to be a resource because it is used up or depleted with increase in GHG accumulations.

If a limit is set to the use of fossil fuels, not by their availability, but by the amount of carbon the atmosphere can safely absorb, the imputed cost (or shadow price) of using them grows at a rate equal to the discount rate until the limit is reached and they can be replaced by non-net-carbon-emitting or backstop technologies. This means that the shadow price to be attached to carbon savings is *undiscounted* in NPV comparisons of costs, at least up to the point where the backstop technologies are widely deployed.

The main point about the rule is that it gives an appropriate weight to those options that are consistent with stabilising (or even reducing) carbon accumulations over the long term. Interestingly, it also means that in NPV comparisons of costs the shadow price on carbon emissions is then undiscounted over time, since its value rises according to a compound-interest law involving the discount rate.

A. CHINA

Global warming has become an issue of great concern and focus of discussions among scientists and the general public the world over in recent years. It is the theme of this year's World Environment Day, which is not accidental. Environmental problems such as greenhouse effect and the depletion of ozone layer have become a major potential threat to the existence and development of the mankind on the earth. This has awakened people and they have become to understand that strategies must be sought and common efforts must be made to deal with these problems.

Facing global warming, what environmental changes will take place in China? What impact will it exert on the economic development and the people's livelihood? What strategies will China take? These are questions the general public of China and all these who are interested in global environmental problems are concerned about.

Data show that in the last hundred years, the global climate has experienced a tendency of gradual warming. 1920s saw the first warming period. In 1930s and 1940s, the global average temperature was 0.3-0.4 Degree Centigrade higher than that at the end of last century. This was then followed by a period with lower temperatures. With the world entering 1980s. The globe began to experience the second warming period which is the hottest decade of the century. The years with the highest temperatures since 1900 are all within this decade. The global average temperature in 1980s is 0.6 Degree Centigrade higher than that at the end of the last century. It can be said that since 1920s, the global climate has been obviously getting warmer.

Various different predictions are available on the tendency of the future global climate change. However, most of the scientists are of the view that the world will continuously getting warmer.

The general tendency of the climate change in China in the past is basically in correspondence with that of the world: It was comparatively cool before 1920s, Warmer in 1940s and cooler in 1960s and 1970s, and rises again in 1980s. However, there are also some differences: First, the temperature came down a little in 1930s; secondly, the recent temperature rise began in the middle of 1970s, which was earlier than the other parts of the world, but the warming in 1980's is not so radical as that of the world. There are two characteristics in China's warming in the last decade: (1) The warming in the North (mainly north of Huai River and Qinling Mountain, particularly in the North-east and North of China) is more obvious than in the South; (2) The warming in winter is more pronounced than in the other seasons. Take the last winter (December 1988-February 1989) for example, the temperature in most parts of China was higher to various degrees than the average temperature of the precious years with the North-east and North being most obvious.

What future climate change will take place in China? According to the predictions by China's meteorologists, the surface temperature in China will tend to rise under continuous fluctuations. The scale of warming will not be smaller than that in the northern hemisphere, and the north will have a higher rise than the south. According to the study by the late meteorologist Mr Zhu Kezhen, in the 5,000 years of civilisation in China, there have been several warm periods and cool periods. The difference in the annual mean temperature between warm periods and cool periods is no more than 3 Degree Centigrade.

It can be seen if the annual mean temperature in 2030s is 1.5-4.5 Degree Centigrade higher than today, it will mean a temperature rise higher than today, it will mean a temperature rise higher than what has happened in the past several thousand years. To put it specifically, if the annual mean temperature rises by 2 degree centigrade, the temperature in Beijing then will be similar to today's temperature in Yellow River watershed; if it rises by 3 Degree Centigrade, Beijing's temperature will be similar to today's temperature in Huai River watershed; if it rise by 4 Degree Centigrade, Beijing's temperature will be similar to today's temperature in Changing (Yangze) River Valley.

Climate change will exert a profound impact on China's economic development and people's livelihood. Though some of the impact is only an assumption, reasons are solid for being on the alert. Climate change will move the northern border of China's sub-tropics from Qinling Mountain and Huai River. The temperature in winter in Xuzhou and Zhengzhou then will be similar to the present temperature in Hangzhou and Wunan. The temperature zone will move to the north by 4-5 degrees of latitude. Most of the frozen soil in the Qinghai-Tibet Plateau and small glaciers in Qilian Mountain and Tai Mountain will disappear. At the same time, the raining belts will also move. There will be more rain in summer in Hainan Island of the Tropics. While in other area, there will be less rain, especially in the north-west which is a dry area and will get even drier in the next century.

Such climatic change will have an obvious impact on China's agriculture. On the one hand, the warming will help to raise the agricultural output on some areas and on the other hand, it will lead to less rain in summer and higher temperature and more rain in winter. This will increase the capital cost of agricultural production. Since China's population accounts for 1/5 of the world population, we should attach great importance to food production. Because China is a developing country with comparatively backward farming methods and low ability to combat natural disasters and climatic change, the warming will have to a serious impact on China.

The rise of the sea level caused by global warming as an other threat to China, has a long coastal line. Owing to the historical reasons, most of China's economic forces are located in the coastal areas with dense population. The delta areas near the river estuaries are mostly rich areas with fairly developed economy. If the sea level rises, apart from flooding part of coastal areas, the area of alkaline and saline soil will increase, erosion and sedimentation in the coastal areas and the downstream of big rivers will become more serious, water way transportation will be affected, difficulties in drainage will increase and the work land of sea dykes and ports will also increase.

As a country with the largest population in the world, China is very much concerned about the trends of global warming because the possibility that global environmental problems will have impacts on China's industrial and agricultural production and people's health is the greatest. Factors causing global warming are complex: There are manmade as well as natural factors. The climate change of 100 years ago was mainly caused by natural factors while human activities are the main cause for the climate change in the recent 100 years.

Now most of the scientists believe that the built-up of carbon dioxide and other greenhouse gases in the atmosphere is the main cause of the greenhouse effect. These gases mainly come from the combustion of fossil fuels. At present, the whole world consumes 7 billion TDE of fossil fuels a year, or 90 per cent of the total energy consumption. The carbon dioxide released from the combustion of fossil fuels accounts for 70 per cent of the total amount of the newly released carbon dioxide. The past century saw the sharp increase in the release of carbon dioxide.

The annual release of carbon dioxide was about 90 million tonnes in the 1960s, but soared to 5 billion tonnes in 1985. The same period, however, saw the sharp decrease of forest—the "sanitation worker" and the "storage" for the atmosphere—at an annual average rate of 9-24 million hectares. This is evident that the greenhouse effect is closely related to the production and subsistence activities of the human beings. To solve the problem of the greenhouse effect and prevent the climate from getting further warm therefore requires efforts to adjust the productive activities and the lifestyle of the human beings. This obviously is an ordous task. Its fulfilment requires various countries in the world to gain a common understanding of the seriousness of the problem and shoulder their international bounden duty.

China's energy consumption amounted to 800 million tonnes standard coal, which is similar to that of Japan. As a major energy consuming power, China fully understands its duty in adjusting energy strategies and in controlling greenhouse effect. China has two problems to solve in the way the energy is utilised: First, fossil fuels make up 95 per cent of the primary energy consumed, most of which are relatively "dirty" energy source-coal. In 1986, in China, coal accounted for 72.4 per cent of the total primary energy production and 76.0 per cent of the total primary energy consumption while it was 23.3 per cent in the United States, 24.8 per cent in the Soviet Union, 18.8 per cent in Japan, 10.0 per cent in France. The world average was 30.0 per cent. Second, the energy efficiency remains low, averaging about 30 per cent which is 50-100 per cent lower than some advanced countries.

Therefore, China has great potential to tap in improving the energy utilisation ways. This is starting point for our participation in the global effort to improve the global environment. The endeavour to control the spread of the greenhouse effect will exert enormous pressure on various countries, but will also provide them with chances to effect changes and make progress. For China, it is a new challenge which will inevitably trigger deep-going reforms to the orientation, scale composition and efficiency of energy consumption. After the decade-long reform and development, China's economy at the crucial stage of the readjustment of the industrial structure. If we can have a good grasp of this opportunity to take an active part in the joint effort to improve the economic development, our effort will pay off to promote the economic readjustment and development.

In participating in the joint action to improve the global environment and prevent the global climate from getting further warm. China will mainly do two things: improving the way the energy is used and strengthening the protection of forest and vegetation.

1. Improving the Way the Energy Utilisation

(1) Striving to control the scale of energy consumption. China is the most populous country in the world. Its huge population has posed a heavy burden on China's subsistence and development. It is predictable that the scale of future energy consumption will widen with each passing day. The countermeasures that could be taken include: First, developing the production of non-fossil fuels whose combustion will not release carbon dioxide and other greenhouse gases, which will be discussed later; second, controlling the growth speed of energy consumption, and advocating and developing the energy-saving production processes and lifestyle. In this regard, we think in this way: although China's per-capita share of energy consumption is low, the total amount of energy consumption is considerably huge. For example, in 1987, China consumed 830 million tonnes of crude coal as energy. And the consumption of coal is expected to hit 1.1 billion tonnes by 1992. This increase in energy consumption brings pressure on both the production and the shipment of crude coal. It also exerts great impact on the environment. Judging from the angle of the long-term development of economy, the way out lies in the development of new industries which consume less energy. However potential must be tapped in improving the energy utilisation efficiency of those energy-guzzling traditional industries. If the thermal efficiency of coal can be raised from 30 per cent at present to 45 per cent. China would be able to save about 300 million tonnes of crude coal, and reduce the pollutants in the atmosphere by 40 per cent. Financially speaking, this will

also save China 150 billion yuan in the production of coal.

(2) Improving the utilisation rate of energy. China's energy utilisation rate is about 30 per cent on the average, compared to 57 per cent in Japan, 51 per cent in the United States and 40 per cent in Britain. As far as the average thermal efficiency of the energy consuming equipment is concerned, it is 35-40 per cent for ordinary thermal power stations, 80 per cent for industrial boilers and 50-60 per cent for industrial furnaces and civilians' stoves in the developed countries, compared to less than 30 per cent, about 60 per cent and 15-20 per cent respectively in China. Generally speaking, China lags far behind the advanced countries, which shows China has great potential to tap in seeking improvement and enhancement. During the economic development in the last few years, China has always attached great importance to saving energy. In 1988, the amount of energy and electricity consumed for the generation of every 10,000 yuan worth of industrial output value dropped by 5.63 per cent and 5.75 per cent, respectively from the previous year. This is a proof of good result. Of course, it is impossible for China to keep the energy saving at such a high rate for a long period of time. There will be more tough difficulties ahead of us. Nonetheless, we'll take energy saving as the orientation of our work for a long period of time to come. Our target is: by 2000 the energy consumption for every 10000 yuan of GNP will be 7.2 tonnes of standard coal (It was 23.73 T in 1985), per-capita energy consumption will be 1 tonne and energy saving rate will be 50 per cent. China is currently extremely in short of energy. About one third of its material production capacity often suffers from power failure and short supply of coal, causing enormous economic loss. Saving energy is therefore the need of the environmental protection and is also of great significance to the economic development.

To save energy, breakthrough must be made in the following two aspects: 1) technical renovation must be made on a large-scale instead of minor changes and modifications to the existing equipment and processes. If China's industrial technologies and energy utilisation technologies still remain at present levels, we will achieve no obvious progress in energy conservation. It means that technological progress is the key link; (2) economic mechanisms for energy conservation should be formulated in energy conservation, general calls and administrative orders will yield little results while economic levers with price mechanisms as the chief tool can achieve doubled success.

(3) Actively developing and using new energy sources: Nuclear energy, solar energy, wind energy, biological energy and other kinds of new energy are accepted as the clean energy which does not endanger greenhouse gases and other atmospheric pollutants. Although new energy sources have their own weak points in technology and cost, they represent the orientation for the world's development of energy; hence they have a bright future. China is making effort to develop and use the new energy.

China has always been attaching importance to the development of hydropower stations in energy production. China is rich in hydropower sources with a theoretical potential of 680 million kw of which 370 million kw can be tapped, 1900 billion kw.hr of electricity can be generated per year. China has built not only the world famous large hydropower works such as the Gezhouba Hydropower station but also numerous small hydropower stations. These small hydropower stations numbered more than 76,000 in 1984, generating a total of 8.5 million kw of electricity. It reached 10.06 million kw in 1986. Construction of the small hydropower stations gained further ground in the last few years. Electricity generated by these small power stations in the countryside has helped reduce coal consumption and, moreover, improve the conditions of farmers in daily life and the conditions for the improvement of the ecological balance.

Under strict safety measures, nuclear energy is a new kind of clean, safe and highly efficient energy. China began to build its nuclear power production bases in the early years of the 1980s. Projects with a total generating capacity of 2.1 million kw of nuclear power a year will go into production soon. By the end of this century, China's nuclear power production will hit 6 million kw. Undoubtedly, the development of nuclear power will help reduce consumption of fossil fuels, contributing to the decrease in the release of greenhouse gases.

In China, the development and use of solar energy have always received encouragement. In cities in north China, many buildings have been installed with solar energy equipment. and efforts are being made to increase the use of solar energy in industrial production. In the rural area, besides encouraging the development of small hydropower stations, China has also made effort to develop methane-generating pits and solar energy stoves to help solve the shortage of fuel in daily life. As a result, energy consumed in the rural areas is diversified, including bilgas firewood electricity gneerated by small hydropowr stations, coal, solar energy, wind energy and plant stalks.

2. Strengthening Protection of Forest and Vegetation

Forests are the 'lung' of the earth. Through photosynthesis, forests inhale carbon dioxide and exhale oxygen, thus playing an important regulatory role to the global climate. China is poor in forestry resources. Due to the incessant wars in history and the wanton falling of trees in the ensuingyears, China's forest areas had further decreased. In order to save the forest resources and expand the areas covered with threes. The Chinese government has promulgated many regulations including the Forest Law and the Regulations. Concerning the Water and Soil Conservation. In the meantime, China introduced the system geared to encourage people to plant trees on a voluntary basis. This system has been followed for ten years. Each year, there were 100 million Chinese citizens who took part in the tree-planting activities. They have planted altogether close to 10 billion trees averaging 1 billion trees, each year. This has increased the greened area both in cities and their suburbs. In the countryside, the reform to the forestry management system and the introduction of the contracted forestry responsibility system have aroused the enthusiasm of the farmers for planting trees. Many barren hills have been greened now. It is China's target to raise her forest cover from 12 per cent at present to 18-20 per cent by the end of this century. So long as we can persist in following the correct policy on forestry and strengthen protection of forest resources in accordance with the law, this target should be able to be attained.

In addition, China attaches importance to the protection of grasslands and prevents them from being reclaimed in a blind way. For lands which are not suitable for reclamation but have already been reclaimed for farming, efforts have been made to turn them back into grasslands. In the meantime, efforts have also been made to avoid overgrazing. The degradation of the grassland eco-system is being reversed.

Efforts made by China to protect the environment have contributed to improved conditions for a better environment and resources, stimulating the development of the economy. Of course, much remains to be done in this tregard, and the task is arduous. We have taken note of the long-term impact of the global environmental problems on the economic and social development. I am convinced that this international workshop will help promote the dissemination of needed knowledge, the exchange of experiences and co-operation among various countries.

B. TAIWAN

For the past 90 years, mean annual temperature increase of about 0.8^{0}C has been observed in a few big cities in Taiwan. Temperature increase would lead to thermal expansion of the oceans and the melting of the Antarctic and Arctic ice sheets. Sea levels would rise, inundating low-lying land areas that include many of the rice fields, fish ponds, and shrimp ponds. Because of the sea level rise, coupled with land subsidence caused by fast groundwater withdrawal rate, the western portion of Taiwan is regarded as the "sinking" coastal region. For the last 80 years, the Peikang Stream delta (a west-bound river system) has been receding at an average rate of over 49.9 m a year. With about 20 million people living on Taiwan island covering only 36,000 km^2, the government is committed to concentrate efforts in slowing down the greenhouse effects and reversing the temperature warming trend. This includes better air pollution controls, protecting current forest resources, promoting reforestation, restricting groundwater use, regulating coastal fishing farming, and strengthening environmental eduction to rouse public awareness.

Although the Industrial Revolution during the mid-eighteenth century is only 230 years old it has allowed the creation of many useful products and has raised the living standard of many people throughout the world and increased the population growth. In 1850, the world's population is estimated at 1 billion; 2 billion by 1930; 3.7 billion by 1970; and 5 billion in 1986. Based on present trends, United nations population experts estimate that world population growth will probably reach 6.2 billion by the year 2000. At the same time it has intensified many existing environmental problems and affected world's climate.

Increasing population cause a rise in resource use, but a rise in the standard of living creates an even greater demand for earth's natural resources. This includes large scale forest clearing to make way for farms, and uncontrolled mining of coal, petroleum, and natural gas to provide for ever-increasing energy demand. These human activities will alter the atmospheric composition, which in turn will affect the regional or even global climate.

Between 1870 to 1983 the total non-human energy consumption from all prime movers (that is, the various devices for harnessing mechanical energy, such as electric motors, and steam and gasoline engines) in the United States. Between 1870 and 1970, the diagram indicates a doubling of non-human energy capacity in almost every decade. Since 1970, the rate of increase reduced slightly. This reduction in energy consumption was triggered by soaring prices and economic recession. In 1870, each American energy consumption was 0.4 horsepower. However, this figure jump to a whooping 100.6 horsepower by 1970. In Taiwan relied entirely (94 per cent) upon non-renewable resources such as oil, coal, natural gas and nuclear for energy generation. This dependence increase to 97 per cent in 1986 and would further increase to 98 per cent by 1996. Nuclear energy comprised only 17 per cent of the total energy consumption in 1986. It would reach 22 per cent by 1996.

Burning fossil fuels and deforestation would increase the atmospheric concentration of CO_2. Since the beginning of the Industrial Revolution from about the middle of the eighteenth century, uncontrolled burning of fossil fuels and widespread destruction of tropical forests have raised atmospheric carbon dioxide levels from 260 ppm to 350 ppm, a rise of 30 per cent. This atmospheric buildup is expected to continue at a faster pace, averaging at least 1 ppm each year. Large quantities of carbon dioxide trap the low-energy radiation emitted from the earth's surface and raise global temperatures, inteisifying the "greenhouse" effects.

According to climate models as well as scientific investigations, these changes in the atmospheric composition have already raised the earth's average temperature by about 0.5^{0}C. By the year 2030, the global temperature would be between 1.5 and 4.5^{0}C warmer as indicated by a number of studies. This is enough to make the earth far warmer than at any time since civilisation began.

Temperature Increases in Taiwan

Meterological observations in Taiwan began as early as 1896. Up till now, there are 4 stations, namely Taipei, Taichung, Tainan, and Hengchun, with over 90 overs of uninterrupted temperature records.

Population for the 4 cities, according to a 1984 survey is 2,570,000;690,000; 640,000; and 35,000; for Taipei, Taichung, Tainan, and Hengchun, respectively. Population density at Taipei, Taichung, and Tainan exceeds 10,000 person'km^2; whereas Hengchun is less than 1,500 person'km^2. Pengchiayu, which is a small island located between Taiwan island and Ryukyu islands, is uninhabited and only use as shelter for fishing boats. The average temperature for the month of January and July as well as for the year are listed for all five stations.

To test the trend of temperature increase in Taiwan, the entire record length is broken up into three segments; the first period (1896-1926), the second period (1927-1956), and the third period (1957-1987). Average summer, winter and the annual temperature were computed for all three periods. Table 2 lists the temperature differences between the second and the first period, and the third and the first period for all five stations.

To test the trend of temperature increase in Taiwan, the entire record length is broken up into three segments; the first period (1896-1926), the second period (1927-1956), and the third period (1957-1987). Average summer, winter and the annual temperature wee computed for all three periods. The temperature differences between the second and the first period, and the third and the first period for all five stations.

Clearly, there is a definite progressive annual temperature increase in Taiwan. Annual temperature increases from 0.38-0.58^{0}C (mean 0.47^{0}C during the second period to 0.67-0.87^{0}C (mean 0.74^{0}C) during the third period. The evidence is pretty strong that the increase will accelerate the future. Among the five stations, Taipei, Taiwan and Taichung are highly urbanised cities. Hengchun is only slightly urbanised; whereas Pengchiayu still remains rural and situated quite a distance away from big cities. Anyone who lives or works in a city knows that it is always warmer there than in nearly suburbs or rural areas. Day in and day out, cities emit vast quantities of heat. Concrete and brick buildings and asphalt pavements absorb heat during the day and release it slowly at night. Tall, closely spaced buildings slow wind velocity near the ground and reduce the rate of heat loss. Thus, it is not surprising that a dome of heat hovers over a city, creating what is called an "urban heat island". It is, therefore, not unreasonable to find temperature increase over the years as urbanised cities like Taipei, Tainan, and Taichung grow and become vast urban regions. However, the fact that Hengchun and Pengchiayu also experienced the same magnitude of temperature increase indicate that this warming trend is no longer a local phenomenon but rather a regional or even a global problem.

The temperature for January, in general, shows a slight increase except in Taipei when comparing the first and the second period, and a slight decrease except in Pengchiayu when comparing the second and the third period. As for the month of July, all stations indicate temperature increase. The average temperature increase during the third period for all five stations is about 0.86^{0}C, which is almost 2.5 times the temperature increase during the second period (1927-1957). The most increase occurs in Tainan and Pengchiayu, the slight and the least urbanised areas in Taiwan, respectively. The trends for January, July and annual average temperature.

As no significant rising trend can be detected in January and July shows a steady temperature increase over the year, it is safe to conclude that the July temperature is responsbile for the annual temperature warming in Taiwan.

The effects of temperature Warming

(A) Sea Level Changes in Taiwan

Gutenberg (1941) studied the world's total records and found that the mean sea level would rise 1 mm each year. Milliman et al. (1988) recently developed three sea level rise scenarios. The first scenario assumes the best case with only 13 cm rise in the year 2050; and 28 cm rise by 2100. The second scenario, which is the worst case, assumes significant rise of 79 cm in the year 2050; and 217 cm rise by 2100. The third scenario is the really worst case that includes groundwater withdrawal faster than recharge causing subsidence or sinking of the ground. Mintzer (1988) also pointed out that for a temperature increase of 1.5 to 4.5^{0}C, the sea level would rise 40 to 120 cm.

Taiwan island is located in between two continental plates, namely the Euro-Asian and the Philippines. The constant collisions of these two plates together with the climatic change would affect the rise and fall of the sea level. Tidal records are the best indictor of the sea level changes. There altogether over 10 tide gaging stations located along the coastal areas of Taiwan. At present there are eight stations still in operation with more than 10 years of continuous records.

In (1987), according to the tidal records at Fukang, reported that the eastern coast of Taiwan is rising at a rate of 27 mm a year as a result of tectonic uplifting. However, all tidal records indicate a sea level rise in the western coast. In the north, keelung experienced about 8 mm rise in sea level; 27.8 mm rise at Taichi in the west; and 0.2 mm rise at Hengchun in the south. Taichi is the most famous fish farming area in Taiwan. Groundwater depletion is a serious problem in this neighbourhood. The existing groundwater level lies more than 10 m deep. Thus, the rapidly rising sea level is due to the combined effect of temperature warming and the excessive withdraw of groundwater supplies, more so due to the latter than the former.

For the last 80 years, the Peikang Stream delta has receded a total of 3.989 km from the original coast line, or an average of 49.9 m a year. Again, this area witnesses a large number of fish farms that withdraw so much water from the underground aquifers. As a result, the water table is much of the region is falling rapidly and reaches at least 20 m below the ground surface-the lowest in the central Taiwan region. This receding action is also due partly to the sea level rise and largely to the land subsidence.

The above evidence indicated that the eastern portion of Taiwan can be considered as the "uplifting" coastal region due to constant tectonic plate collisions. The western portion can be regarded as the "sinking" coastal region. If the problem of land subsidence caused by groundwater depletion and the receding river delta due partly to sea level rise cannot be kept under control, it may pose unpredictable threats to Taiwan's agriculture, water resources, land use, and the environment. A more predictable effect of the sea level rise is on agricultural production. Where rise is usually produced on low-lying river delta and floodplains in Taiwan, without heavy investments in dikes and sea walls to protect the rice fields from seawater intrusion would markedly reduce harvests. In addition, rising sea level will destroy fish ponds, shrimp ponds, and salt pans. Water resources in coastal areas will be at risk. Demand for irrigation would increase. Most water supplies for human consumption would probably not be threatened but would become more expensive, partly because water quality will be a bigger problem.

(B) Other Possible Impacts

At first thought a slightly warmer climate might seem desirable, resulting in longer growing seasons and possible increases in crop productivity in some parts of the world. The milder winters would also save fuel, although some of these savings would be cancelled by increased use of air conditioning during the hoter summers.

Unfortunately along with benefits for some parts of the world there could be serious consequences in other areas. The projected change in average global temperature could modify the distribution of rainfall and snowfall over much of the earth, and reduce the length of crop growing seasons in some parts of the world.

The most likely threat from global warming over the next 50 to 100 years would be shifts in food-growing regions and capacity during a time when the world's population is expected to at least double. Rainfall and snowfall patterns would shift, and some presently productive cropland would have to be abandoned and replaced with land of unknown quality. Existing irrigation and drainage systems and food storage and transportation systems that cost many billions of dollars would have to be rebuilt to reflect new patterns of precipitation, But for several decades there probably would be instability in global food production and distribution because many of the new crop-growing area should have poor soils and would not be organised to farm, irrigate, store, and distribute large amount of grain.

The climatic changes associated with global warming could redistribute the world's water resources. The devastating effect would be most apparent in agricultural regions. Drought and excessive heat in an once productive country would then have to depend on other countries for its supply of the basic commodity. Adaptation to such economic changes would probably cause shifts in the global political structure. However, discussion of the social, economic, and political consequences is obviously tentative at this point.

Another of the principal effects of global warming may be an increased in sea surface temperature, which in turn could increase the frequency of tropical typhoons and other storms. The end result would be stresses on tropical animals, the loss of substantial amounts of coral, and decreased biological diversity among marine organisms. With less coral and fewer coastal mangrove swamps, damage to beaches and waterfront property from wave action and typhoons is likely to increase.

The warming effect will not be uniform everywhere. Temperature increase at the poles is likely to be two to three times the global average, while increase in the tropics may be only 50 to 100 per cent of the average. This increased warming the poles will reduce the thermal gradient between the equator and the high-latitude regions, decreasing the energy available to the "heat engine" that drives the global weather machine.

As the thermal gradient is reduced, global patterns of winds and ocean currents, as well as the timing and distribution of rainfall, will change. Ocean currents that now moderate the climates of certain high-latitude countries may shift, causing the climate of such countries to cool even as the rest of the world warms. The global hydrological cycle is expected to intensify by 5 to 10 per cent if the world warms by 1.5 to 4.5^{0}C. Global rainfall will probably increase somewhat, but the timing and distribution of regional rainfall is likely to change substantially. Some areas will become wetter and others drier.

Temperature increase of a few degrees of Celsius would amplify extreme seasonal effects. Early melting of snow in late winter in some area would flood rivers and streams; summer droughts would be more common. Evapotransiration would generally be higher under warmer conditions. This transfer of water to the atmosphere by plants and soil would make less water available as runoff to replenish surface and groundwater reservoirs.

Government Interventions to Reverse Warming trends

evidence accumulating over the past indicates the earth's temperature has begun an upward climb. Temperature warming as a result of carbon dioxide increase and intensified greenhouse effects has reached a global scale. Now, the need is to address this environmental devastation through international cooperation, information and technology exchange, government intervention and community support. These concentrated efforts would reverse the warming trends and restore the natural support systems of the "spaceship" earth.

Taiwan is a small island with a total area of 36,000 Km^2. According to a 1984 survey, the total population is about 20 million and the population density already exceeds 550 persons'km^2. Island resources are limited. Therefore, based on the consumption of energy resources, emission of "greenhouse" gases is only negligible when compared to other developed countries. The carbon dioxide emission data is lacking. However, this information can be deduced from the total fuel consumption records. In 1987, Taiwan consumed 33,073.9 kiloliters of oil equivalent of fuel. This only comprises less than 0.02 per cent of the global energy consumption which stands at 180,976,988 kiloliters.

Chlorofluorocarbons (CFCs) are not only destroyers of the stratosphere's ozone layer but powerful greenhouse gases as well. CFCs productions in Taiwan in 1987 averages 15,100 tonnes/year; whereas the world's production totals 2,250,000 tonnes. According to import statistics, Taiwan consumes only around 0.5 per cent of the world's CFCs production.

Despite the fact that Taiwan contributes only an insignificant fraction to the global pollution problem, the government hads devoted efforts to counteract such environmental stress. This includes organising environmental protection institutions, promoting energy conservation programmes aimed at emission reduction of greenhouse gases, restricting forest clearing and encouraging tree planting, and strengthening environmental education. Although these measures are done on a local or regional level, they also will help alleviate global environmental problems.

(A) Environmental Protection Organisations

The Environmental Protection Administration (EPA) was established in August, 1987. The Executive Yuan passed the "Guidelines for Environmental Protection Policy of the Present Stage" during September of the same year. The Taiwan Provincial Department of Environmental Protection was formed in January, 1988. The systematic establishment of county and local protection agencies began in July 1988. Since then, the EPA has initiated ten large environmental protection programmes. One of the programmes, the environmental Monitoring Network Project, is working on setting up 189 air quality monitoring stations by the year 1994. Another programme, the Clearn Air (Blue Sky) Project is working on improving air quality on the island's most polluted regions, with an aim to reduce the Pollution Standard Index (PSI) value from the highest recorded value of 240 to less than 100 by the year 1992.

(B) Emission Control Measures

The present EPA air pollution control measures for curtailing emission of greenhouse gases includes:

1. Controlling state-run and private pollution enterprises.

   These enterprises will be required to report on their emissions annually. Any violations will be fined and required to take corrective action within limited time periods. It is estimated that 50,000 metric tons of particulates will be reduced by such measures.

2. Motor vehicle and aircraft air pollution control

Motor vehicle pollution control laws and regulations, emission standards and testing methods will be amended and enacted alongwith the promotion of low pollutant fuels.

3. Promotion programme for solar energy water heating

By the end of 1987, there wee 10,300 units with a total heat collecting area of 52,600m^2.

4. Promoting use of natural gas

using natural gas can reduce CO_2 emissions by 30 to 50 per cent and causes virtually no pollution as compared to burning coal and oil.

5. Ban on the use of CFS as aerosol propellants

Since November 1987, the government froze the licence for manufacturing eight CFCs substances and urges companies to expedite development of a CFCs' substitute and the recovery or decomposition of CFCs.

(C) Reforestation programmes

During the Japanese occupation, forest management is dominated by tree-felling as a major source of revenue. As a result, the natural forest resources is diminishing. By 1984, the total forest land is estimated at 1,876,000 ha, or 52.1 per cent of the total area. Since 1970, the government began encouraging reforestation in the public and private sectors. The total acreage being reforested between 1974 and 1985.

Besides promoting reforestation, the government also passed three forest management policies:

1. Forest should be managed for land protection purposes not for economical reasons.
2. Deforestation should be curtailed, soil and water conservation work should be strengthened and expanded to include forest land.
3. National forest should be managed properly, should not be rented out, and should be considered as one of the protected resources.

(D) Environmental Education

Environmental education is one of the most important partner in environmental protection work. If more people are aware of the environmental needs, environmental protection can be easily implemented with the support and cooperation provided by the people.

Currently, the following activities have been implemented in Taiwan:

1. Promote environmental education in the schools by strengthening their environmental education resources.
2. Help elementary school teachers prepare environmental education teaching tools.

3. Commission the development of a series of television programmes and introduce environment protection in other countries.

4. Organise environmental education centres and publish environmental education journals and magazines.

5. Fund environmental education research projects through the national Science Council of the Executive Yuan.

6. The EPA in cooperation with the Bureau of Industry hold seminars to educate directors and managers of companies with high pollution levels.

Global temperature warming appears to be an undeniable fact along with increases in carbon dioxide, strengthening of the greenhouse effect, and the sea level rise. This is further supported by the significant average annual temperature increase, over the last 90 years from a number of meteorological stations, representing both highly urbanised and rural uninhabited areas in Taiwan.

Taiwan is only a small island with limited resources and extremely high population density. It, therefore, needs to devote more research efforts into what mechanisms are influencing such climatic changes. Although carbon dioxide, CFCs, nitrogen oxides, and methane are probable candidates, the direct linkage between these factors and the global warming is still vague. Thus, further comprehensive studies are necessary to better understand what the major causes of change area. If the culprit can be singled out, more efficient and effective steps can be taken to counteract global warming. Although the process cannot be stopped entirely, it can at least be slowed down. In addition, the effects of global warming are not well understood. Nobody can tell exactly what will happen and to what degree. It is then worthwhile to look into or even quantify, if possible, the effects of global warming and sea level rise on crop, fisheries, land use, and water supplies.

At present, the government is committed to concentrate efforts in slowing down the greenhouse effects and reversing the temperature warming trend. This includes better air pollution controls, protecting current forest resources, promo-ting reforestation, restricting groundwater use, regulating coastal fishing farming, and strengthening environmental education to rouse public awareness.

C. MALAYSIA

Concern in malaysia over global climate change has, over the last 2-3 years, been given greater attention as evidence from various studies confirmed the threat to the global environment due to effect of various "greenhouse" gases. Evidence presented by various national and international organisations obtained through systematic monitoring and research, and presented at various forums have been the catalyst for creating awareness of the problem.

Concern on the subject has increasingly become a topic of discussion not only among scientists and environmentalists but also among the public. This awareness has no doubt been created through the media including local and foreign newspapers and magazines, local radio and television, non-governmental group forums as well as the education efforts of the department of Environment. This year, the Department of environment, in conjunction with World environment Day, is organising an exhibition on the theme which was adopted by the United Nations Environment Programme, entitled "Global Warming: Global Warning."

Concerns of Climate Change

Although no clear consensus has been made on the magnitude of environmental changes due to global climate change, Malaysia, nonetheless, is concerned even if the changes are smaller than predicted. This concern is due to some of the following factors:

(i) Malaysia is the largest producer of rubber and palm oil in the world and any significant shift in weather conditions is likely to affect crop production. Reduction in yields due to climatic changes will reduce earnings from these two commodities which contribute significantly to the economy of the country. Similarly, other crops such as cocoa (of which Malaysia is the world's fifth largest producer) will similarly be affected.

(ii) Malaysia is a coastal state with a coastline stretching more than 2000 km. Much of the coastline consists of low-lying mangrove areas which play an important role in coastal fisheries and aquaculture, mangrove timber supply as well as coastal protection. Inundation, due to sea level rise, of the coastal mangroves and induced changes affecting coastal ecology and stability will have serious implications to coastal fisheries; timber supply (particularly for the piling and charcoal industry), as well as coastal protection works along the coast. Coastal aquaculture, especially those using pond aquaculture systems, will likely be eliminated or moved inland. Coastal funds constructed along most of the mangrove areas and normally protected by a mangrove buffer, will have to be strengthened at great cost.

(iii) The two largest rice growing areas in Malaysia, the MADA in Kedah and the Tanjung Karang area in Selangor, are located on flat coastal plains. Erratic rainfall distribution and low rainfall in recent years have affected water supply for wet rice irrigation. In addition, sea level rise would raise the spectre of salt water intrusion and salinisation of soils which will affect crop production. Engineering measures to strengthen coastal protection works will have to be undertaken.

(iv) Major stretches of coastal areas are presently experiencing severe coastal erosion. The effect of sea level rise which is likely to increase coastal erosion is an area of concern since many of these areas support growing tourism and associated tourism related activities and facilities. In addition, major coastal towns (almost all state capitals are coastal) will likely face inundation or increased frequency of flooding.

(v) Climate changes resulting in higher rainfalls and increased intensity of rain have implications not only to soil erosion particularly in the cultivated hilly areas where major agricultural cash and food crops are grown, but will increase the intensity and frequency of floods. Current flood protection works are based essentially on past records of rainfall and do not include those projected due to climate change. With regards to soil erosion, silt in rivers is the most important pollutant affecting almost all rivers in the country. Soil erosion, originating from land disturbances due to logging, agriculture, urbanisation and others, is difficult to control even presently, given the high rainfall and intensity of rains in Malaysia.

Responses of Malaysia

Several actions are being taken by Malaysia to address the issue of global climate change. These actions, initiated over the past two years, are aimed at:

(i) Improving an understanding of global climate change such as the causes, the process, and the likely affects. Information based on findings by local, other national and international organisations will be used;

(ii) Participating and supporting international meetings and forums on global climate change; and

(iii) Developing local policy responses initially and action subsequently to address problems due to climate change.

International Response

Malaysia has participated actively in international meetings related to global climate change. Malaysia has been represented in all recent meetings such as the "Montreal protocol on substances that Deplete the Ozone layer" and the "Vienna Convention for the Protection of the Ozone Layer." As measure of Malaysia's commitment to the problem the First meeting of the Parties to the Montreal Protocol and the First Meeting of the Conference of the Parties to the Vienna Convention were attended by the Malaysian Minister of Science, technology and the Environment in which the following commitments were made:

> that the Malaysian Government agreed to subscribe to the Vienna Convention and to ratify the Montreal protocol.

Malaysia is presently collaborating with Thailand and Indonesia on a UNEP sponsored project on "Socio-Economic Impacts and Policy responses Resulting from Climate change." The study which was initiated in 1988 will examine climate impacts for each of the three countries and consists of three main stages:

(i) Estimation of the direct impacts of climate change, at selected locations, on agriculture water resources and coastal resources;

(ii) Aggregation of these direct impacts to the "national" level; and

(iii) Formation and testing of policy scenario alternatives.

The project is assisted by three consultants appointed by UNEP. Results of the study will be used to guide the government on policy responses that may be taken in anticipation of global climate changes.

The above UNEP project while it is regional in intent, is also applicable as a national study on the problems due to global climate change. The studies being undertaken or to be undertaken include the following:

(i) Climate change effects on the productivity of major agricultural crops. Specific study on the major rice growing are of MADA in Kedah will be carried out;

(ii) Changes to water resources availability and related flood implications using the Kelanten River basin as a model;

(iii) Coastal changes affecting coastal protection works, mangroves, urban centres, tourism and recreational beaches, and coral reefs are to be studied in different areas.

For the purposes of the above studies a National Steering Committee and various technical committees have been set up with participation from no less than 15 government agencies, research institutions and universities.

In relation to issues on the use of ozone depleting substances the Ministry of Science, technology and the Environment has:

(i) Initiated dialogue with industry and other government agencies on the implications of the Montreal Protocol to discuss various strategies to reduce consumption of CFCs and to seek alternatives;

(ii) Collect official statistics on the import and export of CFCs and halons through scheduling these chemicals as controlled substances under the Customs Act.

Other actions taken, though not primarily directed towards the problem of global climate change, include the development and adoption of a National Forest policy directed at conservation of remaining forest resources to ensure sustainable supply of timber and other forest products, use of forest for various beneficial uses, and a reafforestation programme for logged over forest areas with poor regeneration capacity.

In summary, awareness of the potential impacts due to climate change is growing though the need for greater awareness of the issue by politicians and policy makers needs to be enhanced.

D. PHILIPPINES

Between 1976 and 1986, developing nations accounted for most of the increase in non-European oil consumption, with transportation the fastest growing sector. Rapidly expanding fleets of vehicles have accelerated local pollution and global warming. An exception to this pattern among developing countries is a 15-year-old Philippine transport programme that combines higher gas taxes, expanded public transport and training. This has resulted in improved transport, lower oil imports and reduced CO_2 emissions.

In response to the growing financial burden of oil imports and worsening traffic congestion, the Philippine government in 1975 initiated a number of measures, including:

- Constructing a light rail transit system for Manila;
- Raising fuel prices by nearly 100 per cent
- Establishing car sale and registration fees on a sliding scale with higher fees for larger motors;
- Building oil recycling centres;
- Training vehicle maintenance workers.

The results are striking. Traffic time on roads in areas served by the rail system declined by about one third, in turn lowering gas consumption. Between 1976 to 1985, total gasoline consumption declined by 43 per cent from 2.35 billion litres to 1.34 billion litres. This occurred during a period of increasing population, expanding travel and rising per capita income.

If gasoline consumption per unit of GNP had been as high in 1985 as in 1976, gasoline use would have been 4.2 billion litres, or about 2.9 billion litres more tan actual consumption. The reduction in projected consumption saved about $500 million in oil imports in 1985. Reduced gasoline consumption translated into a reduction of CO_2 emissions by 8.2 million tonnes annually by 1985. Without this reduction, Philippine CO_2 emissions from fossil fuel burning would have been almost 25 per cent higher.

A series of studies by the Indonesian Directorate of Public Buildings and California's Lawrence Berkeley Laboratory have shown that energy usage in new Indonesian buildings can be cut by half using measures with average payback times of two years or less. A large savings potential in buildings is important because residential and commercial buildings consume 58 per cent of electricity generated by PLN, Indonesia's national utility.

The rate of growth of Indonesian electricity production was 15 per cent annually between 1980 and 1989, or more than four times the rate of the country's GNP growth, indicating a very wasteful pattern of electricity usage. If, as PLN now projects, this growth rate is unchanged, financing electricity expansion will constrain economic growth and lead to accelerated CO2 emission. This financial and environmental burden could be greatly reduced by implementing basic improvements in energy usage in buildings. These improvements and the consequent reduction in electricity demand include:

- Use of tinted rather than clear glass, cutting electricity usage by 4 per cent.
- Installation of efficient lightbulbs and use of lighting control systems, reducing electricity demand by 10 per cent to 23 per cent.
- Employment of daylighting (adjusting lighting level to allow for daylight), lowering electricity consumption by 7 per cent to 17 per cent.
- Adding variable speed fans and more efficient chillers, shaving 6 per cent off electricity demand.

Based on the huge financial savings potential shown in these energy efficiency studies, the Indonesian Directorate of Public Buildings seems likely to establish energy standards in 1990 or 1991 for all new commercial buildings. These standards would set maximum electricity usage levels that are almost a quarter lower than average current shock.

Additional steps being considered by the Indonesian Directorate of Public Buildings include:

- Energy efficiency test procedures for appliances.
- Appliance labelling.
- Establishment of appliance standards.
- Energy audits and retrofitting of existing buildings.

A combination of these and other measures to improve building and equipment efficiency offers to cut electricity use in new buildings by half, with average payback time of under two years. Similar savings potential exists in other Southeast Asian countries.

Detailed technical studies indicate that it is cost effective to reduce electricity use in refrigerators by about 80 per cent, mainly by improving the compressor and fan motor and adding insulation and a second compressor. These steps are estimated to cost ¢2/kWh saved. Over the past 15 years, improvements in energy efficiency have cut electricity consumption in new refrigerators by up to 70 per cent. Recent US national refrigerator standards are projected to cut present refrigerator power consumption by at least half. Investing $12.5 billion in improving refrigerator efficiency will save an estimated $60 billion in avoided construction cost in 30GW of new electricity generating plants. Assuming the avoided electricity would have been generated by burning coal, this gain in efficiency will cut annual CO_2 emissions by 160 million tonnes, equal to about 4 per cent of current US CO_2 emissions from fossil fuels.

Use of appliances, such as refrigerators, is growing rapidly in developing nations. A particularly striking example occurred in Beijing, where the proportion of households with refrigerators rose from 1.5 per cent in 1981 to 62 per cent in 1987. Appliances in developing coutries tend to be energy inefficient, imposing high electricity bills on consumers, requiring large national expenditure on building power plants and releasing unnecessarily large amounts of CO_2 into the atmosphere. Making refrigerators more efficient would reduce consumer bills, cut required national investment in power plants and sharply limit growth in CO_2 emissions.

As an example, let us assume that 40 million refrigerators manufactured in or exported to developing nations each year, drawing an average of 800 kWh annually. Assume that manufacture of these inefficient refrigerators is modified to incorporate efficiency improvements that cut electricity usage 75 per cent. After 10 years, annual electricity consumption would have been cut by about 240 billion kWh, reducing the required generating energy by about 46 GW. The cost of expanding energy supply by 46 GW would be $80 to $90 billion. Achieving refrigerator efficiency gains of 46 GW would cost only about $16 billion, or about one fifth the cost of expanding power supply. Reducing refrigerator electricity demand by 240 billion kWh would also cut CO_2 emission by about 240 million tonnes/years, or about 4 per cent of the 5.75 billion tonnes of CO_2 emitted by the burning of fossil fuels in developing nations in 1987.

Lighting is one of the largest and fastest growing uses of electricity in developing countries. A recent study of the potential of efficient lighting in Brazil and India shows that a shift to efficient lighting would reduce electricity costs to consumers, cut required capital investment in unnecessary power plants and, in addition, would eliminate the CO_2 emissions that would have come from these plants.

Installation of efficient fluorescent bulbs in India would cost one sixth of the cost of building new power plants. If India switched half of its current inefficient fluorescent lamps for efficient lighting over the next decade, it would cut CO_2 emissions by 17 million tonnes/year, equal to about 3 per cent of Indian CO_2 emission in 1987.

Savings offered by efficient lighting are even greater in Brazil, where higher plant costs mean that efficient bulbs cost about one tenth as much as building and operating new power plants. By replacing 36 per cent of inefficient flouorescent bulbs with efficient fluorescent lamps, Brazil would cut CO_2 emissions by 5-10 million tonnes/year (assuming replacement of fossil fuel fired electricity generation).

These efficient fluorescent bulbs give off a more balanced, comfortable light at the same intensity, but draw about one fifth as much electricity (16 watts instead of 75 watts). Considerable potential remains to further improve efficiency of existing 'efficient' fluorescent bulbs. Additional improvements in bulb efficiency will allow even greater savings in avoided power plant cost and avoided CO_2 emissions.

A detailed analysis by the American Council for an Energy Efficient Economy and Brazil's Companhia Energetica de Sao Paulo details electricity savings achievable in Brazil through improved efficiency. The study determined that investment of $10 billion of improving energy efficiency over the next 15 years could save $34 billion by substituting for a $44 billion investment in 21.3 GW of new electricity generating capacity, or about half the planned expansion by the year 2000. Reduction in CO_2 emissions from expanded efficiency would be limited because the bulk of projected electricity growth will come from hydropower. However, there are other clear environmental benefits associated with avoiding construction of additional dams.

The "greenhouse effect" is currently considered as the most pressing and complex environmental problem in the world. With the onset of the industrial age, man's activities have increased the quantity of the greenhouse gases such as carbon dioxide, methane, nitrous oxide, ozone and chlorofluorocarbons near the surface of the earth. Thus, when the sun's heat penetrates the atmosphere, these greenhouse gases help to retain some of the warmth that would otherwise be reflected back into space, thereby warming it up.

Scientists initially blame the greenhouse phenomenon to all forms of combustion which produce carbon dioxide. But further studies showed that on a molecule for molecule basis, methane, nitrous oxide, ozone and chlorofluorocarbons (CFCs) are likely to produce worse effects than carbon dioxide. It has been calculated that one molecule of CFC is likely to have a similar effect on the climate as 10,000 molecules of carbon dioxide. CFC is the chemical compound found in styrofoam cups and fastfood containers and in the freon used in air-conditioners and grocery refrigeration cases. CFCs float up into the stratosphere and break down the layer of ozone gas which serves as a shield protecting the earth from much of the sun's harmful radiation.

Indeed, the greenhouse effect poses a very complex environmental problem since if global warming continues, scientists predict that "the polar ice caps will melt thereby threatening the water supply and the existence of productive farmlands in the lowlands. Moreover, with the continuous thinning of the ozone layer, ultraviolet radiation can increase thereby causing more cases of skin cancer and disruption of plant growth or photosynthesis." To date, combatting the greenhouse effect is a global problem and the Philippines is not an exception in recognising the state of the environment.

This paper will discuss the measures being adopted by the Philippines to combat the greenhouse effect beginning with a discussion of the government units in the Philippines responsible in the protection of the environment, the rules and regulations governing environmental conditions including the penalties imposed for non-compliance of these rules and future legislations concerning environmental protection. A separate section is devoted to the five-year energy plan as well as the mitigating measures being implemented by the National Power Corporation, the Philippines' bulk power supplier for the protection of the ecological balance. The final section discusses policy recommendations affecting plant operations and other areas for improvement in order to minimise if not completely eradicate the greenhouse effect.

Government Agencies Responsible for Philippine Environmental Protection

The environment is the sum total of conditions and influences that make up the surroundings of living organisms. Among the factors of environment that regulate the nature of living things are sunlight, water, atmospheric gases, temperature, pressure and the wind, the tides, mineral resources and the substrate. These factors determine the availability of food and shelter needed to sustain life and they regulate the interaction of living things.

In man's quest for a better life, man began overdoing things resulting to the degradation of his environment. Gradually, man realised that there is little that can be done to increase the capacity of nature to disperse the pollutants he is creating. Thus, to prevent more damage to health and property, the concept of controlling polluting materials at their source is introduced. It is in answer to this need that the Philippines' Department of Environment and Natural Resources (DENR) was created in order to pursue a programme of pollution control and improvement of the environment.

The DENR is headed by a Department Secretary who is a member of the President's cabinet. Under him are six staff bureaus, namely: (a) Mines and Geo-sciences Bureau; (b) Forest Management Bureau: (c) land management Bureau; (d) Environmental Management Bureau; (e) Ecosystem Research and Development Bureau and (f) Protection Area Wildlife Bureau. For purposes of environmental management, it is the Environmental Management Bureau (EMB) which acts as the lead agency.

The EMB prepares all legislations, policies and programmes for environmental management and pollution control. It also formulates environmental quality standards and recommends rules and regulations to protect the Philippine environment. The EMB's functions are incorporated in its four divisions such as: (a) Legal; (b) Research and Development; (c) Environmental Quality and (d) Environmental Education. The functions of these divisions are as follows:

(a) *Legal*—Formulates rules and regulations concerning environmental protection. It also conducts continuing legal studies and researches on laws and jurisprudences. This division assisted the Pollution Adjudication Board.

(b) *Research and Development*—Performs ecological studies on pressing environmental problems. It develops analytical test procedures to identify and measure major pollutants, characterise individual effluents and analyse samples from environmental monitoring networks.

(c) *Environmental Quality*—establishes environmental quality indices and guidelines for environmental protection; formulates environmental surveillance, air and water quality monitoring programmes and review of environmental quality standards; responsible for implementing the country's environmental impact statement system.

(d) *Environmental Education*—Formulates and coordinates the implementation of the national programme on environmental education. It designs training modules and compile information packages on environmental management. It also promotes the incorporation of environmental concepts in all levels of education.

Philippine Environment Laws

Presidential decree 984 or the National Pollution Law was enacted in 1976 to formulate a reasonable standard for the air and water quality of the country. The Decree also empowered the Environmental management Board to establish the maximum permissible emission standards for visible emissions and particulate matter and specific air pollutants from stationary and mobile sources. The implementing rules and regulations of this presidential decree likewise regulate fugitive particulates, emission of volatile compounds or solvent, nuisances related to atmospheric air disturbances and open burning.

The salient provisions of the implementing rules and regulations of Presidential Decree 984 are as follows:

Chapter II. Article 8, Section 58-Maximum Permissible emission Standards for Visible Emissions and Particulate Matter.

A. For Fuel-burning Equipment and Industrial Plant

(a) The concentration of air impurities emitted from existing sources at the point of emission shall be such that when compared in the appropriate manner with the Ringlemann Chart or an equivalent method approved by the Board, impurities shall not appear to be darker than shade 2 of the Chart.

(b) The concentration of air impurities emitted from new sources under the above regulation shall be such that impurities shall not appear to be darker than shade 1 on the said Chart.

B. Boilers or Incinerators emitting Solid Particles

(a) For existing boilers and incinerators, concentration of solid particles at the point of emission adjusted to a basis of 12 per cent carbon dioxide shall be such that the total mass of such solid particles does not exceed 300 mg/scm.

(b) For new boilers or incinerators (new sources), the maximum permissible emission limit shall be 300 mg/scm.

C. Any trade, industry, process, industrial plant or fuel-burning equipment emitting solid particles, except those referred to in (b).

The concentration of solid particles at the point of emission of existing sources shall not exceed 500 mg/scm.

For new sources, the maximum permissible emission limit for particulate matter shall be 300 mg/scm.

Chapter II, Article 8, Sector 59-Maximum Permissible Emission Standards for Specific Air Pollutants from Stationary Sources

Chapter II, Article 8, Section 61-Maxmimum Permissible Standards for Mobile Sources

The following standards of concentration and rate of emission of air impurities in respect of the following classics of motor vehicles are prescribed:

(a) Any motor vehicle which has been registered and which is propelled by a gasoline engine:

The concentration of air impurities at or near the final point of emission from the exhaust pipe into the atmosphere is such that air impurities shall not be visible for a continuous period of more than 10 seconds.

(b) Any diesel vehicle which has been registered:

The concentration of air impurities at or near the final point of emission from the exhaust pipe into the atmosphere is such that air impurities shall not be a shade or density greater than 20 per cent capacity for a continuous period of more than 10 seconds where chapter II, Article 8, Section 62-National Ambient Air Quality Standards.

For the purpose of protecting public health and welfare and reduce damage to property as well as providing air quality management control strategy for emission limitation for mobile and stationary sources, location of commercial, industrial and residential facilities and to assist in the promotion and use of an improved transportation systems, the following maximum limiting levels, not to be exceeded more than once a year are hereby established:

The methods for measurement of four of the pollutants are World Health Organisation (WHO) selected methods; sulphur dioxide by acidemetric; photochemical oxidants by potassium iodide; nitrogen dioxide by diazotising coupling and carbon monoxide by nondis persive infrared absorption. Total, suspended particulate matter is measured by light scattering, the comparability of these methods with others may be judged by easing World Health Organisation-specified comparison methods.

The maximum air quality standards of other specific pollutants are set at values given in the table below:

Chapter II, Article 8, Section 66-Air Quality Indices

The following describe the three levels of air quality for particulates, carbon monoxide and sulphur dioxide anywhere in the Philippines:

(1) Suspended Particulates

Good—0 to 100 micrograms/scm

Fair—101 to 250 micrograms/scm

Poor—251 to 350 micrograms/scm

Serious or alert conditions-greater than 340 micrograms/scm

Alert 1 (White)—350 micrograms/scm

expected to hold or increase for 12 hours or more

Alert 2 (Blue)—550 micrograms/scm

expected to hold or increase for 12 hours or more

Alert 3 (Red)—750 micrograms/scm

(emergency) expected to hold or increase for 12 hours or more

(2) Carbon Monoxide

Good—0 to 10 ppm (o to 12.5 milligrams/scm)

Fair—11 to 20 ppm (12.6 to 37.5 milligrams/scm)

Serious or alert conditions-greater than 30 ppm

Alert 1 (White)—30 ppm, expected to hold or increase for about two hours or more but not to exceed four hours; otherwise, next alert is issued.

Alert 2—60 ppm, expected to hold or increase for about two hours or more but not to exceed four hours, otherwise, next alert is issued.

Alert 3—80 ppm (emergency), expected to hold or increase for about two hours or more.

(3) Sulphur dioxide

Good—0 to 0.1 ppm (0 to 285 micrograms/scm)

Fair—0.11 to 0.8 ppm (286 to 850 micrograms/scm)

Poor—0.31 to 0.5 ppm (851 to 1430 micrograms'scm)

Serious or alert conditions—greater than 0.5 ppm

Alert 1 (White)—0.5 ppm, expected to hold or increase for 12 hours or more

Alert 2 (Blue)—0.8 ppm, expected to hold or increase for 12 hours or more

Alert 3 (Red)—(emergency) 1.0 ppm, expected to hold or increase for 12 hours or more.

The implementing rules and regulations mentioned above are only those affecting air quality. Other provisions of the Pollution Control Law which are not included in this paper are those pertaining to water quality and noise management.

The non-compliance of the implementing rules and regulations of the Pollution Control law shall result in the imposition of the following penalties:

(a) Any person found violating or failing to comply with any order or decision of the Board on these rules and regulations for the control or abatement of pollution shall pay a fine not exceeding five thousand pesos (P5000) per day for every day during which such violation or default continues.

The fines so imposed shall be paid to the Government of the Philippines through the Board and failure to pay the fine in any case within the time specified in the above-mentioned order or decision shall be sufficient ground for the Board to order the closure or stoppage of the operation of the establishment being operated and/or managed by said person or persons until payment of the fines shall have been made.

(b) Any person who shall violate any of the provisions of Section 8 of PD 984 of these rules and regulations or any order of the Board shall be liable to a penalty of not to exceed one thousand pesos (P1000) for each day during which the violation continues or by imprisonment of from two years to six years or by both fine and imprisonment and in addition, such person may be required or enjoined to discontinue such violation.

(c) Any person who shall refuse, obstruct or hamper the entry of duly authorised representatives of the Board into any property of the public domain or private property devoted to industrial, manufacturing, processing or commercial use during reasonable hour for the purpose of inspecting or investigating the conditions therein relating to pollution or possible or imminent pollution shall be liable to a fine not exceeding two hundred pesos (P200) or imprisonment of not exceeding one month or both.

(d) Any person who violates the provisions of or fails to perform any duty imposed by PD 984 or these rules and regulations or any order or decision of the Board promulgated pursuant to PD 984 thereby causing the death of fish or other aquatic life, shall be liable to pay the government for damages for fish or aquatic life destroyed, in addition to the penalty described above.

(e) In case the violation is a juridical person, the person shall be imposed on the managing head responsible for the violation.

(f) The remedies provided herein shall not be a basis to nor shall affect any the remedies provided in PD 984 but shall be cumulative and additional to such remedies.

Future Legislations

The law-making body of the Philippines is vested upon the House of Representatives and the Senate. It is through these branches of the Government that bills and resolutions are prepared and acted upon to become laws of the country. The succeeding discussion will present the various bills and resolutions on the environment awaiting action by the House of representatives and the Senate.

A. Environment-related Bills and resolutions pending in the House of Representatives

1. House Bill 7179—increases the penalties for violation of the law on anti-pollution amending Section 9 of PD 984, known as the National Pollution Control Decree of 1976.
2. House Bill 7815—increases the penalties for violation of the air pollution law amending PD 1181, providing for a fine, for the first offense of not exceeding P500, for the second offense to a fine of not exceeding P1000 and for, the third and succeding offense to a fine of not exceeding P2000 plus the suspension of the owner's vehicle's certificate of registration until all requirements of regulations have been complied with.
3. House Bill 6971—creates a regional Environmental Pollution control and Sanitary Waste Disposal Board in each region of the Philippines to be responsible, among other, in the maintenance and protection of the ecological and environmental natural balance of the region.
4. House Bill 17222—requires a monthly inspection for pollution control of all commercial, industrial, manufacturing and processing firms.
5. House Bill 16190—increases the penalties for violation of the law on anti-pollution to a fine not exceeding ten thousand pesos per day for every day during which such violation or default continues. For those violating the provisions of Section 8 of PD 984, a penalty of not to exceed five thousand pesos for each day during which the violation continues is applicable or imprisonment of from six years and one day to twelve years or both.

B. Environment-related Bills and resolutions pending in the Senate

1. Senate Bill 981—provides incentives to organised citizens' environment protection groups and establishes an environment defense fund which shall come from 50 per cent of the total amount derived from fines in violation of the National Pollution Control Law, gifts, donations and grants.
2. Senate Bill 994—regulates the importation of substances that deplete the ozone layer. The bill defines the controlled substances as those listed in the Montreal Protocol.
3. Senate Bill 960—restructures the Pollution Adjudication Board amending Section 19 of executive Order 192.
4. Senate Resolution 919—amends and expands the scope of PD 984. This bill transfers to the Pollution Adjudication Board and the Environmental Management Bureau of the Department of Environment and Natural resources the enforcement of the proposed Pollution control Law and expands the

coverage of the ban to those engaged in industrial, commercial, manufacturing, processing, transportation and agri-business purposes.

Philippine Five-Year Energy Plan and Mitigating Measures Being Implemented in the Power Sector for Environmental Protection.

The energy development programme of the Philippines is prepared by the Office of Energy Affairs which is asked with the formulation of energy policies and coordinating energy programmes. The present five-year energy plan of the Philippines still consists of sustained efforts in the development of indigenous energy sources and a reduction in oil import dependence. The plan defines the government's thrusts as follows:

(a) Promotion of energy self-reliance

(b) Nationalisation of energy prices to reflect the true costs of production and distribution

(c) Encouragement of energy conservation measures to promote efficiency

(d) Participation of private sector in energy projects.

(e) Maintenance of environmental and safety measures for energy projects.

In support of these thrusts of the government, the power sector's strategies are as follows:

(a) Continuation of indigenous energy resource exploration and development

(b) Completion of committed energy projects

(c) Continued departure from oil-based generation except in more efficient systems such as cogeneration

(d) Review of sector efficiency in all stages

(e) Continuation of viable energy conservation programmes and projects

(f) Restructuring of power rates recognising long-term marginal costs

(g) Pricing of domestic petroleum products consonant with international oil prices.

(h) Streamlining and improvement of operations of electric cooperatives and utilities

(i) Maintenance of evaluation and implementation procedures on environmental aspects.

(j) Reduction of public investments by inviting private capital into construction and operation of energy projects.

In view of the fact that the Philippinee energy plan still maintains to tap indigenous energy sources and fossil fuel for its power requirements due to the continuous mothballing of its 620 megawatt nuclear power plant, the Government through the National Power Corporation, the country's bulk power supplier, implements measures to mitigate the environmental impact of the operation of these types of plants. These mitigating measures are:

For Geothermal Plants

Gaseous and liquid emissions from geothermal plants are composed of different non-condensable gases such as carbon dioxide (78-95 per cent), hydrogen sulfide (1-17 per cent), methane (0-12 per cent), ammonia (0-1.7 per cent) and trace amounts of argon. These gaseous emissions may cause acid rain which increases the acidity of soild and leaches valuable nutrients needed for plant growth. To minimise the environmental impact of geothermal plants.

For Thermal Plants

Fossil oil emits carbon monoxide in the air during combustion. Water condensates from the power plants also deposit chemicals into the water. They raise the temperature of surrounding water bodiest to the detriment of aquatic life. This type of plant contributes tremendously in degrating the air.

Policy Recommendation

Philippine legislators can explore various areas for policy formulation to protect the ecological balance of the country.

First, Philippine energy policies should include environmental aspects. Energy planners should make sure that energy supply as well as environmental quality standards should be met simultaneously. Moreover, the electric power industry should try using and adapting new technology in the operation of power generating plants to minimise destrction of the environment. Specifically, the following steps should be taken for the different power plants to protect the environment:

(a) pollutants emitted by coal plants during combustion must be reduced.

(b) reinjection technolgy must be fully applied in geothermal plants.

(c) a more comprehensive impact study for hydro projects must be conducted to reduce environmental degradation.

(d) strict control measures must be instituted by thermal plants to limit chimney emissions to allowable levels.

(e) land allocation for food and energy forests for dendro plants must be judiciously studied.

Second, Regional Offices of DENR should redouble their efforts to ensure that all anti-pollution control measures are carried out effectively. They should efficiently perform the following:

(a) monitor smoke belching vehicles and collaborate with various agencies with police power to fully enforce smoke emission standards.

(b) promulgate stringent rules and regulations for employing efficient air pollution control devices accompanied with strict enforcement as well as discriminate imposition of deterrent penalties.

(c) update air quality standards that will be relevant to the present state of the environment.

(d) conduct comprehensive information and educational campaign.

(e) provide more incentives for industries to put up control facilities.

(f) provide more disincentives for those non-complying with pollution control facilities like restrictive measures ranging from withholding or withdrawal of credit and infrastructure support.

(g) commit more administrative support in the form of additional appropriation for enforcement agencies.

Finally, other aspects that policy makers can look into are:

(a) embark on massive ecological programme.

(b) use of unleaded fuel.

(c) use of smokeless fuels in households and factories.

(d) minimise concentration of motor vehicles in urban areas.

(e) strict regulation on the importation of cars which do not meet standard in pollution control equipment.

(f) relocation of heavy and medium industries away from urban centrers must be pressured.

On top of all these policies, the Philippines should strictly implement existing laws for proper environmental conservation. Further, leadership in this area of responsibility must be clearly identified plus the present government should ably support the implementation in terms of funding and personnel allocation. And more significantly, the Filipinos must be properly educated and involved in environmental protection so that exploitation of the environment can be minimised in exchange for profit.

E. SOUTH PACIFIC ISLANDS

Knowledge about past and present South Pacific island environments varies considerably in both detail and calibre depending on numerous factors including location (for example, proximity to main centres) and resource potential. Such imbalances hinder the production of meaningful (locally-applicable) accounts of recent environmental changes for the region and reduce the credibility of region-specific predictions concerning future changes. For example, no tide-gauge records extending back more than thirty years exist for South Pacific islands so direct (data-equivalent) comparison with secular changes of sea level recorded elsewhere is not possible.

Similarly, region-specific predictions of future sea-level rise commonly depend on high-calibre local information which is not yet available for the South Pacific region although a series of modern gauges was promised to the region by Australia at the 1988 South Pacific Forum meeting. Interpolation between control points across the South Pacific provides an unsatisfactory basis for short-term predictions compared to that used elsewhere.

This paper assesses the evidence for environmental changes in the South Pacific since 1900 which are attributable to contemporary global warming then goes on to evaluate the future impacts of continued warming on a series of typical islands and selected urban centres in the region.

It is widely accepted that the environmental changes which were experienced on many South Pacific islands since 1900, when European settlement was well established in the region, were caused directly by human activity (Farrell, 1972; Barrau, 1980). Increases in the area of degraded land on most islands has been linked to an increasing demand for cash crops, especially for export, although subsistence cropping on resource-poor islands with rapidly-increasing populations has led to a similar situation. On Niue, for example, an isolated, predominantly limestone island 259 km^2 in area, the proportion of degraded land increased from 20 per cent to 45 per cent between 1949 and 1971 (Dahl, 1984).

It is contended, however, that the casual connection between people and environmental change has not been demonstrated with sufficient rigour and that several key studies are open to alternative explanation. An example with relevance to global warming is provided by the Waibau area of Viti Levu, Fiji. The largest river on Viti Levu, Fiji, is the Rewa, the catchment of which loses 35 tonnes/hectare/year. This comprises 24 t/ha/yr from the Wainibuka, 30 t/ha/yr from the Wainimala, 69 t/ha/yr from the Waindina, and 79 t/ha/yr from the Waimanu (Hassan, 1986). The latter catchment includes an area named Waibau which since 1956 has developed into one of the major suppliers of market crops to the nearby urban centre of Suva. In 1988 the total soil loss from Waibau was 138 t/ha/yr.

A major difficulty with attributing soil loss from areas such as Waibau and the Waimanu catchment is the lack of time-series data. Analysis of recent aerial photographs across one-third of the Waimanu catchment showed 620 landslides which had been initiated in April 1986 during a major rain storm. These landslides represented 570,000 m^3 of lost material or 92 t/hr/yr over the whole Waimanu catchment in a few weeks, a figure in excess of that for the whole of 1988 which included anthropogenic soil loss from waibau (all data from D. Glatthaar, 1989, personal communication). It is significant that most landslides in April 1986 were clustered in undeveloped areas outside the Waibau sector which suggests that natural events are the dominant control or erosion in the Waimanu catchment and that soil loss from Waibau, although a serious problem, may have been overestimated in terms of erosion from the whole catchment.

Data referring to the erosional effects of tropical storms, especially those associated with tropical cyclones, are few but it nevertheless clear that the Waimanu/Waibau example is not unusual in that the amount of material released into the river system during such events greatly exceeds that released under 'normal' conditions. Given this and the increased frequency of tropical cyclones experienced recently in the South Pacific, it is rash to insist that anthropogenic causes are the primary contributors to soil erosion and land degradation in the region.

It is clear that climatic changes manifested as secular changes in meteorological variables may have had a part to play in land degradation at Waibau and elsewhere. Such changes include catastrophic events, many resulting from tropical cyclones, and long-term secular changes, the potential effects of which can be noted but not demonstrated given the calibre of data available for the region. In assessing the potential role of climatic change on the environment of Pacific islands over the last ninety years or so, variations in meteorological variables are examined.

For the last hundred years or so, it is considered that a global warming has taken place at the Earth's surface of around 0.4°C but criticisms have been levelled at the northern-hemisphere continental bias of such global syntheses and the data from which they derive. One attempt to overcome these problems was by using radiosonde temperature data for the lower troposphere which show an undoubted recent temperature rise in the South Pacific.

Tropospheric warming in the South Pacific is probably matched in broad terms by ground and sea-surface temperatures in the region. Although insufficient long-term data are available from the islands, the former trend is supported by ground-temperature analyses for New Zealand, "a rise having been experienced throughout the whole region during 1935-1970, temperatures climbing by 1°C over these years". Sea-surface temperatures in the South Pacific have risen by 0.5-1.00°C since 1912.

Tropical tropospheric warming in the southern hemisphere has led to an intensification of the hydrological cycle in the period 1949-1979 which may be expected to have been manifested by increases in suspended fluvial load, evaporation and precipitation. Most authorities regard future climatic changes resulting from global warming to include increased precipitation in low latitudes, especially the oceanic realm where increased seawater evaporation will result from an increase in sea-surface temperatures.

Sea-surface temperature is a critical factor in the development of tropical cyclones which occur when the South Pacific Convergence Zone (SPCZ) juxtaposes places where the sea-surface temperature exceeds 27-28°C. In Fiji, for examples, this latter condition applies to Rotuma Island (12.5°S) all year on average, to Viti Levu (18°S) from January to April inclusive, and never for Ono-i-Lau (20.5°S-Basher, 1986). A recent summary by Folland *et al.* (1984) concluded that the average sea-surface temperature in the South Pacific had risen by 0.5-1.00°C since 1903-12 and that this latitude zone had experienced a rise of between 0.1-0.20°C since 1951-60.

It is therefore reasonable to assume that the oceanic 28°C isotherm has moved polewards recently in the South Pacific thus increasing the area over which tropical cyclones have developed. This in turn may be expected to have resulted in increased tropical cyclone frequency in the region. Tropical cyclones with hurricane-force winds (hurricanes) have indeed been experienced with increased frequency in the South Pacific recently. Only twelve such events affected Fiji between 1941 and 1980 (3.1 per decade on average) yet six occurred between 1981 and 1985 (15 per decade on average-Thompson, 1986).

It is clear that one possible explanation for the recent increase in tropical cyclone frequency on South Pacific coasts is that it is a consequence of global warming. If this is so, then careful studies of what has happened to different islands under conditions of increased tropical cyclone frequency are essential preparation for the future which is likely to see such conditions maintained or intensified.

Coral reefs are highly-sensitive indicators of environmental changes. Periods of recent coral death in the South Pacific have been variously attributed; fish poison, Acnthaster infestation, landfill de\redging and increased lagoonal sedimentation are among the common explanations. Many case studies of dredging and sedimentation reveal an undoubted connection such as the death of coral colonies around the mouth of the Quraniqio River, Viti Levu, Fiji, caused by sedimentation associated with the construction of the Pacific Harbour resort just inland. In other cases, is it not clear why there has been increased sedimentation recently.

Much of the sediment responsible for smothering coral reefs reaches them via rivers. Increases in the amount of suspended sediment load is linked to increases in sediment loss from the catchment. This in turn could be the result of forest clearance and land degradation and/or be related to an increase in precipitation. As seen above, the latter could have been occurring recently as the result, in part at least, of global warming, as suggested by the increase in tropical cyclone frequency.

The clearance of mangroves for coastal development in the South Pacific islands has undoubtedly led to greater sediment mobility in nearshore areas and has probably seen responsible for widespread coral death but, as with other explanations, it is difficult to prove rather than simply to infer a connection.

Sea-level Changes

The behaviour of sea level during the late Holocene is becoming increasingly better known for the South Pacific but little is known specifically about that since 1900, principally because of the lack of long-term tide-gauge data distributed equably throughout the region and the paucity of temporal successions of aerial photographs and coastline maps.

Several investigations have been carried out on recent sea-level changes by interviewing elderly inhabitants of long-established coastal settlements in tectonically stable areas about how the shoreline has changed since they were children. On the central east coast of Viti Levu, Fiji, it was demonstrated that sea level rose 10-30 cm in the last 70 years resulting in the inundation of coastal settlements. At Naloto, for example, around 60 per cent of the village that existed in the 1940s was underwater in 1987.

A major project, as yet unpublished, extended these investigations throughout the South Pacific. In all the places listed, except for the tectonicality arising parts of Vanautu and Tonga, the results show that a sea-level rise has been in progress for at least the last ninety years in the South Pacific region.

The amount of lateral inundation which this sea-level rise has caused varies depending on the nature of the coastline; in these data, it has a maximum of 70 m at Satalo in Western Samoa which represents an annual inundation rate of over 1 m.

Figures relating to the vertical sea-level rise were obtained by measuring the depth at the old low-tide level at the time of present low tide. Operator error, tectonic subsidence, swell effects, and sea-floor scour since this level was abandoned are all likely sources of error in the measurement of actual sea-level rise. The measured magnitude of sea-level rise varies from 40 cm to 7.5 cm giving annual rates of sea-level rise of between 7.27mm and 110.29 mm. These figures are considerably higher than the changes recorded from tide-gauges in the Pacific which vary from 0.7 mm/yr at Honolulu, Hawaii, 1.6 mm/yr at Wellington, New Zealand, to around 2 mm/yr at Pago Pago in American Samoa yet confirm that sea level has been rising recently.

The present long-term trend of regional sea level is believed to be a falling one, a transgressive maximum having been recorded on most South Pacific coasts around the time of the Holocene climatic optimum. Recent sea-level rise is as yet a tiny variation in this trend yet one which, because of its persistence, would be expected to have caused noticeable changes in shoreline processes and character. The changes in shoreline processes associated with a recent sea-level rise on South Pacific coasts include an increase in water depth offshore, usually in a reef-bordered lagoon, and consequent increase in lagoonal/nearshore sediment mobility. This factor combined with the Bruun effect resulting from sea-level rise on equilibrium coasts and the reduction in effectiveness of the offshore reef as a protection against wave erosion, especially during storm surges, has led to an increase in the amount of shoreline erosion along most island coasts. This has been especially noticeable where the coastline comprises loosely-consolidated material such as beach or aeolian sand ridges.

This condition is shared by 70 per cent of the world's sandy coastlines, less than 10 per cent of which have been prograding recently, the remainder of which (20-30 per cent) have been stable. Beach erosion is not necessarily an indication that sea level is rising but, where other factors suggest that this is so, as in the South Pacific, coastal erosion provides useful corroboration. The changes in shoreline character resulting from increased erosion and decreased natural protection vary depending on factors including initial shoreline profile and composition, type and density of coastal vegetation, and the presence or absence of artificial structures to protect the shoreline. From a study in Fiji, Nunn reported that around 15 m of shoreline retreat, representing a loss of 3,500 m^3 of material, has occurred at the school at Ucunivanua in the last fifty years despite the planting of deep-rooting, saltwater-tolerant trees.

Nunn also reported that most sea walls in rural parts of Fiji were neither of sufficient durability nor appropriately designed for the changing shoreline environment. This situation seems common to most South Pacific islands, especially along non-urban coastlines. From the recent survey of the region from the Solomon Islands to the Cook Islands, only 1 out of 63 rural settlements which had experienced a sea-level rise over the past 90 years or so had a concrete seawall, while only 19 per cent had used cement in seawall construction and 49.2 per cent had no remaining/effective artificial structures for coastline protection.

Summary of Past Effects of Global Warming on South Pacific Islands

1. Global warming has probably caused increased precipitation in the South Pacific over the last few decades which has been responsible, in part at least, for increased soil erosion and land degradation in the islands.

2. Increased sea-surface temperatures have led to an increase in tropical cyclone frequency recently.

3. Likely increases in suspended fluvial sediment load have been a major contributor to increased coral and

mangrove death along South Pacific coasts recently.

4. Sea level has been rising along South Pacific coasts for probably the last ninety years.
5. Coastal erosion has worsened recently as the result of reef overtopping, the Bruun effect and increased nearshore sediment mobility.

Predictions of environmental change are not wholly convincing because of the number of unknowns in the interactive complex on which such predictions are based. Credibility must be stretched further in the South Pacific where base data used to fuel these predictions elsewhere are not available. The best that can be done is to apply a range of globally-generalised scenarios to the region for the purpose of impact evaluation. This was the subject of a major study carried out by the Association of South Pacific Environmental Institutions which assumed the following temperature and sea-level parameters.

Where adequate base data existed, maps were made of the coastlines of selected urban and rural sites within the region for the medium and high scenarios in 2025 and 2100. In most cases, sufficiently precise data to construct the 2025 shorelines were not available. Although the detailed results are given elsewhere, a summary and selected examples are given below.

The case studies used and the situations they typify are as follows.

Apia, Western Samao— large urban centre to be considered a case by itself.

Beqa, Fiji— small volcanic island with narrow coastal plain, highly variable offshore reef configuration (similar to many small islands in papua New Guinea, Solomon Islands, Vanuatu, Tonga and French Polynesia).

Cicia, Fiji— small volcanic island with high periphera *(maketea)* limestone, only a fringing reef offshore (similar to Lakeba and many Lau islands in Fiji, Rurutu and may other Austral islands, Mangaia and Aitutaki in the Cook Islands).

Labasa, Fiji— town at delta head threatened both by shoreline retreat and impounding of sediment in narrowly-constricted valleys (similar to many smaller settlements on Viti Levu and Yanua Levu in Fiji, on Upolu and Savai'i in Western Samao, and throughout Vanuatu and the Solomon Islands).

Moala, Fiji— large volcanic island with narrow coastal plain and both fringing and barrier reefs offshore (comparable to large volcanic islands like Viti Levu in Fiji, some of the Hawaii group and islands in French Polynesia).

Nafanua Harbour, Tonga— artificial harbour threatened from rear but afforded considerable protection already by naturally cliffy shoreline (similar to cliffed limestone coasts elsewhere on Tongatapu and vava'u in Tonga, in the southern cook Islands and eastern Fiji).

Nuku'alofa, Tonga— large urban centre considered a case by itself.

Rarotonga, Cook Islands— high volcanic island with extensive coastal plain fringed by beach ridge with only a fringing reef offshore (similar to Moorea and Tahiti Nui in French Polynesia, Savai'i in Western Samoa, and Kadavu in Fiji).

Savusavu, Fiji— town developed along narrow coastal plain threatened by impounding of steep narrow valleys at its rear as well as shoreline retreat (similar to many settlements on Viti Levu and Vanua Levu in Fiji, and throughout Solomon Islands, Vanuata and French Polynesia).

Suva, Fiji— large urban centre to be considered a case by itself.

Tarawa, Kiribati— ribbon atoll similar to most others in Tuvalu and Kiribati, the northern Cook Islands, and the island *motus* of French Polynesia).

Vatao, Fiji—moderate height limestone island with upland agriculture and variable offshore reef configuration (similar to 'Eua and Niuatoputapu in Tonga, and the Loyalty Islands in New Caledonia).

Vatulele, Fiji—low limestone island with lowland, often wetland, agriculture, variable offshore reef configuration (similar to many true atolls and almost-atolls, also Vava'u and Tongatapu in Tonga).

For each of the above case studies, land loss was calculated for the medium and high scenarios for the years 2025 and 2100 in order to illustrate the likely effect on most islands in the South Pacific and also to provide base data for planning and economic impact assessment. Most environmental issues associated with future sea-level rise can be illustrated by three case studies.

Case Study I—Beqa Island, Fiji

Beqa is a small (36.3 km^2 in area), high (reaching 440 m) volcanic island similar to many in the South Pacific and considered to be typical of the commonest building block of larger volcanic islands in the region. Beqa is wholly the product of subaerial volcanism and formed from a single volcano, the old centre of which lies near the southwest corner of Malumu Bay. The major eruptive phase was 3.1-4.2 million years ago. The island's interior is moderately dissected and rock dips closely follow ground surface dips. The nine settlements are located on a narrow discontinuous coastal plain cut in bedrock. Mean annual temperatures are around 250^0C varying only slightly throughout the year. Precipitation reaches a maximum of 3000 mm annually in the southeast quadrant of the island.

In 1976, the population of Beqa was 1143, a density of 31.5 persons per km^2. Most of the population are engaged in subsistence agriculture although significant amounts of seafood, *yaqona* (Piper methysticum), copra and staple crops are taken by boat for sale at nearby markets on mainland Viti Levu. construction of circum-insular road has been temporarily suspended.

Land loss on Beqa as sea level rises will be approximately in the following proportions; agriculture (18.2 per cent), coconut plantations (13.2 per cent), mangroves (37.3 per cent), settlements)12.3 per cent), coastal sand flats and beaches (19.1 per cent).

The coastal lowlands are generally intensively used either for settlement or for coconut plantations but the soil is not generally of sufficient depth or fertility to sustain intensive annual subsistence cropping which is mostly practised on the lower upland slopes. The shortage of lowland, which is being felt already, will be exacerbated as sea level rises. Some villages, like Dakuibeqa, have expanded upwards from the coastal plain but most are backed by steep cliffs and/or narrow valleys so will eventually be forced to relocate to hither ground.

Given the paucity of level ground in higher parts of Beqa, it seems likely that existing communities will become spatially fragmented in the process of relocation. Although the shole island is owned by the inhabitants ("native land"), the traditional system of land ownership, as elsewhere in the Pacific, greatly restricts the movement of particular communities and thus the options for relocation.

Coastal erosion is likely to increase as offshore reefs are overtopped and it will become more difficult to maintain the coconut plantations which exist along much of the coast. It would be sensible to start planting coconuts on higher ground if only to ensure a source of income from copra in the future. Mangroves exist around the coasts of Malumu Bay, the deep north-south embayment into which Lalati village faces. Mangrove acreage will certainly decrease in the future thus reducing its potential as a food source and as a form of protection against coastal erosion.

It is likely that the potential of the offshore reef as a source of seafood will deteiorate as sea level rises but it is less clear what will happen to numbers of lagoon fish, on which Beqans presently depend daily for food, but changes in lagoon circulation and accelerated currents are certain to cause changes. It is clear that a decreased reliance on seafood could be replaced by an increase in terrestrial food production which should be encouraged. Agricultural diversification, especially from lowland to upland crops should also be encouraged. The possibility of fish farming inland or in sections of coastal embayments should also be explored.

Water on Beqa is mostly supplied to the villages from upland catchments and few problems will result from the effects of global warming described. The circum-insular road will only be affected by inundation in a few places since it has been cut mostly above coastal plain level.

Case study II—Labasa town, Fiji

Labasa is the third largest town in Fiji, the largest on the island of Vanua Levu, and is the centre of an extensive and important area of sugar cane, Fiji's principal agricultural export. Labasa is located at the head of a delta formed by the Wailevu, Labasa and Qawa rivers which flow through comparatively straight, steep-sided valleys from their sources in the mountainous backbone of the island in the south. Most of the flat land and lower slopes is fertile and farmed intensively but the coastline is fringed by mangroves. Although no data are available, the coastline fronting the Labasa lowlands has been actively prograding over the last few thousand years as the result of large amounts of sediment being deposited on the shallow and gently-sloping delta front. Progradation has undoubtedly been aided by the sheltered location of the delta (front) and rapid mangrove colonisation. Parts of the delta which have become liable to periodic inundation recently are protected by sea walls.

The economy of Labasa and its surrounds depends mostly on sugar cane, which is milled on the outskirts of the town and shipped from a deep-water wharf at Malau Head, a high volcanic promontory 8 km northeast of the town. Timber is becoming increasingly important but little of this derives from lowland areas in the vicinity. Labasa has a population of around 16,500. Rainfall is 2445 mm annually, tropical cyclone damage is above average as on all leeward coasts in Fiji.

The Labasa area and the shorelines at 1.5 m and 3.5 m which would apply for the medium and high scenarios in the year 2100. Most of the delta would be inundated but would undoubtedly become unusable long before as seawater penetrated the water table. Increased seawall construction is not an option because of the thickness and permeability of the deltaic sediment carpet and the alluvial flats with which it merges, and also because of the high cost-benefit ratio of such structures over such large areas which would have to include not only the delta front but also the major river channels.

In the recent past, the three main rivers in the area have brought down sufficient sediment load and water to enable coastal progradation to take place. As sea level rises and the shoreline moves inland, much of this water and sediment will be deposited on the lowland floodplains. If sea level rises as predicted, this effect will probably be manifested as a sharp increase in large-magnitude flood frequency which would cause increased annual failure of the sugar-cane crop and short-term build-up of floodplain alluvium.

The site of Labasa town would probably become isolated: 90 per cent of it would be inundated by a 3.5 m sea-level rise. Land loss by land use would be in the following proportions; residential (16.9 per cent), industrial (2.2 per cent), port-related (34.8 per cent), commercial (11.2 per cent), services (21.3 per cent), open space (4.5 per cent) and educational (9 per cent).

The only feasible long-term option for Labasa town seems to be relocation but since at present the town depends largely on sugar cane for its economic well being, perhaps such an option should not be too readily embraced until alternative economic strategies for the area are drawn up. Many lower parts of the uplands, including river terraces and denuded interfluves would support sugar cane but the sole longer-term option is diversification of the area's agricultural base. Decreased dependence on sugar cane is necessary at the same time as increased utilisation of upland areas.

Case Study III—Nuku'alofa, Tonga

Nuku'alofa is the capital city of Tonga and occupies the central part of the north coast of the limestone island of Tongatapu. Part of the coastline of Nuku'alofa borders the inland Fanga Uta lagoon which although connected directly to the sea experiences moating effects causing differences in elevation between tides on ocean-facing and lagoon coasts. Nuku'alofa is home to about 18,500 people who live mostly in areas bordering the port area and city centre which is the commercial and cultural heart of the island kingdom. Three areas of Nuku'alofa were examined in order to give contrasting pictures of impacts of future sea-level rise on various land-use categories.

Area N1 is the city centre. Although sea-level rises of 0.2 m and 0.5m would not cause substantial direct inundation because of a continuous seawall, most of this area would be rendered uninhabitable by about 2025 if sea level rises as predicted because of a rise in the surface of the water table. Land loss through direct inundation by land use would be in the following proportions; residential (2 per cent), commercial (75.6 per cent), Royal Palace and Tombs (17.1 per cent) and open space (5.4 per cent).

Area N2 is the port area of Nuku'alofa and includes many hydrocarbon storage facilities and is the site of much light industrial manufacturing. Land loss would be in the following proportions; agricultural (15.9 per cent), mangroves (1.1 per cent), residential (40.6 per cent), industrial (11.3 per cent), commercial and ports (7.7 per cent) and others (23.4 per cent). A continuous seawall may prove ineffective even to the effects of a 0.2 m sea-level rise. Wharf facilities will clearly need building up if they are to remain viable.

Area N3 is a representative area of suburbs bordering the Fanga Uta lagoon. The shoreline here is not seawall-protected but fringed by mangroves and the debris of human occupation, Nearly half the area would be directly inundated were sea level to rise 1.5 m. The effects of moating would amplify the effects on ocean-facing coasts by about +0.15 m in the lagoon. By land use, land loss would be roughly in the following categories; agricultural (0.5 per cent), mangroves (2.5 per cent), residential (96.2 per cent), commercial (0.2 per cent) and others (0.6 per cent).

Much of the present site of Nuku'alofa would be rendered uninhabitable within 30-60 years if sea level rises as predicted and, as with Labasa, relocation seems the only appropriate long-term solution. Higher ground in the south of Tongatapu would be an appropriate area but existing constraints on relocation will probably mean that the present concentration of people and activities is split up into a number of smaller centres.

Inhabitants of limestone islands such as Tongatapu invariably depend heavily on the underground freshwater lens for drinking water. As sea level rises, this lens will shrink in size and its domed surface will rise leaking fresh water at the ground surface in low areas. Alternative sources of fresh water should be developed.

Continued global warming will push the 28^{0}C sea-surface isotherm further polewards resulting in increased tropical cyclone frequency which will not be countered by local effects in the South Pacific as it may elsewhere. An increase in the frequency of tropical cyclones, presently one of the main causes of seasonal variations in annual precipitation additional to that resulting from an intensification of the hydrological cycle and increasing sea-surface temperatures.

The amount of erosion caused by heavy rainfall during tropical cyclones is disproportionately large compared to that which occurs at other times. On high islands, the amount of soil and regolith moved in landslides adds appreciably to the increased fluvial load at such times and will contribute to increased land degradation on many islands. Sand cays and low limestone islands may suffer substantial additions or losses to their coastlines during tropical cyclones. For example, a coral rubble bank 19 km long and up to 4 m high formed on the ocean-facing side of Funafuti Atoll in Tuvalu following the passage of Tropical Cyclone Bebe in 1976. The land area of Funafuti was increased by one-fifth as the result of this bank forming.

Effects like these will undoubtedly become more common in future and it is imperative that problem areas (in space) are identified and appropriate measures taken.

Summary of Future Effects of Global Warming on South Pacific Islands

1. *Land loss.* As sea level rises, land loss will occur through direct inundation and through a rise in the groundwater table, an effect likely to be abetted by an increase in precipitation. The effect will be felt most on inhabited atolls such as in Kiribati, Tokelau, Tuvalu and the Federated States of Micronesia. Many atolls may become uninhabitable within thirty years if the high scenario for sea-level rise becomes reality. Coastal settlements on other islands will be affected to varying degrees. Relocation is the only feasible long-term option for many settlements.
2. *Coastal erosion.* This will increase drastically in intensity as offshore reefs are overtopped, nearshore sediment mobility and lagoon floor scour increase, and intertidal vegetation such as mangroves are destroyed.
3. *Agriculture.* Rises in sea level and the surface of the groundwater table will result in flooding and saltwater pollution of many lowland areas, particularly large river deltas, effectively destroying their agricultural potential. Many lowland river valleys used for agriculture will experience similar effects but will also be affected by increased flood frequency.

Existing Policies

South Pacific nations appear to be in urgent need of straightforward and balanced advice on global warming and the effects it is likely to have on their islands over the next hundred years. Most governments in the region depend for professional advice of this nature on external or extra-governmental bodies such as the United Nations sponsored CCOP/SOPAC (Committee for the Co-ordination of Offshore Prospecting in the South Pacific) and SPREP (South Pacific Regional Environment Programme). Yet, probably since global warming is not perceived to be central to the mandate of such bodies, the public messages they have issued have been superficial and subdued (e.g. Matos and Tiffin, 1988; Tiffin and Matos, 1988). Although articles by academics in some regional publications (e.g. Roy and Connell, 1989; Nunn, 1989) have been balanced, those reports which command the widest audiences have appeared in the daily press. Such reports are generally noteworthy for their imbalance and sensationalist message. Added to this is the way such articles commonly contradict each other so that the educated readership, which undoubtedly includes most prominent government personnel, gets a thoroughly confused picture and understandably concludes that so much disagreement among 'experts' surely means that there can be nothing immediate to worry about, let alone to use scarce financial resources to plan for.

There is clearly a moral duty incumbent upon responsible regional bodies to produce coherent and reasonable statements about global warming and the effects which it is likely to produce. Such statements should be devoid of personal prejudices and preferences and should be based on a wide reading and understanding of available literature.

It is clear that regional advisory and consultancy bodies in the South Pacific have failed markedly to provide regional governments with a clear picture of what is likely to happen in the future as a result of global warming. Probably reason of for this failure have to do with the preferences of the scientific leaders of regional advisory bodies, their existing briefs and the range of expertise represented by their staffs. In CCOP/SOPPAC, for example, most of the staff of the technical Secretariat are earth scientists without interest or expertise in climatology. Thus, even though, at the Sixteenth Annual Session of CCOP/SOPAC, representatives of all South Pacific nations expressed great concern at the potential impact of the "greenhouse effect", no significant response was made by the technical secretariat.

Most of the Pacific island atoll states (Kiribati, Tokelau, Tuvalu and the Federated States of Micronesia) have economies which depend largely on foreign aid and money sent home by relatives overseas. Other income is generated through favourable trade agreements such as SPARTACA (South Pacific Trade and Economic Agreement) with larger richer countries such as Australia, New Zealand and the United States of America on the region's periphery. The responsibility of such countries towards the atoll nations is likely to be felt more keenly in the future as conditions deteriorate and options for increased migration are likely to be implemented (Committee to Review the Overseas Aid Programme, 1984; Roy and Connell, 1989).

Migration seems the only long-term option for most atoll inhabitants but many, worried at the potential for loss of cultural identity in more cosmopolitan societies may opt for migration within the region rather than outside it. Such moves would be encouraged by the overall success of the resettlement of Banabans from Kiribati on Rabi in Fiji and of Vaitupu (Tuvalu) islanders on Kioa in Fiji, although the Atolls resettlement Scheme in the North Solomon islands was less successful.

1. Options for the resettlement of atoll inhabitants should be explored.
2. Options for the relocation of coastal settlements in rural parts of the Pacific should be explored and implemented.
3. Options for the preservation of larger (urban) settlements should be explored and solutions implemented.
4. Subsistence agriculturalists in lowland areas should be educated about the potential of higher ground.
5. Options for the transference of cash-crop agriculture to higher ground should be explored.
6. Advice about the most suitable forms of artificial structures for shoreline protection should be made widely available.
7. Maps should be made to show the susceptibility of particular parts of the South Pacific islands to certain processes,

such as land degradation, flooding and coastal erosion, which are likely to increase in magnitude causing problems of increasing severity. This might be done as part of a hazard mapping exercise in the islands.

8. Data on suspended sediment load and other related variables for many of the large rivers should be collected in order that coherent systems for river management under a rising sea-level regime can be advised.

9. A series of state-of-the-art tide gauges should be set up throughout the region.

10. Several coastal sites should be selected for monitoring shoreline changes through the use of regularly-repeated satellite and/or aerial photograph and/or ground survey.

F. BANGLADESH

The implications of global warming for Bangladesh including sea level rise, increased frequency of floods and cyclones and other changes in weather patterns are considered along with possible responses and actions that may be taken.

The possibility of global climate change, particularly global warming due to the so-called Greenhouse Effect has increasingly been raising attention recently both amongst scientists and the general public. Bangladesh is one of the countries which is likely to be most critically and adversely affected by the consequences of such a global warming and subsequent sea level rise which will affect the low lying coastal areas as has been pointed out by a number of authors. However, the consequences for Bangladesh are likely to be far more complex and will require a much better understanding of the complexities of each sub-ecosystem in order to make reasonably accurate predictions and plan appropriate responses. Some of the possible impacts, the areas for further research and the responses being undertaken are described below.

Implications of Greenhouse Effect on Bangladesh

Bangladesh is a low lying deltaic region of South Asia with a substantial coastline and three major rivers systems (Ganges, Brahmaputra and Meghna) flowing through it in a generally north to south direction carrying approximately 1.5 to 2.4 billions tonnes of sediment from the Himalayan mountains every year while being hit by monsoon precipitation from clouds which come from the south to the north, often preceded by violent cyclonic storms and tidal waves. The combination of enhanced river water flow and monsoon precipitation often results in flooding of large parts of the country causing widespread distress and damage.

The Bangladesh coast is geologically one of the youngest and most dynamic deltas. In addition, much of the coastal area, parts of which is itself an active delta, still being formed, lied within less than 1 m above sea level and will be susceptible to inundation if (and when) the sea level rises.

It is evident that whatever the impact on Bangladesh it will involve a complex interplay of the different ecosystems, particularly the coastal systems and will have an effect on all the components of that ecosystem, namely the Terrestrial, Aquatic and Human systems. Various authors including Milliman *et al.* (1988) and Broadus (1987) have postulated an inundation of upto one third of the country due to 1 to 3 m sea level rise by taking the contour line of 1 and 3 m above sea level of the country and then calculating the habitation, land, production, etc. within that area to estimate the scale of damage. It is extremely unlikely however that the effect will be so straightforward as it will take a long time (30-50 years) for the sea level to rise allowing much time for a dynamic interaction of the coastal system further complicated by the incoming silts in huge quantities.

It is also difficult to distinguish the sea level rise from land subsidence of which there is also some evidence. The country consists of a number of geologic formations in active interaction and these interactions also need to be taken into consideration. Brammer has taken a more ralistic approach based on a detailed examination of the different agro-ecological zones and predicted the possibility of the primary effect being increased inundation and flooding in the depressed basins upstream rather than that at the coast or the areas within the tidal limits only. It is obvious that much more needs to be known about the different ecosystems and sub-ecosystems (e.g. mangroves, estuaries, islands, cities, habitation beaches etc.) before any reasonable impact scenario can be established.

Some of the likely impacts on the various sub-ecosystems would include the following:

(a) *Mangrove forests:* These are in the Sundarbans and Chokoria Sundarbans and are already suffering from the effects of saline intrusion due to reduced water flow through the main rivers. The rate of saline intrusion is likely to be increased due to any increase in sea level.

(b) *Estuary and tidal rivers:* These are likely to be affected by any rise in sea level which will interact with the water flowing through the rivers, the sediment load and the surrounding land formation. The dynamics of the estuary which is already undergoing considerable erosion and acretion needs to be considered with respect to the possible sea level rise.

(c) *Cities:* The two main cities likely to be affected are Chittagong and Khulna and to a lesser extent Cox's Bazar, Patuakhali, Barisal and the new post city of Mangla. The situation with respect to height of harbours and ports and possible effects of sea level rise need to be studied.

(d) *Islands:* A number of offshore islands including some with substantial populations and habitations (e.g. Bhola, Sandwip and Hatia) and a coral island (St. Martins) are likely to be threatened by the prospect of a rise in sea level. The likely effects on the people, habitation, infrastructure, erosion and natural coral reefs need to be studied.

(e) *Beaches:* One of the world's longest natural beaches is located at Cox's Bazar and will definitely be threatened by a sea level rise. This aspect needs to be considered.

(f) *Fisheries:* Fisheries is very important activity in Bangladesh both as a source of protein in the diet of the people (particularly of the poor) as well as for the employment it generates (there are an estimated 10 million full & part time fishermen in Bangladesh).

The main capture fisheries are the coastal (marine) fisheries (e.g. for shrimp) and estuarine fisheries (e.g. Hilsa) while major shrimp farming also takes place in the coastal districts of Satkhira and Cox's Bazar. The possible effects of sea level rise on these needs to be considered.

(g) *Depressed Basins:* There is a sizeable number of depressed basins located around the country, particularly the "Haors" of is quite likely that these will be affected both by increase of flood water as well as drainage congestion due to rise in sea level. This will result in lengthening of the flood season and related environmental and human impacts. These need to be studied.

(h) *Agronomy:* The increase in temperature may affect the crops being grown in different parts of the country, particularly in the higher temperature zones. The effects of such increase in temperature needs to be considered. Further, sea level rise will inundate several of the agriculturally most productive areas.

(i) *Wildlife:* Some of the country's most important wildlife reserves and sanctuaries, including the habitates of such mammals as tigers and elephants lie in the coastal zone and will be threatened.

(j) *Habitation:* The threatened coastal zones are inhabited by a large part of the population which will be threatened by sea level rise, flooding as well as cyclones. Appropriate measures for their protection must be devised.

(k) *Natural Hazards:* The likelihood of increased flooding has been reported due to increased precipitation (from more intense monsoon rains), increased snow melt (from temperature rise) and blocked drainage (from sea level rise). This will effect the entire floodplain.

Also increased frequency of cyclones and storms can be expected causing increased damage to life and property.

(l) *Salinity Intrusion:* Any increase in sea level will necessarily increase the already rapid rate of salinity intrusion into both surface water as well as ground water aquifers. This will affect both agriculture and human health.

(m) *Biodiversity:* Many indigenous plant and animal species are likely to be threated with extinction from the coastal and inundated in land areas particularly the Sundarbans. Efforts must be directed to inverntorise, identify and save these for continued biodiversity.

Responses

The responses that are being made and need to be made may be divided into the following three levels:

(a) Scientific

The state of knowledge, both about the global climatic change and the sub-ecosystems of Bangladesh as described above need to be increased. This will require efforts of scientists, both within Bangladesh and internationally, as well as strong cooperation between such Bangladeshi and international scientists. Amongst a number of initiative being taken the Bangladesh Centre for Advanced Studies, a private, non-profit, scientific research institute with a major focus on Environmental and Resource Management issues has already started studying the specific problems of the sub-ecosystems of Bangladesh as well as linking up with the leading scientific research groups in the world. It is also sensitising the policy planners on these issues.

The scientists also need to raise these issues at appropriate national and international meetings.

(b) National Policy

The national policy of the Government of Bangladesh needs to take cognizance of the possible effects of global climate change and to begin to plan appropriate responses. The formation of National Committee has been advocated and the Government of Bangladesh has begun to form such a committee under the aegis of the Planning Commission.

It is also essential to raise the issue and participate in all appropriate international fora such as the Intergovernmental Panel on Climate Change (IPCC), UNEP and other international scientific meetings. This is being done to some extent by Bangladeshi scientists and government representatives who attended recent meetings in New York, Delhi, Paris and London. The government has also supported the formation of an International Institute for Environmental Studies and Disaster Management in Dhaka.

The government also needs to support scientific research and manpower development and training in appropriate fields to deal with the problems likely to be occurring over the next few decades.

(c) People

The knowledge about global warming and its possible consequences needs to be disseminated amongst the people by both the scientific community as well as policy planners to enable people to take appropriate action of their own. There has been considerable media attention to the issues of global environment in Bangladesh recently and also a number of national and international seminars and conferences on various aspects of the problem which were widely reported.

People in Bangladesh are generally extremely poor, malnourished, and prone to natural hazards and a changing ecosystem. They have developed some coping strategies and resilience to encouner such changes. These also need to be studied and built upon.

Bangladesh being one of the poorest and most densely populated countries with a low resource base has real difficulties in taking a long-term planning perspective. The competition for allocation of the limited resources in intense. Despite these, there is an emerging awareness amongst some of the policy-makers, scientists and institutions to do their best to encounter the devastating effects of the highly probable global warming, climatic changes and the sea level rise.

What is needed to address these questions are the following:

(a) An in-depth study of the vulnerable ecosystems, both coastal and in-land;

(b) Greater cooperation between Bangladeshi and international scientists to study these issues;

(c) Planning and development of appropriate skilled manpower to tackle the problems likely to affect Bangladesh as a consequence of the Greenhouse Effect;

(d) Develop policy options to mitigate the effects;

(e) To study and enhance people's coping capability and

(f) To promote and support international actions at the global level to reduce the impact on most vulnerable areas, such as, Bangladesh.

G. INDIA

The composition of certain gases in the atmosphere, such as carbon dioxide, methane, oxides of nitrogen etc. play a crucial role in determining the earth's climate. Levels of these so called "greenhouse gases" are particularly important because they play a vital role in preserving the balance between incoming and outgoing energy from the Sun and consequently maintain an even temperature on earth. Short wave radiation from the sun reaches the earth's surface unhindered, but the outgoing long wave radiation is partially trapped or retained by the atmosphere. Thus, like glass in a greenhouse, the atmosphere abosorbs some of the long wave radiation emitted by the earth, and radiates energy back to it. This downward flux warms the earth and is known as greenhouse radiation. The gases responsible for this phenomenon are known as greenhouse gases (GHG). If the atmosphere were not present, temperatures on this planet would be much lower than they are, about-20 degree Celcius. It is due to greenhouse effect that the average temperature of the earth is about 15 degree Calcius; and life as we know it, is possible on the earth.

Since the 1930s, however, scientists have considered the possibility that anthropogenic variations of some of the greenhouse gases might bring about perturbations to the natural greenhouse effect.

Scientific Update of GHG and Climate Change

There are two ways human activity

increases atmospheric concentrations of GHG: (i) by increasing the strength of sources (processes that produce GHG) and (ii) by decreasing the strength of sinks (processes that destroy and/or absorb GHG). The main source of carbon dioxide is from the use of fossil-fuel (coal, petrol etc.). Most estimates suggest that over two-fifth of the carbon dioxide being released at present is absorbed by the oceans due to their thermal inertia. Forests also play a crucial role as sinks. Greenhouse gas concentrations are determined by the balance between sources and sinks. Carbon dioxide is the most important greenhouse gas, followed by chloro fluoro carbons (CFCs), methane, nitrous oxide and tropospheric ozone.

Atmospheric abundance of these gases is increasing respectively at 0. 5, 4, 0.9, 0.3 and 0.5% per year. CO_2, CH_4, N_2O, CFCs and tropospheric O_3 have contribited 55, 15, 4, 11 and 110%, respectively, to climate change forcing since pre-industrial times. Global circulation models predict 4. 5, 5.5^0 C increase in global temperature due to greenhouse warming by the year 2030 which will in turn cause widespread changes in precipitation pattern and monsoon circulations as well as in sea level rise. Most climatic models predict that dobuling carbon dioxide would reduce current food production by between 3 and 17 per cent. Global wamring will also afect natural biota as species ranges are expected to shift poleward by 700 to 900 km within the next century. Significant changes, in the distribution of major biomass of the world, is also suggested. The atmospheric life-time or residencoe period of CO_2 is 50-200 year and of methane is about 8-12 year but each incremental molecule of methane is about 25 times more effective as GHG than each additional CO_2 molecule in the atmosphere. N_2O has long tropospheric life time (about 150 years) and as a GHG each aditional molecule of N_2O is about 250 times more effective than every additional molecule of CO_2. Quantities of carbon dioxide (atmospheric life time 50-200 years) are increasing in the atmosphere because of industrialisation and land mangement activities. It has increased from a concentration of less than 280 ppm a few hundred years ago to the present level of about 353 ppm. The principal cause of increase in atmospheric carbon dioxide in recent yards has been combustion of fossil fuels which releases about 5.6 billion tonnes of carbon in the atmosphere annually. Industrial nations contribute about 75% of these emissions though contributions from developing nations are also growing. Of the total sources of CO_2 biotic portion contributes about 20-30 per cent whereas land use related and soil-borne sources contribute 10-115 per cent. The other known major source of carbon dioxide is deforestation which is estimated to account for 15-30% of annual carbon dioxide emissions. Every year about 0.6 per cent of the total global forest area is cleared for permanent use. Forest clearing releases 0.3-2.6, Gt of carbon annually.

The Indian economy is largely dependent for its energy on fossil fuels. As for the present utilisation pattern, coal accounts for about 60% followed by liquid petroleum with about 30% and the rest is made up by natural gas. Steel, power, cement, and railways are the major industrial consumers of coal and account for over 70% of its production. Petroleum caters to the energy needs of transport sector including railways, and also industries. Natural gas finds major utilisation in power and fertiliser industries. Coal is the biggest single source of CO_2 emission in India. According to the current trends, about 60% of the commercial energy consumption in the country is derived from coal. The consumption of coal in the power sector alone shows a steady increase from about 49% in 1986-87 to 60% in 1989-90. Unlike pertoleum, the use of coal in industry is governed by its quality which is highly inconsistent. Thus the apportioning of coal-as-mined is done through the intermediate operation of coal a beneficiation wherever needed, say, in the steel sector. In the steel sector, coal of three categories, viz., cleans from coal washery and raw coal and dome imported coal are used. In the power sector, on the other hand, the middlings from the washery and raw coal are burnt. A small amount of coal is also exported and is thus not to be included in the calculation for CO_2 emission.

A doubling of the concentration of greenhouse gases with linger residence periods in the atmosphere is projected to occur towards the middle of the next century. Various 1-d and 2-d climate models indicate that the most significant change will be a global warming. Some of the potential consequences of this change, such as sea level rise, will depend directly on how the surface temperature responds. Many other important effects, such as changes in rainfall and soil moisture, may take place well before there is any detectable warming. Although uncertainities remain, particularly regarding the regional impacts, there is a global consensus that the risk of climate change is justifiably real and serious in a long-term perspective.

Making a significant reduction in atmospheric concentrations of greenhouse gases requires international co-ordination and agreement. The First World Climate Conference held in 1979 recognised climate change as a serious problem. The Inter-governmental Panel on Climate Change (IPCC) was created by the United Nations Environment Programme (UNEP) and the World Meteorological Organisation (WMO) in 1988. The IPCC has brought out several assessment reports which have become basis for further discussions on the subject. An inter-governmental Negotiation committee (INC) was set up in 1991 under the UN auspices to draft a global treaty on climate change.

The First World Climate Conference held in 1979 recognised climate change as a serious problem. The Inter-governmental Panel on Climate Change (IPCC) was created by the United Nations Environment Programme (UNEP) and the World Meteorological Organisation (WMO) in 1988. The IPCC has brought out several assessment reports which have become basis for further discussions on the subject. An Inter-governmental Negotiation Committee (INC) was set up in 1991 under the UN auspices to draft a global treaty on climate change. The Convention remained open for signature at the Treaty Section of the UN Office of Legal Affairs in New York until 9th June, 1993.

As per the existing norms, the F.C.C.C. will enter into force on the 90th day after the date of deposition of the 50th Instrument of ratification. The required 50 ratifications have since been received and the Convention would come into force from March 21, 1994. The ratification of F.C.C.C. by ratified the Convention. Notable among them are United States of America, Canada, Australia, China, Mexico, Japan, Algeria, Sweden, Norway, New Zealand and Czech Republic.

Developing Member parties who ratify the Convention will be eligible for funding available under the aegis of F.C.C.C. but will also have to comply with commitments as applicable under F.C.C.C. especially under Articles 4 and 12.

Main Features of the Convention

The Convention sets out obligations that are binding on the party States. Three new legal principles or approaches in international environmental law have emerged from the Convention.

- "Change in the Earth's climate and its adverse effects are a common concern of humankind". All countries now have a requirement to protect the atmosphere from serious harm, and to include measures to deal with its adverse effects in their national policies.

- "Common but differentiated responsibilities and respective capabilities". According to this concept all States are responsible for preventing further damage to the atmo-sphere. This responsibility is in direct proportion to their contribution to the causes and to the means at their disposal.
- The "Precautionary Principle" This principle supplements the traditional imperative 'react and correct' with 'forescast and prevent'. Certain activities are to be permitted only if it is certain that they do not cause serious damage to the atmosphere.

Steps that need to be Taken in India

The Framework Convention on Climate Change imposes a commitment on the developing countries to incorporate measures related to the impacts of emissions and the enhancements of sinks in national policy related to development. For meeting this commitment, the following steps will need to be taken in India on the basis of available project options and within our means, keeping in focus that the eradication of poverty and alleviation of hunger are the principle challenges and overriding priorities before us.

- Prepare national inventories of sourcs, and sinks of anthropogenic emissions not controlled by the Montreal Protocol (which is concerned with CFCs and aerosols responsible for ozone depletion).
- Exchange information related to implementation by publishing inventories and description of steps taken to mitigate the likely adverse impacts of climate change.
- Conservation and enhancement of natural sinks of GHGs.
- Take climate change considerations into account in the planning process generally, including the annual and Five year Plans, particularly in Energy, Agriculture and Forestry sectors.
- Implement national programmes containing measures to mitigate climate change.
- To reduce the scientific uncertainties regarding climate change through research, systematic observation and development of data achieves.
- Promote education, training and public awareness, and encourage public participation in the process.
- The developed countries will provide new and additional financial resources and facilitate transfer of technology. While a sustained financial mechanism will evolve in scope and magnitude upon ratification of convention, the interim arrangement is as follows:
- IPCC, for updating the body of scientific knowledge available on the subject.
- GEF, for interim funding for developing countries.

Assessment by the Conference of the Parties

- Progress of implementation of Convention by Party States.
- Consideration of projects for financing.
- Formulation of eligibility criteria and modalities related to the financial mechanism.
- Amendments, if any, with mutual agreement among parties.

Glory became India's, albeit in limited sense, when it carved out a leadership role at the Climate Summit in Berlin. Supported by China, the Indian proposal of a 20 per cent cut in carbon dioxide emissiosn of the industrialised countries by the year AD 2005—a revised form of the earlier Alliance of Small Island States (Aosis) draft protocol—broke the deadlock created by the American strategy of blocking the negotiations on specific cuts on emissions, and the time frame.

The Indian stand was based on the pragmatic concern that the reduction emissions must be initiated by the industrialised countries to provide the necessary space for the developing countries to increase their emissions. The 1st meeting of the Conference of the Parties opened in Berlin of March 20, to establish a fundamental basis to implement the United Nations Framework Convention on Climate Change (UNFCCC) signed by 153 nations at the Earth Summit in June 1992. The 10 days climate summit discussed substantive issues, such as the Adequacy of Commitments, Joint Implementation and the Financial Mechanism.

The industrialised countries (IC) are mandated by the ICCC to reduce their carbon dioxide emissions to 1990 levels by 2000. However, the national reports filed to date by the ICs to the Interim Secretariat on Climate Change reveal that they are actually increasing their emissions by an average of 16 per cent annually. Prior to the Summit, Trinidad and Tobago; on behalf of the AOSIS, had submitted a draft protocol to the Interim Secretariat. Haunted by the spectre of sea-level rise and submergence, the small island states highlighted the ominous truth that the existing commitments of the ICs were grossly inadequate to meet the objectives of the convention.

In political terms, even though the 20 per cent carbon dioxide reduction demanded by AOSIS is considered ambitious, yet, it is just not enough in terms of what is needed to avoid to expected climate change damage. The Inter-governmental Panel of Climate Change (IPCC) which provides scientific inputs to the policy-makers, predicts that 60 per cent cuts in greenhouse gas emissions from the 1990 levels are needed to stabilise our climate and prevent global warming.

But the AOSIS draft protocol failed to generate any enthusiastic support from the G-77 countries. It was stumped by the German proposal, made in the same context, which sought to create a separate category of the more developed among the developing countries, such as China, India and others, and committing them to reductions in the emission of greenhouse gases. The German proposal enviasged working "towards balance commitments on the part of industrialised and developing countries, for example, by means of further reporting the commitments for non-Annex 1 parties and commitments to limit the rise in emission in the case of certain more advanced developing countries."

Initially, countries like India and China were opposed to a reductions protocol, fearing it might open up negotiations for including the developing countries in reducing or stabilising their emissions. But later, the stand was modified when Germany decided to drop its proposal, veritaly torn apart by the developing countries and the environmental NGOs.

That the ICs should shoulder the main responsibility for the current climate problems, and, therefore, lead the rest in cutting emissiosn, was reiterated by India's minister of environment and forests, Kamal Nath. In a letter, on the eve of the Berlin meet, to the leaders of both, industrialised and developing countries, Kamal Nath asserted that the "developing countries have no—or indeed negative—responsibility for causing global warming.

To the extent that the Convention specifically requires, under Article 4(2)(b), a review of the adequacy of the commitments undertaken by the developed-country parties and the industrialised East European countries (called Annex-1 Parties), the German gameplan only served as s diversionary tactic to ward off the potential criticism of their commitments. In the absence of legally binding targets and timetables, the Convention fell short of being an effective instrument for combating climate change.

G-77 agreed to support a working group to negotiate a COP mandate based on the elements of the revised protocol as presented by India. As a 1st priority, the working group will aim to set specific, legally binding targets for the reduction of carbon dioxide and other greenhouse gas emissions by the ICs. However, at the conclusion of the Summit, all the emerged was weak and vague agreement; nearly 130 participating nations agreed to work towards a protocol by 1997, which would set legally binding targets in time for the 3rd climate summit to be held in Japan. On the issue of the II, countries like China and India made a major concession, and perhaps took a risky path, by agreeing to a 5-year pilot phase programme. The II pilot phase will, however, present an immense challenge to climate negotiators, as the mechanism and modalities for the sharing of credits for emission reductions is yet to be established.

Energy Efficiency and Fuel Substitution

The production and use of modern energy in India generated 115 million tonnes of carbon in 1985, or 10 per cent of all carbon emissions emanating from the developing world. From a global perspective, India will account for 21 per cent of the increase in carbon emissions produced from energy use in developing countries between 1985 and 2025. Between 1970 and 1990, the intensity of India's primary modern energy use increased by 38 per cent. In various sectors of the Indian economy, the intensity of energy use per unit of value addedd exceeds that found in most industrilised and many developing nations. The increasing intensity in India reflects a replacement of human and animal draught power by mechanical and electrical devices. In particular, agriculture electricity use is increasing much faster (at almost 20 per cent annually) than the value added derived from this sector. The high intensities also reflect the underlying inefficiency of energy use in industry, transport and power generation. For example, energy intensities of cement production for wet, dry and the wet-dry processes were as much as 20 per cent and, in some cases, 50 per cent above international norms. Similar figures have been docoumented for steel, aluminium and other major energy intensive industries.

A recent study has shown that direct and indirect carbon emissions may be higher from construction activity than any other component of India's final demand. Energy-intensive materials such as glass, cement, bricks, steel, aluminum, and asphalt constitute the bulk of the components of a building or any other type of structure. Thus, emissions associated with construction are large. Reducing construction activity to decrease emissions is not a viable solution since a growing infrastructure is necessary to maintain the pace of economic development in India. Using materials more productively through improved designs of buildings and other infrastructure would be the better approach.

India's opportunities for curtailing emissions of carbon include rectifying its currently uneconomic allocation of fuels and inefficient energy-use patterns. Abundant but carbon-intensive coal resources satisfy almost 50 per cent of India's modern energy demand. In recent years, the costs of coal production and transportation have risen primarily because of changes in technology and low productivity. Arguably, more efficient fuel options (primarily oil, natural gas and renewables) could serve as economically viable substitutes for coal in the future.

Determining which alternatives provide the most cost-effective means for India to restrain the growth of CO_2 entails a through economic evaluation of the available options. In the industrialised countries, economic evaluations of reducing carbon emissions have focused on taxation policies. However, in most developing countries where fiscal and technological resources are scarce, any effective emissions abatement strategy must go beyond evaluating the impact of changes in domestic taxes on levels of carbon emissions. Carbon conservation efforts must identify the types of energy-supply and energy-use technologies needed to restrain the growth of carbon. They must also assess the capital investment and foreign exchange requirements needed to acquire less carbon-intensive technologies and fuels.

Through the 1980s, imports of crude oil and petroleum products constituted the largest single commodity groups in India's import bill. The share of oil imports in total import in each year between 1980-81 and 1990- 91. It also relates oil imports in this period to export earnings. Although we find no secular trend, we can discern a clear pattern in which oil imports were linked to the growth in consumption of petroleum products and the trends in the domestic production of crude oil. The higher international oil price in 1980-81 boosted India's oil import bill. In response, India accelerated the exploitation of the offshore Bombay High Reserves. Thus, production of Indian crude tripled to 28.9 million tons in 1984-85.

The consumption rate accelerated in the second half of the 1980s. Consumption increased at an annual average rate of 5.5 per cent during the Sixth Plan (1980-85), but it increased at 6.8 per cent per year during the Seventh Five-Year Plan. As Bombay High was exploited fully in the mid 1980s-and no new discoveries on the same scale as in the mid-1970s were made, oil imports rose inexorably. Crude oil imports, therefore, increased in 1985-86, and petroleum product imports followed suit in 1987-88. As we show later, the nexus between oil demand and import payments has an important bearing on India's capacity to reduce emissions.

Generation, transmission and distribution of electricity are the most capital intensive of all energy activities. In the growing Indian economy demand for electricity has increased faster than that for other forms of energy. Driven the Sixth (1980-85) Fifth-Year Plan, for example, the government allocated 27.2 per cent of its plan outlay to the energy sector. This share increased to 30.6 per cent during the Seventh Plan. Even this increased share was insufficient to provide adequate power to the growing Indian economy. Power shortages are now common throughout the economy and 'brown-outs' in many cities during peak periods are the norm. The government is pursuing private sector power generation to alleviate power shortages. But the government has not yet set priorities in selection of technologies. Moreover, the management and operation of the power sector itself must be improved prior to seeking additional financing.

A recent analysis of the growth of carbon emissions and the economics of abating emissions from modern energy use in India highlights the relation-ship between abatement strategy on the one hand and capital investment and foreign exchange or hard currency requirements on the other.

The analysis assumes annual average rates for GDP and population growth. A linear programming model couples these rates to energy demand growth by sector and end-use. The model minimises the cost of providing energy services to the Indian economy. Energy service may be provided by new energy supply or higher efficiency of supply and/or use. The model computes the investment and foreign exchange requirement for meeting the estimated demand for energy.

In Scenario 1, energy intensity is frozen at 1985 levels. Energy sector investment as a proportion of GDP increases to 7.0 per cent by 2005. Foreign exchange requirements increase from 1.9 per cent in 1985 to 4.4 per cent by 2005. In each case, the sharp increase will require that financial resource be transferred to the energy sector from other sectors which will also demand more capital and foreign exchange.

Reducing the intensity of energy use as illustrated in Scenario 2, restrains the growth of carbon emissions to 340 million tonnes. Since this reduction is achieved primarily through cost-effective efficiency improvement, the cost of abatement is negative. The Indian economy benefits from restraining carbon emissions growth. The scenario captures the many opportunities available to use electricity more efficiently, which reduces investment requirement. However, the opportunities for reducing petroleum products demand are limited, and more difficult to implement, and this is reflected in the 4.2 per cent foreign exchange to GDP ratio, which changes little from Scenario 1.

Switching to less carbon-intensive fuels (Scenario 3) can reduce emissions further than in Scenario 2. This result is achieved through either the import of natural gas as illustrated in Case 1 or through the use of renewables as illustrated in Case 2. Renewables include the use of wind, solar, hydro and biomass resources. Natural gas imports increase the ratio of foreign exchange to GDP to 4.8 per cent while reducing the investment needs. Increased use of renewables drives up the investment requirement sharply to 6.2 per cent of GDP. Restraining emissions beyond that might be achieved through efficient fuel allocation and use would increase either capital investment or hard currency requirements.

The unit cost of conserved carbon rises from Scenario 1 to Scenario 3 as more expensive approaches are used to curtail carbon emissions. By conducting alternative runs of Scenario 2, and placing progressively tighter constraints on carbon emissions, the cost of conserving carbon at levels between those in Scenarios 2 and 3 were determined. For example, the cost of conserved carbon is US$0.02 per kilogram when emissions are reduced from 340 million to 300 million tonnes.

The Scenario 3 costs of restraining carbon emissions are lower than those for Scenario 1. This implies that India can reduce carbon emissions at a net benefit to the economy, and the energy sector would therefore not require any new resources to reduce emissions. Scenario 3 costs compared to Scenario 2 are between 1 and 11 billion dollars higher. Many energy efficiency improvements are embeded in Scenario 2. Thus, if India's energy sector were to become more efficient, this analysis implies that it would need additional resources to reduce emissions. Which of the two paths (Scenarios 1 or 2) India takes to provide energy services will determine whether resources need to be transferred to or away from the erergy sector, and consequently whether the nationa would be justified in seeking resources from the world.

Restraining emissions from energy use in India will require the fuel allocation and energy efficiency be improved to their maximum potential. The scenario analysis includes many opportunities for improving efficiency of electricity use and supply. Additional measures to improve oil use efficiency would have further reduced the FE/GDP ratios. What approaches might have yielded a more efficient use of oil in India? Four important shifts in strategy could be implemented:

1. a shift in long-haul freight movement from road to rail in order to reflect the economic cost of transportation for each mode;
2. a shift in cooking fuel from kerosene to LPG to encourage the use of more efficient stoves;

3. electrification of non-electrified households which would reduce kerosene consumed for lighting;
4. replacement of diesel pompsets with electric pompsets.

These four strategies would have reduced demand for oil in the scenarios between 1980 and 1990 by 14.4 per cent. More importantly these steps would have reduced by 5 million tonnes the demand for kerosene and diesel, the two critical middle-distillates which are imported. The result of four strategies would be a reduction of the 1985 FE/GDP ratio of 1.9 per cent to 1.3 per cent. If we assume that similar improvement may be achieved by 2005, then the ratio for Scenario 2 would fall to 2.9 per cent from 4.2 per cent. A lower share of GDP allocated to importing oil would make the fuel import payment more manageable.

Strategies 3 and 4 would increase demand for electricity which would add to investment requirements for 2005. As we pointed out above, there are many options to improve efficiency of electricity use that are easily implementable. The options for improving efficiency of oil use involve a diverse set of actors which make them difficult to implement. For some end uses, such as lighting, electricity use is more efficient than oil use. Shifting to electricity in such selected end-uses would improve a nation's energy efficiency. Further, since improving system efficiency for electricity may be easier than for oil, increased electricity demand from strategies 3 and 4 could be better controlled.

In light of the cost, capital investment and foreign exchange parameters, to what extent can India restrain carbon emissions from modern energy use? Stabilisation of emissions or limiting their increase to 20 per cent over a 20-year period has been discussed for the industrialised countries. Analyses show that this goal could be achieved without a net loss of GDP in some of the countries. The Energy Modelling Forum-12 in its deliberations on scenarios assumed that restraining emissions growth to a 50 per cent increase over a 20-years period has been discussed for the industrialised countries. Analyses show that this goal could be achieved without a net loss of GDP in some of the countries. They Energy Modelling Forum-12 in its deliberations on scenarios assumed that restraining emissions growth to a 50 per cent increase over a 20 years period was plausible for the developing countries. In contrast, Scenario 3 at best, emissions for 2005 could be held to 155 per cent (280 compared to 110 million tonnes) above the 1985 level. Further reduction in emissions would reduce annual GDP growth from the 4.9 per cent assumed. Indeed, we estimate that to achieve the 50 per cent limit suggested by EMF-12 would require that the current pattern of unsatisfied energy demand continue in the future and that India's GDP growth be limited to 3 per cent. Most probably, these requirements would be unacceptable to India.

New Perspectives on Sectoral CO_2 Emissions and Incremental Costs

Global negotiations on climate change have generated considerable interest among environment policy makers, economists, atmospheric scientists and various other interested parties. There are four major greenhouse gases (GHG): carbon dioxide (CO_2), methane (CH_4), chloro fluorocarbons (CFC), and nitorus oxides (N_2O). Among these, CO_2 receives the most attention, because:

- CO_2 emissions can be most reliably estimated from the energy balances of each country because they are directly related to energy consumption and cement manufacturing.
- CO_2 is the most important greenhouse gas emitted by the developed countries in terms of magnitude (as well as ratiative forcing).

- Increases in CO_2 emissions are measures regularly and global emission levels for CO_2 are increasing more rapidly than, say, methane.

For these reasons, various authors have carried out analyses of the possibilities of reducing CO_2 emissions at global and national levels.

While global approaches highlight the policy direction and the magnitude of the efforts required. Blizer *et al* have pointed out that the possibilities of GHG emission reduction needs to be discussed using *country-level* models with sufficient structural detail. Their model for Egypt shows that if Egypt curtailed GHG emissions in 20 years by 20% over base year CO_2 production levels, it would reduce the GDP growth rate by 3.13%. A reduction of 40% would slow GDP growth by 32.4%. Alternatives at *country-level* depend *inter alia* on:

- Energy resources and technologies used (i.e., coal, oil, gas, hydro or nuclear).
- Development patterns (whether agrarian, industrial or service-oriented economy).

India and China are considered to be major players in global climate change because of their likely increase of CO_2 emissions due to increases in income and level of population. Moreover, both depend on coal. Several authors, for example Mathur and Sathaye and Ketoff, have carried out CO_2 accounting exercieses by considering energy supply and energy end-use activities such as manufacturing steel, fertilisers, cement, and so on. While this simple and quick method could show that India's emissions in 1985-86 were 115 million tonnes and will rise to 615 million tonnes by 2025, it is not adequate to give policy guidance or to assess the impacts on the economy of alternative emission reduction policies, which may depend not only on technologies but also on fuel substitution, fuel consumption, patterns of development, etc. Indirect CO_2 emissions can also impact substantially. Thus, an overall economy wide exercise can provide additional insights and policy guidance. This article reports on such an exercise, using the input-output table for the year 1983-84 that was used for the Seventh Plan, 1985-90.

A case study of India is of particular interest, because India may be the most populated country in the world by 2025 and its GHG emissions may rise substantially. Depsite the fact that in 1986 it emitted only 0.2 tonnes per capita of fossil CO_2 (compared to 5.4 tonnes per capita in the USA), it is the sixth largest CO_2 emitter in the world. Parikh *et al* have shown that, to accommodate even a modest rise of emissions by only India and China, the developed countries would have to reduce their GHG emissions by 30% by 2025, to keep global emissions in 2025 at the same level as in 1986. India's emissions are projected to increase fourfold compared to 1986, but even then would be only 0.36 tonnes per capita below the world average of 1.2 tonnes per capita in 1986.

There is yet another concern about India's future fossil CO_2 emissions. India's major energy source is coal, with resources of more than 186 billion tonnes, compared to production of 221 million tonnes in 1990-91. Proven coal resources are estimated at 56 billion tonnes, while oil and gas resources are only 756 million tonnes and 686 billion cubic metres respectively. Thus, coal will continue to provide more than 60% of India's energy needs. Coal has the highest CO_2 emission coefficient per primary Giga Joule (GJ). Coal in India has nearly 35% ash content. Standard Indian coal in calibrated to have 20.9 GJ per tonne as compared to the United Nations standard coal at 29.7 GJ/tonne. This difference in the Indian energy data system has not been kept in view and has created confusion in international data systems.

Thus, the combination of high population, coal predominance in energy-use and high potential growth due to currently low income levels cause concern about India's carbon emissions in international cricles. On the other hand, India's viewpoint is different. From its perspective, India's fossil-based carbon emissions were only 115 million tonnes in 1985 compared to 5.4 billion tonnes for the world and 1.3 billion tonnes for the USA. Even in 2025, its total emissions are likely to be only 0.6 billion tonnes—a per capita emission of 0.36 tonnes compared to a world average of 1.2 tonnes per capita. These emissions would not be a problem if the developed countries' present and past use of fossil fuels had not resulted in an excessive build-up of CO_2 in the atmosphere.

India has to tackle basic problems of health and nutrition and needs to work on basic environmental problems such as increasing access to safe drinking water and sanitation and reducing the use of bio-fuels which harm the health of women and children. Under the "Poluter pays principle', those above the world average should pay money to those below the world average. Under certain conditions, a tradeable emission quotas system is recognised as economically efficient in allocating resources to the reduction of CO_2 emissions. 'Debtor' countries who find it costly to reduce CO_2 emissions will transfer resources to 'creditor' countries by purchasing emission rights. Creditors can then use these resources according to their own priorities.

This is the best option in international negotiations. However, until such an international system is in place, there are possibilities for other countries to invest in India to reduce CO_2 emissions in those projects which are of interest to India while others can claim credit for the reductions in CO_2 emissions. These are known as 'offsets' in other countries. This may be a second-best solution from debtors to creditors is restricted to emission-reducing investment, and funds may not be directed towards the environmental priority of their choice (e.g. drinking water). However, such measures may be negotiated even bilaterally by, say, one of the EC countries in response to the CO_2 emissions-reduction policy of the EC.

Regardless of these negotiation issues, the structure of India's CO_2 emissions is of interest to policy analysts in India and elsewhere. We try to quantify emissions here by economic sector, using the input-output approach described in the next section.

Recent years have witnessed and growing concern regarding the accumulation of greenhouse gases (GHG) in the earth's atmosphere. This concern has led to many in-depth studies of the phenomenon. Individual country studies have ranged from simple (albeit data-intensive), inventories of GHG to evaluation of policy options of stabilise or reduce emissions in some future year. Impact studies have focused on better understanding the effects of gases on atmospheric temperature, monosoon patterns and sea level rise.

The Inter-governmental Panel of Climate Change estimates that the global termperature would increase up to 3.5^0 C by 2100 under the most likely scenarios. The results of these models and other studies prompted calls for an international treaty which nations could adopt to restrict the growth of emissions. Such a treaty was put forward at the 1992 UNCED meeting in *Rio de Janeiro*. The text of the treaty was debated for many months prior to the convention. One of the critical divisive issues was the sharing of burden among the various parties. This issue lies at the heart of the debate among nations on climate change. Burden sharing is difficult to resolve since the emissions burden that attention shoulders is different for each gas. And, it depends on the historical cumulative emissions a country may have emitted thought its use of various fuels. The burden varies by the type and extent of impact that a national may have to bear as well.

The cumulative share of carbon dioxide emissions from the developing countries between 1970 and 1986 is estimated to be only 15 per cent but, with 76 per cent of the world's population, their share of energy-related carbon dioxide emissions in 1986 was about 27 per cent. This share is increasing since their modern energy growth is faster than for other countries. IPCC scenarios of future emissions indicate that worldwide emissions of carbon and other gases would continue to increase even if the developed countries were to reduce emissions from their current levels. These shares suggest that the developed countries should shoulder a greater responsibility for historical emissions but the opportunities for reducing incremental emissions may be more abundant in the developing countries.

Unlike other airborne pollutants, such as S and N, which are emitted in trace amounts, are very reactive and can be scavenged at the source, GHG are emitted in far more diluted and much larger amounts, which poses a problem in attempting to physically remove the gases. The main alternatives consist of eliminating or reducing the emissions of these gases, or growing biomass to sequester carbon dioxide. Much of the current deforestation occurs in developing countries, and reversing or slowing this process would aid in reducing future emissions.

Given the above background, developed countries have insisted that reducing GHG emissions from the developing countries is vital . If necessary, such emissions reduction could be accomplished with assistance from the Global Environment Facility (GEf). On the other hand, the developing countries have argued that the growth of emissions is an unavodiable consequence of economic growth, as was the case with the industrialised countries during their past. Hence, global environmental protection should not be allowed to penalise development. Is this difference reconcilable?

For a signatory to the Climate Convention, the adoption of policies and strategies to restrain emissions growth is an important goal to pursue. A nation would find it easier to follow such options to the extent that these are adoptable without hindering its current or anticipated development trends. Many studies have argued that alteration of growth patterns will lower social welfare and add to the cost of future socio-economic development. We will cite several studies for India to illustrate the opposite view that the adoption of such policies need not reduce social welfare. Indeed, accelerated adoption of certain energy of forestry policies, some of which are already being promoted and implemented, will lead to reduced carbon emissions and/ or increased carbon sequestration at no additional cost to the nation. The pursuit of such policies will shift the business-as-usual growth to basic-needs-oriented development.

Adoption of such policies may be slowed or thwarted by many barriers in developing economies. In the case of India, scarcity of capital and of hard currency are twin dilemmas which often limit adoption of the most efficient policies. Lack of institutions to facilitate the adoption of high energy efficiently technologies is another barrier. Reddy (1991) lists the many barriers to improving energy efficiency.

If adoption of such policies were to increase social welfare and achieve reduction of GHG emissions at no extra cost, then is a nation justified in seeking international support of implementation of these options? As we illustrate for India, even if the life cycle cost of abatement projects is less, the up-front costs of these projects may make them prohibitively expensive to pursue. As provided for in the Convention, India could justifiably seek support for projects meriting such assistance through the Global Environment Facility.

This chapter addresses three main issues dealing with these topics: emissions inventory and the uncertainty of estimates, energy efficiency, fuel substitution and the economics of GHG abatement; and emissions and sequestration from biomass growth. The Climate Change Convention calls for each Party to prepare an emissions inventory. GHG included in the inventory are carbon dioxide, methane, carbon monoxide, nitrous oxide and CFCs. Carbon dioxide and methane are emitted in larger amounts than the other GHG, and CFCs are regulated under the Montreal Protocol. Several other different inventories have been prepared for India by experts, both national and foreign. Emission estimates of CO_2 from energy sources are in relatively close agreement. Estimates of CO_2 from non-energy sources, and of methane from all sources, vary widely and have been debated in many fora.

We show the estimates of annual greenhouse gas emissions from anthropogenic activities in India for 1986. Carbon, emitted as carbon dioxide, emissions total 164 teragrams. Of these, emissions from forestry and land use changes amount to 20 teragrams. A more recent estimate places carbon emissions (as CO_2) from fossil fuel use at 133 teragrams in 1988 compared on 139 in 1986.

Methane emissions are estimated at 55 teragrams. Boden *et al.* estimate emissions of this task from livestock and rice paddies to be 10 teragrams, compared to 48.2. The lower figure is the result of new emissions measurements which reflect smaller rice biomass from Indian paddy fields and the fact that areas emitting high methane flux are a fraction of total paddy area; and above-surface biomass weight is reportedly smaller than elsewhere.

The emissions inventory is for a single year and does not provide guidance on trends or future growth of these emissions. Oak Ridge National Laboratory has tracked historical emissions for several countries, including for India. Carbon dioxide emissions from India have increased at 5.7 per cent annually since 1950 as India climbed from thirteenth to fifth place in the world as a national contributor. With increased shares of oil and gas, the share of CO_2 emissions from coal has declined from 87 per cent in 1950 to 71 per cent in 1989.

We have selected future scenarios from two authoritative reports for carbon dioxide emissions from energy and forestry sources. We show emissions estimates for commercial energy sources for 1985 and 2025, and for biomass sources for 1986 and 2011. Emissions from modern energy sources increase at a rapid pace in these scenarios but those from biomass increase much more slowly in either scenario.

The concept of Joint Implementation is new and seems to represent different views to different people. Our understanding has been more in the nature of appreciating the various aspects as to what is really meant by it, and its applicability in the overall context of the Convention itself. Before we actually go on to assign a definition to Joint Implementation it is necessary that we assess its implications. Humankind is today in an unprecendented-and, unenviable position and climate change is one of the challenges facing humanity. For millions of years the greenhouse effect has maintained the earths climate at a stable average temperature, permitting life to flourish. However, human activities on land, water and soil have gradually grown unsustainably, and have experienced a quantum leap during the last century or so. In the span of a single human generation the earth's life-sustaining environment is now expected to change more rapidly than it has in any comparable period of human history. Much of this change will be of our own making, and we have it in our power to control it to some extent.

Climate change itself was virtually an unknown issue 15 years ago, but even small changes in our gaseous environment have big effects. The impact of climate change measured in terms of increase in an average global temperature would result in sea level rise, extreme weather patterns, regional disruption of water supply, declining food production, damaged forests, wildlife and biodiversity, and harmful, unpredictable effects on public health. A UN report says that the world faces a 3 degree centigrade rise in average temperature within the next century, and if this is to be contained, we would have to go in for at least a 70 per cent cut in human generated carbon-dioxide and other greenhouse gases urgently.

The bulk of the earlier CO_2 build-up is on account of industrial countries, which are even today the largest sources of gaseous emissions. In spite of this, in a spirit of global cooperation, the developing countries, who have done little or nothing to create the problem, entered into negotiations and a large number of them, including India, signed the Framework Convention on Climate Change at the Earth Summit in 1992. It has now received the required number of ratifications and the Convention is expected to come into force on the 21st March 1994, (coincidentally but appropriately on the day of the Spring Equinox, regard since pre-historic times the world over as a day of natural regeneration).

The Convention acknowledges that the Change in the earth's climate and its adverse effects are a common concern of humankind; it at the same time recognises that all Parties, taking into account their common but differentiated responsibilities, and their specific national and regional development priorities will formulate, promote and cooperate in the implementation of the objectives of the Convention. The objectives deal with reduction and sequestration of greenhouse gas concentrations in a manner that would prevent danger to our ecosystems and reduce interference with the climate system, while ensuring that food production is not threatened, and enabling economic development to proceed in a sustainable manner.

These goals have to be achieved by reducing emissions through control of emission sources and sequestering harmful gases by enhancing carbon sinks. Perhaps the mechanism of Joint Implementation has been flagged in this context. Ever since the concept of Joint Implementation was mooted, divergent views have been voiced. But whatever interpretation is ultimately agreed upon, I am sure that all of you would agree that there can be no mechanism which encourages the needless growth of emissions, on the one hand, and creates hurdles in the path of swift development of economically poor societies on the other.

One cannot allow legalistic wrangling to side-track the fundamental issue of international equity. We have to recognise that sustainable development globally is unachievable without the removal of obstructions of equitable relations between and within nations. Central to this discussion is an absolute need for a thorough socioeconomic assessment of 'global benefit', *what* it is and *who* provides it; and equally important an assessment of 'global disbenefit', *what* it is and *who* provides it. In making this assessment, we would have to accept the fact that people who live sustainably (and have been doing so for centuries) in terms of globally equitable and efficient use of energy per capita, provide global benefit, whilst people with unsustainable lifestyles, in terms of globally equitable resource use per capita, create global disbenefit. The real tension in this debate is that between 'costs' and 'benefits'. Cost/Benefit Analysis is being used to distract attention from the kernel of the debate. We cannot allow Joint Implementation to be given an interpretation which would result in an attempt to 'pass on the obtainment parcel' to the *wrong* address, without even putting on the proper postage

One would like to draw the attention of this workshop to the fact that while change in the earth's climate and its adverse effects are admittedly the common concern of humankind, the Convention envisages that the actions to meet the challenge are to be based on *differentiated responsibilities* and are *dependent* on the *capabilities* of the countries. While all States are responsible for preventing further damage to the atmosphere, it is inescapable that this responsibility is in direct proportion to the contribution of each to the harmful gases, and to the technological and financial means at their disposal. The 'precautionary principle', supplements the traditional imperative of 'react and correct' with 'forecast and prevent': activities are to be permitted only if it is certain that they do not cause serious damage to the atmosphere. But this cannot be intrepreted to mean that the socioeconomic development of the South can be thwarted, while the prodiguous lifestyles of the North can remain untouched, by juggling around with averages and semantics.

Conspicuous lifestyles have harmed our environment irreparably. Deprivation and poverty, are not going to help the environment in any way, either. Everyone is vulnerable to the unmitigated continuation of start inequalities in living standards. If disaster strikes our planet we all suffer. A cocoon of luxury will not cushion us from the reverberations of unjust and skewed patterns of development. There are a host of areas which revolve around this question, and which would be resolved only with fundamental structural change in the international economic order.

One would not like to prejudice the deliberations of this international workshop with a strait-jacketted view in the inaugural session itself, but it would certainly like to underscore the need for standing on its head any presumption about the utility or otherwise of any interpretation of Joint Implementation. We would have to address a variety of questions:

- To what extent would Joint Implementation facilitate the transfer of technology to developing countries?
- To what extent would the overall objectives of emission reduction be achieved?
- Do we have reliable methodologies to determine the base scenario of emissions which could be used to calculate the amount of emission actually avoided through Joint Implementation?
- What are the financial and economic implications of Joint Implementation?
- What are the political implications in terms of sovereignity over natural resources?
- Will Joint Implementation adequately facilitate the quantum reductions envisioned by scientists, or would it result inside-stepping core issues?

Emissions and Sequestration from Forest Biomass

Forests, defined as woodlands with more than 10 per cent crown cover, occupy about 20 per cent of the Indian land area. This proportion may be compared with 57 per cent for heavily forested Indonesia and Malaysia. Deforestation has led to a major decline in forest area in most countries; India is no exception. About one-fifth of the forests standing today are exteremely degraded. Half a million hectares or 0.8 per cent of total forest area was deforested in 1986. The primary conversion activity in India is agriculture followed by pasture and harvesting. The primary conversion activities vary by region. In the state of Karnataka, conversion to agricultural land accounts for 33 per cent, but submergence and resettlement due to power and irrigation projects account for 42 per cent, and mining for 21 per cent of the lost forest area between 1956 and 1984.

The carbon emissions from India associated with deforestation. These are divided into three categories: inherited, prompt and delayed. Inherited emissions are caused by past forest activities and occur in the base year. Prompt emissions are those that are generated immediately as a result of a forest conversion activity. Delayed are cumulative emissions that take place over time as decomposition of biomass occurs. Releases of carbon dioxide due to changes in soil organic carbon, both from forest conversion activities and areas under harvesting and afforestation programmes, are also included here.

The committed emissions displayed have been much debated. An earlier World Resources Institute estimate placed these emissions at 140 million tons. More recent estimates from other sources place the committed emissions closer to the figure. In the section above on energy, we discussed the potential and economics of restraining emissions from that sector. Growing biomass offers an opportunity to sequester carbon which would reduce net emissions from India. What is the potential for such offsets and to what extent might it be achieved while simulataneously pursuing or accelerating currently planned socio-economic development?

The development of agro-forestry and tree plantations on previously unforested lands can provide India with options to reduce its net emissions. The biomass density and carbon accumulation of new forests can exceed those of the initial natural vegetation depending on the silvicultural practices. In addition, afforestation projects can provide gainful employment to rural peasants who formerly earned their living through activities resulting in deforestation.

India has a strong, rapidly growing afforestation programme. India's afforestation process was accelerated by the enactment of the Forest Conservation Act of 1980 which aimed to stop forest clearing and degradation through a strict, centralized control of land-use rights. The afforestation activities resulted in a oral of 11.5 million hectares as of 1986. Another 5.6 million more hectares were afforested between 1986 and 1989, raising the total planted area to 17.1 million hectares.

We show the carbon uptake, the annual carbon balance and the net committed emissions from India. The committed carbon uptake is almost twice as large as India's committed emissions. Over the years, the country would sequester 56 million tonnes of carbon from past afforestation activities. This figure is sufficiently large to make a dent in the emissions from the energy sector. The annual carbon balance shows a net uptake of 5 million tonnes from forestry activities. Combining this figure with emissions from energy would reduce emissions from modern energy and forestry to 105 million tonnes of net carbon emissions. These figures have important implications for future carbon emissions from India. The carbon sequestration occurred and is occurring through programmes with the main goal of promoting sustainable rural socio-economic development. Carbon sequestration is an unintended consequence of these actions. Given the potential for reducing net carbon emissions from India, the various factors that have contributed to this sequestration are worth noting. They include:

1. establishment of the Forest Conservation Act of 1980.
2. preparation of an environmental impact statement required with the beginning of the Fifth Five-Year Plan of 1975-80;
3. reduction of subsidies to forest-based industries beginning in the late 1970;
4. increased industry-farmer links which have encouraged production of tree crops;
5. decentralised political decision making-village and district level authorities have been established in West Bengal and Karnataka, for example, that are far

more motivated to ensure the prudent use of local resources;

6. growth of strong environmental movements in different parts of the country;
7. biomass fuel conservation programmes in all the States.

Strengthening these programmes would enhance carbon sequestration and accelerate rural development. As Saxena points out in a recent article, implementation of programmes will benefit some groups at the expense of others within India. Most of the programmes proposed in the article, such as welfare forestry on forest lands, social security plantation, farm forestry for the poor, etc. will benefit the rural poor and the rich at the expense of lower level officials. However, not pursuing such programmes will make everyone lose in the long run.

Favourable scenario have projected a net carbon uptake from forests in 2011 to be 121 million tonnes. Interpolating between the base year, 1986, and 2011 gives a net uptake of 57 million tonnes in 2005. Thus, forests could offset India's modern energy related carbon emissions shown in Scenario 2, by 17 per cent in 2025. India's net emissions from these two sources would be reduced to 283 million tonnes in 2005. This outcome would be achieved at no additional cost to those shown in Scenario 2. Further, given the exponentially higher sequestration potential for 2011, forests could offset as much as 25 per cent of the energy emissions in 2011.

Carbon is stored away or released when the biomass from a tree is utilised. The type of use and its duration determine the net carbon emissions. A tree burnt for the purpose of land clearing will release most of its biomass immediately. A tree providing lumber used to make building will store carbon away for decades. What is the best use of trees in order to sequester carbon?

Hall, Mynick and Williams point out that while sequestering carbon in forests is a relatively low-cost strategy for offsetting carbon dioxide emissions from fossil fuel combustion, substantially greater benefits can be obtained by displacing fossil fuel with biomass grown sustainably and transformed into useful energy using modern energy conversion technologies. Biomass substituted for coal can be as effective as carbon sequestration, per ton of biomass, in reducing carbon emissions. However, fuel substitution can be carried out idefinitely, while carbon sequestration can be effective only until the forest reaches maturity. Also, greater biomass resources can be committed to fossil fuel substitution at any given time than to carbon sequestration because biomass (such as sugarcane bagasse) can be obtained from sources other than new forests. Thus, biomass can play a much larger role in reducing greenhouse warming by displacing fossil fuel than by sequestering carbon. Moreover, biomass energy is potentially less costly than the displaced fuel energy in a wide range of circumstances, so that the net cost of displacing carbon dioxide emissions would often be negative.

India's carbon emissions are likely to grow in the future because of the increasing energy and food consumption needed to support a growing economy. However, strengthening the ongoing afforestation programmes, increasing energy efficiency, and prudent use of renewable options in selected applications have potential to offset a significant portion of the GHG emissions.

Implementing the three types of options will not be easy. Energy and forest products consumption and supply patterns, and forest land use are shaped by a large number of actors at various levels. Improving resource allocation and use patterns will require action at the national and international levels. Reddy has outlined many barriers to energy efficiency improvement. Similar barriers exist to increased afforestation and renewable energy use. Energy consumers are often uninformed, first-cost sensitive, indifferent and helpless to improve efficiency. National institutions are supply-biased, with little incentive to innovate. The government is uninterested, is short of capital and skills, has inadequate training facilities, and limited access to hardware and software. Energy efficiency agencies are relatively powerless compared to their supply counterparts or they are part of the supply agency and there fore have no incentive to reduce demand for their product.

Further, bilateral and multilateral aid agencies target the supply aspects of energy systems with inadequate attention to demand-side measures. Other issues, such as an anti-innovation attitude, the large-is-convenient under and project-mode sponsor contribute to the lack of attention to the three options. Many of the barriers listed above arise because there is no incentive for the various actors to behave differently. Concern about climate change can provide this incentive. The establishment of the GEF and the growing attention being paid to environmental issues at the World Bank is a positive sign which will alter future lending practices of multi-later institutions. Increased attention to environmental issues holds out the hope that these and other similar institutions will begin to address the concerns of the poor, and not just afforestation schemes are being discussed and addressed. Concern about climate change can improve on this dimension by explicitly developing projects which provide sustainable solutions to meet the energy food, water and other needs of the poor. These projects will halt deforestation and/or lead to increase greening of rural areas in India.

Our analysis suggests that if India pursues basic-needs oriented development with emphasis on end-use efficiency, decentralised renewables and its economy will improve more rapidly. Simultaneously, it is in the interest of the developed countries to fund India's incremental costs of switching to less carbon-intensive technologies. Such technologies represent the most cost-effective path to economic development. For perhaps the first time in history, the interests of the developing world are aligned with those of the industrialised countries creating an unprecedented paradigm for future human development. More importantly, many of the measures of implement the three options have the potential to improve the condition of the poor in the developing countries. Efficient energy use and selected renewable options have been successfully demonstrated as necssary means to provide better water supply, lighting and fertiliser, which has fostered rural development. Afforestation in India, through natural regeneration programmes, directly aids rural villagers.

Concern about the shared global problem of climate change offers a unique opportunity to align the interests of the developed and developing countries, rich and poor. While competition and dissimilar goals have often frustrated and defeated co-operative ventures, climate change offers a common incentive for collective action. Pursuing the socio-economic development goals of the South is consistent with the environmental goals of the North, and provides joint benefits to economy and ecology that are in the shared interests of all.

5. Global Warming and International Co-operation

GLOBAL WARMING is a phenomenon that will have profound implications for the study of international relations. The traditional interpretations, concepts and frameworks that we use to study the discipline will necessarily have to be reconsidered. The purpose of this paper is to present a theoretical construct that identifies the factors that promote international cooperation on general issues of collective goods. While being introduced, this framework will be applied to the global warming issue in order, not only to examine the record thus far, but also to suggest what barriers and openings may lie ahead in the international negotiations.

The Earth's atmosphere is constituted so that it allows most sunlight to stream in uninterrupted. After striking the Earth's surface, this solar energy is reflected as longer-wavelength infra-red radiation. Some of this radiation is subsequently trapped in the atmosphere by clouds and 'greenhouse gases' (which include carbon dioxide, methane, nitrous oxide and chlorofluorocarbons). Without this greenhouse effect, the surface of the Earth would be about 33^0C cooler than it presently is. Therefore, the phenomenon is necessary for life, as we know it, on the planet. Since the Industrial Revolution humankind has pumped more gases into the atmosphere and has, in effect, thickened the greenhouse blanket that surrounds the Earth—thus, more heat has been trapped near the surface. As a result, it has been estimated that the average global temperature may rise by between 1.5 and 4.5^0C within 40 years—an occurrence also known as 'global warming'. Such modifications could induce a number of world-wide physical changes, with the most significant being: rising sealevels, changing patterns of precipitation, shifting zones of flora and fauna, and a harsher and more severe climate. The ramifications that these physical changes would have upon social systems are potentially large—more displaced peoples, weakening economies, and increased conflict over scarce resources are just a few illustrations.

'Cooperation' is a term that is used freely in the study of IR—it seems to be able to encompass a wide range of international associations. Since the word is central to this paper's investigation, an accepted definition is required. Therefore I would like to adopt Robert Keohane's explanation:

> Cooperation requires that the actions of separate individuals or organisations—which are not in pre-existent harmony—be brought into conformity with one another through a process of negotiation . . . (Thus,) cooperation occurs when actors adjust their behaviour to the actual or anticipated preferences of others, through a process of policy coordination. To summarise more formally, intergovernmental cooperation takes place when the policies actually followed by one government are regarded by its partners as facilitating realisation of their own objectives, as the

result of a process of policy coordination. (Keohane, 1984: 51-52)

By accepting this definition, a couple of key restrictive points are evident. First, it is clear that cooperation is only a concern in issue-areas that can be characterised by positive-sum games. Furthermore, the benefits from such games must be contingent upon the actions of other players. Therefore, if cooperation is present, independent decision-making will not be undertaken because the definition stipulates that communication and consultation take place. It also specifies that, for the most part, the cooperating states attempt to coordinate their actions so as to obtain desirable outcomes—in other words, it is anticipated that all players will accrue benefits.

Cooperation is of course not always present. States do not necessarily have to cooperate in order to attain their desired goals—even with regard to environmental issues. A country might be able unilaterally to pursue and realise its set objectives. For example, it might want to clean a lake that is within its territorial boundaries. Because the lake's health will be determined by the actions of that country and that country alone, there is no need to coordinate policy with other states. Furthermore, even the attainment of international environmental goals may not necessarily involve cooperation. For example, diplomatic, economic or military coercion—without negotiation—may allow a state to achieve any of a variety of international environmental objectives—stopping dangerous river dumping by a neighbouring state's or obtaining another state's strategic natural resources are only two examples.

But there is at least one problem with which unilateral state action will be unable to deal effectively: global warming. Instead, some form of cooperation will be necessary. This is clear once we recognise that the issue fits neatly into a traditional 'commons' framework. In other words, the global warming issue can be modelled as a 'dilemma of collective action'.

Though global warming presents an unprecendented difficulty, it is not the only international phenomenon that has ever been represented as a dilemma of collective action. Since the beginnings of our present international political system in 1648, states have entered into mechanisms for international cooperation (e.g., conventions, understandings and treaties) in order to realise some sort of order in the anarchical system and in order to escape sub-optimal results. Such issues include: human health standards, mail service, transportation regulations and trading practices. Furthermore, there have been a number of environmental issues that present comparable scenarios—regional pollution problems (including above-ground nuclear testing) and most recently, the issue that some have seen as a 'dry run' for global warming, the case of ozone layer depletion. It is not my intention here to analyse each of these, because that in itself would be an enormous task. Instead I would just like to identify them and to suggest that they are able to provide some help in the study of this subject—in attempting to aid our investigation into the factors that affect international cooperation on the issue of global warming.

The Methodological Attack

It is interesting to note that this approach, especially given its significant magnitude, has received relatively little attention recently. There are two major reasons for this oversight. First, much of the work in International Political Economy (IPE), the 'natural birthplace' of this field of study, has concentrated upon the questions of regime 'preservation'—i.e., what to do, as Keohane has asked, 'After Hegemony'?. And, second, the taboo of international organisations and multilateral cooperation during much of the 1980s meant that investigation into such issues was not in demand. It has only been in the last year or two that this question has resurfaced and once again began to occupy a more important position upon the IR agenda. In my work, I have reviewed the fields of regime analysis, international institutions and cooperation, game theory, economics, international political economy and analyses of empirical recounts of similar dilemmas of collective action (some of which I listed earlier) and I have attempted to synthesise this material in order to hypothesise the major factors for cooperation.

This structure consists of ten elements—elements which loosely can be called 'factors for cooperation'. The framework has been constructed in response to the call for the identification of specific conditions under which cooperation takes place. The two sections that follow—the four preconditions for cooperation and the six catalysts for cooperation—form an 'if-then' statement coupled with an 'as-then' statement. In other words, if 'the first four preconditions for cooperation' are realised, then 'cooperation will occur'. Furthermore, as the 'six catalysts for cooperation' are present in greater and greater concentration, then 'the process by which the preconditions will be achieved will be accelerated'. Furthermore, for. total methodological accuracy, I should point out that these factors are not independent; they feed among themselves in a variety of directions. Thus, some overlap necessarily exists among the ten categories.

Nevertheless, I feel that such an approach allows us to make sense of the past and it also permits us to further our insight while we investigate the future. There are currently only two major international initiatives concerned with tropical forests, and both have significant problems. The Tropical Forestry Action Plan (TFAP) was lunched in 1985 by the World Bank, United Nations Development Programme, the Food and Agriculture Organisation of the UN (FAO) and the World Resources Institute, as an emergency response to the tropical forest crisis. The response involved a massive increase in development assistance to the forestry sector. The total funding envisaged over the five-year period of the plan was US$8 billion, up from approximately US$500 million per year being spent on forestry assistance at the time. Projects would be developed as a result of national planning exercises, which would involve a review of the problems facing the forest sector in the country concerned. This review would identify both policy changes needed at the national level and projects for development assistance funding.

Although the level of development assistance to the forestry sector has increased to US$1 billion per year since TFAP began, TFAP has not delivered acceptable results in most countries. Firstly, most national plans have not included adequate recommendations on changing government policies which lead to deforestation. Underlying causes such as North-South inequalities, land distribution and tenure issues, etc., are rarely addressed. National Forestry Action Plans (NFAP) have often becomes simply a catalogue of projects. However, even as a list of projects they are not satisfactory. Insufficient recognition is given to the fact that many development projects in the past (such as those funding logging or plantation establishments) have led to deforestation in the name of economic development.

As a result many of these destructive projects find their way into National Forestry Action Plans. A more general comment is the NEAP are mostly developed in a "top down" manner, with inadequate participation of people who live in and depending tropical forests, and of NGOs. The nature and even the existence of TEAP is currently under review and there is some hope that the most serious failings.

The International Tropical Timber Organisation (ITTO) was set up in 1986 to implement the International Tropical Timber Agreement (ITTA), a commodity agreement negotiated under the auspices of the UN between countries which consume and produce tropical timber. The ITTA has a number of objectives including promoting sustainable forest management, the expansion of the tropical timber trade and improved market transparency.

Although sustainable forest management for commercial timber production occurs only to a tiny fraction of forest areas, the ITTO has made the promotion of sustainable forest management its primary goal. With a small staff and a project budget of only some US$20 million per year, ITTO's ability to influence the timber trade is limited. Its most effective role is probably that of a forum bringing together producing and consuming countries, the timber trade and NGO. Recently ITTO assessed forest management in the States of Sarawak in Malaysia, where there have been serous conflicts between loggers and local people. Although the ITTO report on the situation contained a number of useful observations, its recommendations were weak. Partly as a result of this the ITTO was unable to get any real commitment from the Malaysian government which might have solved the problems.

In review, neither ITTO nor TFAP has been able to make a significant impact on the crisis facing tropical forests. One can conclude from this that traditional development assistance approaches, particularly in the form of aid projects are insufficient to solve the problem,. Countries from the North and the South must address together the broader issues of trade and economic development that underlie the destruction of tropical forests. Issues such as the terms of trade and commodities produced in tropical countries, trade barriers which discriminate against locally produced products derived from tropical forests, and the net flow of resources from South to North, define a marco-economic framework in which poverty and resource degradation are inevitable.

Those who advocate putting off actions that are proposed in anticipation of global climatic change often hear that waiting will make the costs of response higher. This position is grounded in pessimism about future technology, future economic resources and the ability to acquire new information. Several lines of evidence argue against this pessimistic assumption. First, the menu of response options may increase through innovations, technological and otherwise. If it were decide today that societies should immediately reduce their greenhouse-gas emissions by 20 per cent, several countries might achieve this reduction by building light-water nuclear reactors. Such investment would further lock these societies in to energy technologies that may be inferior to alternatives, nuclear or non-nuclear, that will be available fairly soon. In short, it is not clear that it is better for the United States to spend, say, $100 billion during the next five years on nuclear and alternative energy technologies to reduce carbon dioxide emissions than to wait and spend the equivalent amount during a future period.

In fact, if there is economic growth in the intervening period, the available and spendable resources should increase. It may then be possible to select choices from the menu that are now excluded because of high cost. In other words, if a future society is wealthier than today's, it may be able to finance highly attractive response options that are unaffordable at present; these options might render superfluous the clumsy interventions that now dominate the set of feasible choices. In the US Environmental Protection Agency's scenario of unimpeded growth in the emission of greenhouse gases, annual per capita global income in the year 2100 is projected to be $35,600. Per capita income in the United States would be $150,000 or more, if current income ratios hold. Such flows of wealth open wide vistas for adaptation to climate change, as well as for expenditures to reduce greenhouse-gas concentrations.

When we shift our perspective away from the world's centers of prosperity, the decisions are different but the conclusions not necessarily so. For parts of the world such as Africa, many urgent questions of immediate survival and economic development must be faced. In these regions it is important to analyse carefully whether now is the proper time to begin allocating resources in anticipation of the uncertain impacts of global warming. And in the USSR and the nations of Eastern Europe, the optimal timing of greenhouse investments must be explored in light of the urgent need to put in place better private and governmental decision-making processes.

The argument about timing is also dependent on the quality of the information now in hand and the probability that it will improve. There are areas in which we have reason to be optimistic that we will have, as time goes on, information useful in making mesured responses to the challenge of climatic change. New information could, for example, allow us to better assess the likelihood of various amounts of sea-level rise and thus guide resources more efficiently into a variety of adaptive responses. In the case of large, lumpy investments such as those needed to protect coastal regions from flooding and seawater instruction or to transfer water between river basins, it is not apparent that the quality of current forecasts is sufficient to trigger action.

There may, however, be actions that are more easily and effectively taken now, or soon. As the world becomes more crowded with people, it is likely to become more difficult to reserve land for corridors that would allow the migration of plants and animals as climate shifts. In some regions, such as coastal zones and floodplains, waiting may further set society into patterns that increase vulnerability and reduce flexibility. A parallel argument applies to the use of fossil fuels: The further they are allowed to penetrate into the fabric of society, the larger becomes the job of substituting for them.

In summary, if society becomes better supplied with tools, money and knowledge, the menu of options for adaptation will be superior in the future—far superior, if the technological and economic progress of the past 100 years continues apace. Because information is easier to transfer than are wealth and technology, it is especially important to acquire better information as quickly as possible. A stitch in time saves nine *if* you know how to make the right stitch. It is quite possible that the optimal present strategy on the greenhouse issue, for many nations, is a continued emphasis on increasing the information available and improving its quality.

The contention that there will be no winners from global warmings seems to rest on three arguments. One is the risk of a shift to a calamitous climate, in which warming would reach a catastrophic extreme—a hot climate as extreme as the cold one experienced 18,000 years ago, when the average temperature was cooler by 5 degrees Celsius and an ice sheet buried North America. A second arguement emphasises the accumulation of global problems. The third is based on the perception that society is becoming increasingly brittle.

The first argument emphasises the non-negligible probability that climatic change will follow a catastrophic course, in which case the "everybody loses" statement is—tautologically—true. This could occur if greenhouse-gas emissions are extremely high, if atmospheric concentrations reach very high leves, and if the expected warming follows.

A scenario of even greater concern is that of a "runaway" greenhouse effect, in which a relatively small change in concentration triggers an unexpectedly large change in the climate system. Such a scenario is physically possible, but so far is not supported by much more than highly conjectural assertions. One set of interesting conjectures relates to shifts in ocean currents, which appear to have occurred quite rapidly and unaccountably in the past. Another relates to warming in high latitudes; this might release large amounts of methane from clathrates in the tundra and continental margins, thereby greatly increasing greenhouse-gas concentrations in the atmosphere. It might be useful to survey experts to ascertain what they think is the probability of such a scenario: 1 in 10, in 100, in 1000? Contingency plans must be developed for high-consequence, low-probability scenarios, just as contingency plans are made for earthquakes, reactor failure and threats of nuclear war.

The second argument for the "everybody loses" hypothesis is that environmental change will cause extreme and comprehensive destabilisation of social and economic conditions world-wide. Like the runaway greenhouse effect argument, this is plausible but entirely conjectural. A scenario has yet to be worked out in detail, beyond popularisations. The essence of the view is that almost everything in the world is going wrong—that all the indicators about population, food, terrorism, radiation and the ozone layers, for example, are moving ominously. The overall system is heading for catastrophe, and climate is part and parcel of this. Climatic changes will add stress by reducing the reliability with which a population of 10 billion can be fed and by adding to the number of refugees as a result of rising seas. In this view, the amount of climatic change matters, with larger being worse, but even a small change could be bad—the straw that breaks the camel's back.

A variation on this argument holds that sociotechnical systems are becoming more brittle. Although for many specific contigencies societies may be insulating themselves successfully from climatic change, this protection comes through increasingly elaborate systems that are more vulnerable to catastrophe from both natural and social hazards that may occur infrequently. In the open and enlarging world system of today, vulnerability may be shared even more widely, and consequences may ripple into previously unrelated areas or societies. This argument is comparable to concerns about electricity blackouts, the implication of which multiply with our growing dependence on a reliable power supply. In this perspective, even small climatic changes could trigger collapse of social and physical structures.

Balanced against arguments that any climatic change is bad or extremely risky for everyone is specific research showing that there could be winners from climatic change. This proposition seems logical. The present climate of the earth is harsh in many, if not most, parts of the globe. It is easy to imagine that several, even many, regions will find themselves in more favourable conditions—milder winters, more abundant rainfall—as climate warms. One study suggests that there may be an increase of from 3 to 11 per cent in global mean percipitation.

Indeed, there is a growing body of literature showing, based on analysis of the widely cited results of general-circulation models, that there would be benefits from global climate change. For example, on the basis of detailed studies of Japanese agriculture, Masatoshi Yoshino and his colleagues concluded in 1988 that in Japan, scenarios based on a doubling of atmospheric carbon dioxide levels "would, *ceteris paribus*, result in an enormous rice surplus." This conclusion is based on analysis of rice yields relative to the long-term baseline productivity for a "dumb farmer," that is, one who takes no adaptive steps but goes about his business in the climate brought about by doubled CO_2 in the atmosphere as if it were the climate of today. Yoshino's group noted that most of their estimates reflected a poor response for a well-managed commercial crop. The surplus would be even greater if farmers were, as is probable, to adapt their cultivation methods to then we climate by actions such as introducing more heat-loving rice varieties with earlier planting dates. Such adjustments might result in increase or rice yields in areas of Japan of up to 25 per cent, relative to recent levels. A similar case study showed that major national surpluses of good grains can also be expected in Finland.

It is also worth noting that migration trends suggest that many people seem to prefer a warmer climater. Within the United States, the migration toward the Sunbelt has been widely analysed. To a large extent this is a climate-induced migration; people even accept lower wages to live in warmer climates. A similar phenomenon has occurred in Canada. The behaviour of populations over the past few decades thus suggests that many people might feel themselves winners to have a warmer climate, other things being equal or nearly so.

Finally, there is also the curious fact that previous, detailed studies showed that on balance, global cooling would be adverse and global warming beneficial. These studies, undertaken as part of the Climate Impact Assessment Programme (CIAP) of the US Department of Transportation in the early 1970s, were primarily concerned with the possibility of global cooling that might be associated with a large fleet of supersonic transport planes operating in the stratosphere. The warming scenarios were explored partly as a reference point. Both warming and cooling produced a blend of benefits and costs. Global-cooling scenarios produced net costs, whereas moderate global-warming scenarios tended to produce net benefits. Studies combining warming and cooling scenarios with both increases and decreases in precipitation yielded essentially the same result, with a warmer and wetter scenario the most favourable. The numbers suggest a logical conclusion: that there are both winners and losers with any gradual climate change.

(The Japanese the Finnish studies take account of an important additional fact: that carbon dioxide is food for plants. Holding other conditions steady, doubling the CO_2 content of the atmosphere appears likely to increase yields by somewhere between 10 and 50 per cent.)

It is tempting to speculate that analysis of any potential climatic change—whether cooling in the early 1970s or warming for the past decade—are somehow biased by a sense of threat. Alternatively, it is possible to argue that human societies are now precariously balanced from a socio-economic viewpoint at a climatic optimum reminiscent of the gradations used by the ancient Egyptians to interpret the readings on their nilometers, which measured the height of the Nile River. It is true that either a vast expansion of glaciers or Venusian heat would render human habitation of earth more difficult, very miserable, and perhaps impossible. But no convincing evidence exists to replace the stone markings used by the Pharaohs to measure the height of the Nile with measures of annual average global temperature. By coincidence, the number 16—the river level signifying "abundance," located not far below disaster—is about the current number for global average temperature, in degrees Celsius.

There are also losers from a scenario in which limitations are placed on emissions of greenhouse gases. In this case the losers are the individuals, groups or nations whose economic development is slowed because of higher energy prices or other barriers to access and use of carbon (or of other resources related to emissions of greenhouse gases). The losers also would include the carbon-rich nations whose resource, in or on the ground, is devalued by restrictions on carbon use. In fact, the biggest economic losers from a reduction in CO_2 emissions might be the less-developed countries and the oil-exporting countries. Carbon is the largest export of the poor nations of the South to the rich nations of the North. If the industrialised nations continue to increase their energy efficiency and reduce reliance on hydrocarbons, markets dry up for Mexico, Venezuela, Ecuador, Gabon, Nigeria and Indonesia. Carbon is also by far the largest export of the USSR

Finally, a discussion of winners and losers is incomplete without mention of the direct beneficiaries of the greenhouse issue—namely, those employed in research and negotiation about it. The "greenhouse industry" will employ at least several thousand scientists, lawyers, diplomats and others. In summary, statements that there are no winners or are only losers from global warming appear biased, unless based strictly on a apocalyptic scenario. It seems more accurate to emphasise that there will be complex distributive issues involved in climatic change, with continually shifting sets of losers and beneficiaries, both from the changes in the climate and the regulation of economic activities that my be causing the climatic changes.

Over the last century—a mere instant of geologic time—the activities of the human species have caused unprecedented changes in the atmosphere. A continuing buildup of certain chemical compounds—most importantly, carbon dioxide (CO_2)—is propelling the environment towards (a) potentially catastrophic shift. . . . Fossil fuel combustion has spewed 150 to 190 billion tonnes of carbon into the air, and forest clearing for cropland and fuelwood has contributed an additional 90 to 180 billion tonnes. . . .

While some climatic change is inevitable, societies can gain precious time to adapt if action is taken now to dampen its ultimate magnitude and slow its pace. The first step requires curbing the use of fossil fuels, now the leading cause of the CO_2 buildup. But there is other step crucial to restoring atmospheric balance: protecting our remaining forests and planting more trees. Forests and woodlands are vast storehouses of carbon, so clearing and burning them—as is now happening on a large scale in the tropics—contributes to CO_2-induced climate change. Because trees remove CO_2 from the air through photosynthesis, planting more of them can be part of the remedy. Therein lies an opportunity to capitalise on the enduring link between earthly life and the atmosphere—by reforesting the earth. . . .

The earth's trees, shrubs, and soils hold about two trillion tonnes of carbon, roughly triple the amount stored in the atmosphere. When vegetation is cleared and burned, or just left to decay, the carbon is contains, along with some of that in the underlying soil, is oxidised and released to join the atmospheric pool of CO_2.

Today the bulk of the CO_2 emitted from the land in this way comes from developing countries in the tropics. Each year, 28 million acres of tropical forest are destroyed through the combined action of land clearing for crop production, fuelwood gathering, and cattle ranching. Commercial timber harvesting degrades an additional 11 million acres. All told, an area of trees slightly larger than New York and Vermont combined is lost or logged each year.

Scientists have spent many years trying to pinpoint how much carbon is added to the atmosphere through forest clearing. . . . In a January 1988 *Science* magazine article, R.P. Detweiler and Charles Hall placed the amount of carbon released by tropical deforestation in 1980 somewhere between 0.4 and 1.6 billion tonnes. Another authoritative study, led by Richard Houghton, associate scientist at the Woods Hole Research Institute in Massachusetts, figured it to be between 0.9 to 2.5 billion tonnes. Since about 5 billion tonnes are emitted each year by fossil fuel burning, the midpoint of Houghton's estimate would attribute one-quarter of the total annual carbon buildup to deforestation.

Houghton and his colleagues. . . . estimate 40 per cent [of these emissions] comes from tropical America, 37 per cent from tropical Asia, and 23 per cent from tropical Africa. Just five countries account for half of all carbon emissions: Brazil, Colombia, Indonesia, the Ivory Coast, and Thailand. Brazil alone already contributes a fifth of the total, a figure likely to rise if deforestation in the Amazon continues. A recent study of satellite data by Brazil's Institute of Space Research suggests that 20 million acres of forest were cleared and burned in the Brazilian Amazon in 1987, some three times the annual rate of clearing that had been estimated for the early 1980s.

A worrisome twist to the forest/climate change link is how the world's remaining forests will behave in a warmer climate and an atmosphere richer in CO_2. . . .

Higher CO_2 levels usually have a fertilising effect on plants, spurring them to grow faster. . . . If trees did indeed grow faster as atmospheric CO2 levels increased, they would remove carbon from the atmosphere more rapidly. This "negative feedback" would help slow the global warming. So far, unfortunately, no convincing evidence suggests that trees in their natural environments would respond this way. . . .

Another possibility results in positive feedback, a worsening of the warming trend. George Woodwell, director of the Woods Hole Research Institute, points out that as temperatures rise, trees and microorganisms in the soil substantially increase their rates of respiration. . . . The danger is that an increase in respiration because of rising temperatures could release more CO_2 to the atmosphere, reinforcing the very buildup that initiated the warming.

If respiration exceeded photosynthesis for an extended period of time, trees would stop growing altogether and ultimately die. . . . Woodwell maintains that a widespread forest die-off could release enormous amounts of carbon to the atmosphere—perhaps hundreds of billion of tons—depending on the speed of the warming. He warns that "the sudden destruction of forests by air pollution, now being experienced in northern and central Europe. . . . is but a sample of the destruction that appears to be in store."

Woodwell's scenario might never come to pass. Ecologists do not yet agree on how forests will respond to a warmer climate, or even on whether that response will add CO_2 to the atmosphere to remove it. Some point out, for example, that higher temperatures would increase rates of organic decomposition, which in turn would release nutrients to the soil and thus potentially boost the productivity of trees. This could cause a helpful negative feedback: since trees would be growing faster, they would remove more CO_2 from the atmosphere, helping slow the warming.

How forests will actually respond looms large in the climate change picture, since the potential for a strong feedback—positive or negative—clearly exists. Protecting forests and planting trees need to be high on the international agenda for several compelling reasons other than stabilising the global carbon cycle. The growing wave of deforestatioin has left in its wake a severe energy crisis in the Third World (wood provides the primary source of energy for more than two-thirds of the people in developing countries), an accelerating loss of the earth's biological diversity, and large areas of degraded land. . . .

Countries are unlikely to invest substantial resources in tree planting solely to ward off global warming. But, in much of the Third World, satisfying fuelwood needs and restoring productivity to degraded ecosystems provide a sound—even urgent-rationale. Expanding forest cover for these purposes would yield the added bonus of slowing the buildup of CO2, which gives industrial countries ample reason to step up support for tree planting in the Third World.

An analysis by the Worldwatch Institute suggests than an additional 320 million acres of trees—an area nearly twice the size of Texas—will be needed by the year 2000 to meet developing countries' growing demands for fuel and industrial wood products and to rehabilitate deteriorating ecosystems. What contribution could that much tree planting make to slowing the CO_2 buildup? Too many uncertainties exist to answer this question precisely, but some back-of-the-envelope calculations suggest. . . . that [an effort of this sort] would cut net carbon releases from tropical forests by nearly half.

Of course, slowing the destruction of existing forests is also crucial. Halving the CO_2 contribution from deforestation in Brazil. Indonesia, Colombia, and the Ivory Coast would reduce net carbon emissions from tropical forests by more than 20 per cent. Together, that achievement and the carbon-storage benefits of 300 million acres of trees would cut releases from tropical forests by two-thirds. The total amount of carbon added to the atmosphere from all human activities—deforestation and fossil fuel combustion—would be cut by 17 per cent.

According to Houghton and his colleagues, natural systems in Europe and possibly Japan and South Korea are already accumulating more carbon than they are releasing. Many European countries have in recent decades abadoned substantial areas of cropland and allowed forests to regrow. . . . Japan and South Korea have purposely planted large areas in trees. Japan now has the fourth largest area of commercial tree plantations in the world, nearly 24 million acres. . . . [South] Koreans planted an area in pines roughly equivalent to two-thirds of the area cultivated in rice, their staple food.

As a result of the Food Security Act of 1985, the United States could also turn some of its land into a carbon bank. The Act created the Conservation Reserve, under which farmers will take at least 40 million acres of highly erodible cropland out of production by 1990 and plant it in trees or gas. Each acre of recovered grassland or woodland will store roughly 16 more tonnes of carbon than when it was cultivated. Assuming the carbon accumulates at an average yearly rate of half a tonne per acre as the land undergoes conversion, the reserve would absorb a total of 20 million tonnes of carbon annually for the next three decades, assuming the land remains in the reserve. . . .

Can the community of nations plant trees on the scale required to improve prospects in the Third World and simultaneously help balance the global carbon cycle? There are reasons for optimism. International development agencies now recognise that rural people form the only labour force large enough to plant trees on the vast scale that is needed. More than ever before, this labour force is being mobilised into action. . . .

International relief agencies, such as CARE in the United States and Oxfam in the United Kingdom, have orchestrated some of the most successful reforestation projects to date. World-wide, thousands of women's groups, peasant collectives, churches and other small, local organisations have taken up the cause of tree planting. . . . Kenya's Greenbelt Movement. . . . has enlisted. . . . more than 15,000 farmers and a half-million school children in setting up 670 community nurseries and planting more than two million trees. . . .

Chinese officials have set a goal of getting 20 per cent of their country's territory in trees by the year 2000. . . . Indian Prime Minister Rajiv Gandhi tripled funding for forestry in his development agenda for 1985-90, gave forestry new prominence within his ministries, and created a National Wastelands Development Board to spearhead a "people's movement" for reforestation. . . .

Reforestation's potential to help avert climate change barely gets mentioned in reports or plans than sketch out forestry's future. But as the consequences of global warming becoming clearer, and their magnitude and cost hit home, tree planting solely for the purpose of stabilising climate could appear on the international agenda. Environmental philosopher Rene Dubos pleaded for people to "think globally, act locally." Perhaps no activity harmonises global interests and local needs better than does reforestation. Reforesting an area equal to twice the size of Texas by the year 2000 entails successfully establishing some 18.4 billion trees annually. That target sounds staggering, but it would only requires each person now living in the Third World to plant and care for five seedlings a year.

A strategy of forest protection and tree planting cannot by itself ward off climatic change. . . . But curbing deforestation and planting trees will help our generation ensure that succeeding ones inherit a habitable earth. In such a low-energy, high-efficiency future, the great energy supply debates, such as coal vs nuclear, which preoccupy us so, lose much of their significance, and pollution problems are knocked down to more manageable proportions.

Large energy efficiency gains, and the consequent reductions in CO_2 emissions, will be essential in addressing what is probably the most serious environmental challenge of all: the global warming, which seems already to have begun. I recognise the uncertainties remaining in characterising the Greenhouse Effect, but give the risks, I would advocate consideration now of a series of international conventions responsive to the various aspects of the problem. First, we need to secure swift international approval for the ozone layer protection protocol signed in Montreal last year. We need this for its own sake and to continue the momentum that can get the nations of the world back to the table so that a complete, swift phase-out of CFCs can be negotiated. The phase-out is fully justified on ozone layer grounds alone, but the fact is that a CFC phase-out is the fastest and cheapest way societies can do something major to contain the Greenhouse Effect.

Second, we need an overall global climate protection convention, the prime goal of which should be to stabilise atmospheric concentrations of greenhouse gases of safe levels. This convention should focus particularly on step needed to secure reductions in CO_2 emissions from fossil fuel use. Two facts stand out in this regard: the United States and the Soviet Union together account for almost half of global CO_2 emissions today, and the United States, the Soviet Union, and China together account for about 90 per cent of the estimated coal reserves.

Third, the time is ripe for an international agreement to protect the world's tropical forests and to reforest the spreading wasteland areas in many developing countries. The industrial nations have a double stake in halting the now rapid clearing of the tropical forests. Not only are these forests repositories for about half of the wildlife and genetic wealth of the planet, but CO_2 emissions from biotic sources such as deforestation are estimated to be about a fifth of CO_2 emissions from fossil fuels. Our stake in the salvation of these forests in sufficiently large that we should be more than willing to help provide financial incentives—incentives that will be necessary if countries of the tropics are to turn their attention to what often appears to be a low priority or even a threat to development and sovereignty. I suggest that we go far beyond the debt-for-nature swaps under way today and consider a global bergains as part of this international convention. This bargain would involve the easing and forgiving of international debts in exchange for forest conservation. Of the top 17 most heavily indebted countries, 12 are destroying their tropical forests at extraordinarily rapid rates, contributing to the world's annual loss of 27 million acres.

And fourth, we need international agreement on the protocol now being developed in limit NOx emissions. Unless capped, increasing NOx emissions will lead to increasing ozone concentrations, and ozone is a greenhouse gas as well as a source of urban and rural air pollution. My concern about nuclear power, as things stand today, is that it probably will not, in the end, provide a major part of the answer to global warming. Its public acceptability is too low and its price is too high. If we try to solve the greenhouse problem by cramming nuclear power down the throats of an unwilling public and unwilling investors, we will be setting the stage for prolonged confrontation and stalemate. Moreover, I believe there are safer and cheaper alternatives for the short run, including the vast potential for efficiency gains in how we generate and use electricity.

In all these areas, in seeking these treaties and in setting an international example by acting on our own, US leadership and EPA leadership could not be more important. The world is not exactly waiting on our leadership, but neither will it get very far without us. Let me conclude with a word about why I am optimistic that the world can indeed be saved. This address, you have doubtless noted, reflects a deep appreciation of the importance of economic and technological forces in the modern world. One reason for optimism is that science and technology are presenting us with answers. We are in the midst of a revolution in industrial and agricultural technology, both with huge potentials in the areas we have been reviewing.

But if solutions are found, they will come from another realm as well, from the hopes and fears of people, from their aspirations for their children and their wonder at the natural world, from their own self-respect and their dogged insistence that some things that seem very wrong are just that. People everywhere are offended by pollution. They sense intutively that we have pressed beyond limits we should not have exceeded. They want to clean up the world, make it a better place, be good trustees of the Earth for future generations. With Thoreau, they know that heaven is under our feet as well as over our heads. Politicians around the globe are increasingly hearing the demand that things be set right. And that is very good news indeed. It would also be completely illogical. Countless studies have shown that investment in improved efficiency, at both process and end-use levels, would be far more cost-effective as a way of reducing CO_2 emissions, than the supply-oriented approach out-lined above. Such investment may, in many cases, provide a better return than investment in new fossil-based power generation capacity, implying that CO_2 abatement may be achieved at zero, or even negative, total; cost when the full energy cycle is considered. The world community would be better off if it used less energy, even without considering the threat of climate change.

What is needed, therefore, is not new energy sources 'plugged in' to the supply side of the existing world energy economy, but a new approach to energy planning and development, which focuses on satisfying the needs of end-users in the most cost-effective way, taking into account the full environmental costs of energy production and use. While much work is still required on identifying the policy options available and establishing their respective economic implications, it must be emphasised now that the transition to a sustainable energy future need not be seen as a brake on economic development. Given the uncertainties surrounding the issue of climate change, recognising this point is essential: less the hard-pressed policymakers of the developing world respond that their countries have more pressing needs than to sacrifice growth for the sake of alleviating a distant problem, for which they are largely not responsible, and whose consequences are still unknown.

The problem is that existing energy markets and national-level institutions may directly oppose the sort of reforms required to control the environmental cost of current patterns of energy development. The immediate response to the Canadian study cited earlier indicates where some of the obstacles lie. The task force itself concluded that, in spite of the consultants' findings, it would be premature to accept the Toronto Conference's recommendation as a formal goal, and proposed that it should remain an 'illustrative target'. Concern was expressed in particular that drastic unilateral action to reduce CO_2 emissions could damage the international competitiveness of Canadian industry.

Current energy development is costing more than it need do, primarily because far more energy is being produced than is actually required to provide for the needs of end-users, and irreparable harm is being done to the biosphere in the process. The reason for this, ostensibly irrational, state of affairs lies in the traditional approach to energy pricing and policy formulation. Present energy policies, in effect, subsidize environmental destruction. As work continues on identifying the possible impacts and response strategies to climate change, the world energy community can make a substantial contribution through identifying and eliminating the obstacles to a more rational pattern of energy development, which should allow considerable reductions to be made in CO_2 emissions at no cost in terms of economic growth.

THE NEED FOR GLOBAL COOPERATION

The key obstacle, at least initially, to reducing global emissions of CO_2 is not the total impact the transition may have on the world economy. Indeed, on balance, the damage may be negligible. The problems stem from the depth of the changes which may be necessary to national energy economics, and from the unprecedented level of international cooperation which will be required to ensure that individual countries are not penalised by adopting more sustainable policies.

At the national level, energy planning has, traditionally, focused on delivering supplies to meet projected aggregate demand at minimum 'cost', narrowly defined as the cost of production and distribution. Comparative analysis of energy options has thus, in general, failed to take into account the full costs of 'capital' resource depletion and the environmental side effects of energy production and use. This has meant that energy options which externalize a substantial proportion of their costs, such as fossil-based power generation, have appeared more attractive than those which internalize almost all costs, such as renewable energy systems.

At the microeconomic level, prices charged to energy consumers do not, in general, reflect the true cost of the energy consumed. In the case of electricity in particular, prices often do not even reflect the full marginal internal cost of supply, quite apart from the external environmental costs. The individual or corporate energy consumer is thus deprived of the opportunity of realising the full value of any energy savings made through investment in efficiency or in alternative energy sources.

> A global commitment to the principle that the energy user should pay the full cost of the energy used, not merely the internal costs of production and distribution, would be a major step towards a transition to rational, sustainable energy development.

Such a commitment will require substantial changes to be made to many countries energy economics. A recent study found that the true cost of coal-fired power generation, including the costs of the damage done by acidification but not including possible costs due to climate change, is more than double the cost calculated on a traditional analysis. In many countries, electricity subsides, as a proportion of the full marginal cost of electricity production, run to over 90 per cent.

Powerful vested interests are likely to oppose the necessary increases in energy prices. A key argument will clearly be that high energy prices may have damaging effect on a particular nation's terms of trade , as in the Canadian example cited earlier. An individual country would be reluctant to impose substantial increases in energy prices unless its trading partners did the same. Close international cooperation will therefore be essential.

An encouraging consideration is that the machinery for regulating the effect of subsidies on international trade already exists, in the form of the General Agreement on Tariffs and Trade (GATT). There is no reason, in principle, why this agreement should not be extended to cover subsidies on the energy content of exported products.

Much work remains to be done, particularly in developing a consensus on how the true cost of energy is to be defined. International agencies and intergovernmental bodies will have a major role to play, both in coordinating research into what the full costs really are, and in identifying the policy options available to contain them. The on-going work of the United Nations Environment Programme's Global Environment Monitoring System (UNEP/GEMS) and the Inter-governmental Panel on Climate Change (IPCC) provide examples of the type of activity which will have to be expanded in future, particularly in the preparation for a global convention on climate change.

International agencies can also assist in the development of new approaches to energy policy-formulation and the introduction of environmentally-sensitive energy planning tools: these are currently priority areas of activity for the UNEP Energy Unit. Individual countries must, however, identify their own strategies for incorporation external costs into energy policy and prices, in line with their traditions, sectorial structure and available resources. A consensus will only be essential where issues of international trade are involved.

It is important to recognise that an increase in energy prices to reflect environmental costs need not involve energy importing countries in any loss of foreign exchange, as the oil price increases of the 1970s and early 1980s did. Prices could be raised through some form of taxation based on the various energy sources' environmental impacts, providing governments with a significant source of revenue with which to promote more sustainable energy development. Energy importing nations, particularly those for whom energy represents a significant proportion of their foreign exchange commitments, may benefit significantly from higher prices leading to reduced domestic consumption.

Although many of these reforms may yield benefits in the long-term, they will not be entirely painless. The international community has a major role to play in facilitating the transition. Industrialised countries in particular have a responsibility to assist developing countries in managing the costs, since they have been responsible for the major part of CO_2 emissions to date.

Developing countries must recognise that it may not be to their advantage to follow the supply-oriented approach to energy development which has left many industrialised countries with such grossly inefficient energy economies. Bilateral and multilateral assistance must be channelled less towards expanding energy supplies and more towards satisfying the needs of end-users, particularly the basic needs of the poor. The proportion of energy-related overseas aid invested in improved efficiency and small-scale, end-use-oriented energy source development has, to date, been pitiful.

The capital cost of more efficient energy technologies and new, renewable energy systems will be a major hurdle to their increased use, particularly in developing countries where capital is scarce and investment horizons are short. Assistance in over-coming this cost must be a priority, and diverting existing development aid for this purpose will be neither acceptable not sufficient. Research and development in end-use efficiency and renewable energy sources has also suffered from underfunding, largely because, in the framework of a traditional energy economy, R&D oriented towards centralised, large-scale supply promised much more attractive returns.

At the end of the 1980s , the world energy agenda is dominated by the environment, just as, at the beginning of the decade, it was dominated by oil prices and supplies. The key issue is the threat of climate change. The risks of continuing on the world's present energy trajectory are appalling. The transition presents enormous challenges, but also offers opportunities. The world community must achieve an unprecedented level of cooperation in the face of an unprecedented problem, but success will mean not only a global crisis averted, but a more rational, sustainable world energy economy in the future.

Whose Growth?

Industrial CO_2 output and GDP growth are closely linked. They have risen together consistenly since the beginning of the industrial revolution. Between 1860 and 1990, 83 per cent of this growth occurred in the North. The pre-emptive nature and timing of this growth has triggered the global growth of poverty, population and political and environmental impact Divergence between beneficiaries and disbeneficiaries of this growth is increasing. This divergence is a growing threat to human development and survival. The Framework Convention on Climate Change makes a commitment to "strong and sustainable economic growth". CO_2 emissions restraint is only provisional. Consequently vested growth interests in the North, remain pre-emptive. Further divergence of benefit and disbenefit is inevitable unless future growth is appropriately differentiated within optimal scale and the disbenefit of past/current growth clearly identified.

By measuring accumulated CO_2 output differntials between countries North and South in relation to climate-derived sustainability criteria, benefit/disbenefit curves can be calculated. The basis for recovery exists in the reconciliation of these curves. Global temperature rise over the past 140 years. The mean rise is nearly one degree Celsius. Accumulated industrial output and accumulated atmospheric build-up of CO_2, measured in gigatonnes carbon. The discrepancy perhaps reflects "constant airbonrne fraction".

The assumption in 1990 (IPCC) was that to stabilise rising concentrations of carbon dioxide in the atmosphere at 1990 levels, an immediate cut of between 60-80 per cent of human source CO_2 output was needed. A "stabilisation" per capital industrial CO_2 global output average could therefore have been reckoned at 0.46 TC.

However, if stabilising concentrations was possible at a 0.46 TC level, why were atmospheric CO_2 concentrations rising already by 1860 when total industrial output was 2 per cent of current output, and when average per capita industrial emissions were only 0.031 TC? The idea of "sink-strengthening" and "constant airborne fraction" (CAF) has been suggested. For the purposes of this paper we accept the CAF proposition, but note that the probability of feedbacks becoming more positive over time, is under-represented in this approach. The global population explosion started in Europe. It maintained lead-population-density and intensity growth rates as a function of this density throughout history. Industrial CO_2 pollution intensity began in Europe and increased there first. These inter-linked with each other, were pre-emptively linked to the emulative growth patterns which began to emerge beyond Europe a century later. Current global political relations are pre-determined by disparities consequent of this primary forcing.

The evolution of industrial growth, impacts and systemic consequences is discussed in more detail in the GCI publication "Climate change, Population and the Paradox of Growth". The next section of this publication looks at the evolution of per capita CO_2 impact since 1860, comparing the trajectories of countries for which data exists or could reasonably be extrapolated. These are then compared with the global per capita average trajectory and with the global per capital average reduced by 60 per cent. This trajectory we have called the benefit'disbenefit threshold (B/D-T). This second comparison reflects the constant airbone fraction observations identified inter alia by the IPCC.

Countries are then allocated the reduced per capital global average of CO_2 emissions by population for each year. This total is then subtracted from their actual recorded output total. This equitably allocated to each country by population, a "sustainable" share of CO_2 output and notes the excess or shortfalls in each country for each year. The final calculation adds these excess or shortfall figures up time, giving CO_2 output accumulations for all countries as credits and debits 1860-1990.

The Power of Primary Forcing

This exercise demonstrates something about pre-emptive and emulative development, and the power of primary forcing in the socio-biosphere. It is consistent to find the countries which achieved the highest population densities and pollution intensities earliest are also those which consistently exceed the benefit/disbenefit threshold. As we demonstrate in other GCI publications, these are the countries whose economic growth rates also carry the highest levels of externality costs.

There are no pre-emptive countries whose trajectories accumulating on the benefit side of the reference actually decrease in real terms. There are very few emulative countries whose trajectories accumulating on the disbenefit side of the reference trajectory decrease in real terms.

Population data is compiled from UN sources and the Penguin Atlas of Population History. The CO_2 emissions data is from Carbon Dioxide Information Analysis Centre (CDIAC) Oakridge, Tenessee. They, using UN and other sources, have reconstructed nationally specific industrial CO_2 emissions data since 1950, and have also presented a global total trajectory since 1860. We retrogressed the national trajectories back from 1950 to 1860 by allocating to each country in each year the same proportion of each year's total global emissions that each was responsible for in 1950. For example the US was responsible for 42 per cent of global emissions in 1950, the UK was responsible for 8 per cent. consequently in our reconstruction, they are each regressed at that fraction of the given global total as a constant for each year between 1950 and 1860. This is imperfect, but defendable in the absence of an authentic record. If anything, this would tend to exaggerate the contributions of developing countries in the first 90 years of this reconstruction.

We have stuck almost entirely to national geographical boundaries as defined in 1990 throughout the period defined. This too is a simplification, but a reasonable procedure in the face of an array of other more arbitrary options. (In fact it is probable that early in the period, 100 per cent of carbon from carbon dioxide emissions remained airborne with this fraction reducing over time as the sink capacity in the biosphere was triggered to greater absorptive capacity. Consequently, for the purposes of this model, using only 60 per cent throughout, favours the first movers as they would most probably have been responsible for all accumulations in the first part of the period under review. However since developing country contributions are pretty marginal during this period even using the retro-gressive allocation (of constant 1950 shares), this kind of inaccuracy does not seriously disturb the integrity of the exercise overall).

(Future trends analysis shows China as an exception-it crosses the B/D-T on to the beenfit side 2010. This is however, in the context of B-A-U emissions and population projections. These projections represent losses in global efficiency terms i.e. those defined by the need to negate emissions output).

(For example India at 1 per cent of 1950 global output and regressed at 1 per cent of all previous years in the reconstruction, becomes responsible for one million tonnes of carbon from fossil fuel burning in 1860. Fossil fuel burning was largely unknown in indigenous India in the middle of the 19th Century. And anyway, such as did take place was almost entirely servicing the needs of the colonial power, and the case could be made therefore for adding this to the British total).

The global industrial per capita average from 1860-2025. It also shows this curve reduced by 60 per cent, ie consistent with the atmospheric "stabilisation" cut suggested for 1990 output levels following the IPCC and the "constant airborne fraction" theory. This is the benefit/disbenefit threshold (B/DT). It also shows the actual per capital curves (expressed in metric tonnes of carbon-TC) for US, UK, India and China for comparison with the benefit/disbenefit threshold and reconciliation accounting. Accumulating benefit and disbenefit in these countries. Benefit is expressed as a positive amount and disbenefit as negative. Positive reflects an increasing share of ecological debt, and negative a declining share of conventional wealth. Emissions trading has been proposed. Emissions quotas would need to be allocated consistent with reconciling these differential "faits accompli" as well as prospects. Trading the quotas is of secondary relevance. It is primary forcing which must be curtaile.

A Global Energy Efficiency Strategy

A recent global energy study, found that improvements in energy efficiency through the year 2020 would allow sustained economic growth with almost no increase in CO_2 emissions from burning fossil fuels. Drawing on in-depth analyses of India, Brazil the USA and Sweden, the study assumes use of the most energy efficient technologies currently available or demonstrated. Under this efficiency scenario, energy-use in OECD nations declines while their economies expand. Meanwhile, energy use in developing nations more than doubles, allowing these nations to reach the same level of consumption and services as Europe experienced in the 1970s. It is interesting to note that if current CO_2 releases came equally from all people, global per capita CO_2 emissions would be about the same as in Spain today.

Because of a shift from coal to natural gas, slightly higher energy use in 2020 translates into marginally lower overall CO_2 emissions. Efficiency costs are not provided, but the study does detail energy efficient industrial processes, transportation, etc. Shifting to these more energy efficient technologies generally costs far less than building new power plants, so the pattern of investment detailed in this study frees capital from expenditure on inefficient energy consumption for more productive investment.

This study does not address the political and institutional obstacles to energy efficiency and in this sense is not a projection but rather a model of what could be achieved if the political will existed. Because energy models are generally based on past trends, the dramatic shift toward energy efficiency envisaged in this study has been dismissed as unrealistic. However, energy scenarios describe the evolution of future energy production and consumption based on plausible sets of future events. The threat of global warming makes it necessary and therefore plausible that we will find the political will to achieve a global shift to energy efficiency. The scenario above, therefore, maybe at least as realistic as any other model for future energy use.

The studies reviewed above demonstrate an enormous untapped potential to promote economic expansion in developing nations—by freeing up capital from unnecessary energy expenditure—and to slow growth in CO_2 emissions. Institutional barriers, such as energy subsidies, clearly can be surmounted either by legislation such as that under consideration by Indonesia's Directorate of Public Buildings, or by construction and training measures like the Philippine transport programme. Probably the largest impediment to developing nations achieving energy efficiency is the minimal financial and technical assistance to date from OECD nations for energy efficiency. Unless industrialised nations shift development assistance funding to support a rapid increase in energy efficiency in developing nations, growth in CO_2 emissions from these countries is likely to far outpace any conceivable reduction in CO_2 emissions by developed nations.

If a research-only strategy is unwarranted, given the high costs associated with waiting for scientific certainty, and an adaptionist focus is flawed, are there preventive measures that could significantly affect the greenhouse trend, that make economic sense, and that seem politically realistic? Each of these conditions is essential, and the criteria for meeting them should be stringent. Because of the scientific uncertainties, any preventive measure should be either inexpensive, given the context, or economically justifiable on its own merits.

There are, in fact, preventive measures that meet all three conditions. As Mintzer's analysis suggests, a policy based on increasing efficiency and phasing out CFCs would significantly slow the warming trend. It could delay the commitment to the equivalent of a doubled atmospheric carbon dioxide level for many decades beyond current trends, thereby buying time for the development of alternative energy supplies and for taking the adaptionist steps that will inevitably be necessary. This last is an important point. The two strategies are not mutually exclusive. Adaptionists point to energy conservation as a prime step, just as preventionists do, and some adapting will be necessary under even a fairly aggressive preventionist approach. There is nonetheless, a significant difference in policy emphasis between the two approaches, and an important choice to be made. Because of its global scope and long time scale, dealing with the greenhouse effect will take an unusual degree of political will. Certainly, a general consensus on the overriding goals for action—either to avert climate change or adapt to it—will be a necessary precondition.

Research in many countries and economic sectors confirms that energy efficiency improvements remain a very large resource, and by far the lowest-cost source of new energy supply at least for the short- and mid-term future. Although generally unrecognised, the potential for energy savings is as great in the developing countries as it is in the industrialised ones; poverty is as great a cause of wasteful energy use as is great wealth.

The poor who light their homes with a wick dipped in kerosene, for example, get one-hundredth the illumination of a 100-watt electric bulb for the same amount of energy. At the other end of the spectrum countries just now building large energy supply systems can capitalise on the insights gained in the industrialised nations over the last decade and save literally billions of dollars. For example, a detailed study of Brazil shows that an investment of $4 billion in high-efficiency technologies (such as refrigerators, lighting and motors) would supplant construction of 21 large power plants for a net saving of $19 billion between now and 2000. Cutbacks in fossil-fuel use that are achieved through strategies to conserve energy will also ameliorate acid deposition and related air pollution problems.

The first step in a global greenhouse strategy, then, is a concerted push toward national energy plans built on energy efficiency and, to the extent possible, fuel switching to the less carbon-rich riels, especially natural gas. At the international level this entails a major shift in bilateral and multilateral assistance programmes, which now do not encourage efficiency investments.

The second step is a prompt agreement to phase out CFC use. Analysis has shown that it would be technically easy and inexpensive to halt the current rapid growth in emissions and actually to cut use by one-third in about five years. The US Government has taken a step in this direction by formally proposing a production freeze at current levels and a phase-out over the next 10 to 20 years. As of this writing, there is considerable momentum for a far-reaching international agreement to reduce CFC use; the motivation in so far, however, has been dominated by the need to protect the stratospheric ozone layer. If the case for slowing the greenhouse warming could also be factored in, a phase-out agreement seems not at all visionary.

It is not too early to begin international discussions on the greenhouse effect and on long-range energy strategies. A scientific exchange on this subject between the Untied States and the Soviet Union is already under way and could rather easily be elevated to the political level. Considering that between them these two nations account for 40 per cent of global carbon dioxide emissions (and 52 per cent of global coal reserves), this would be a significant step. The United States should also seek to engage China in discussions of that country's energy options. China currently sees few alternatives to a heavy reliance on coal, but the government has also shown itself to be receptive to major policy shifts—if they will save money. Two other regional discussions should also begin—among the Organisation for Economic Cooperation and Development nations and among developing countreis. A conference exploring alternative energy strategies in developing nations would be a sensible first step, especially if it explored the foreign exchange implications of the various alternatives. Energy efficiency improvements generally, have the triple advantage of costing less, generating more jobs, and requiring less foreign exchange than new supply options. Given many coutries urgent need to ease their foreign debt burden, this last is a major advantage.

Finally, greater efforts must be made to slow tropical deforestation. Although the size of the contribution of deforestation to the warming trend is the subject of much debate, the possibility that the biota are a significant source of carbon dioxide, combined with the overwhelming human and ecological reasons for saving the tropical forests, makes a sufficient case for immediate, high-priority action. In the research realm, a far more vigorous research and development programme on nonfossil energy alternatives should be launched in the United States and elsewhere. Solar energy options need consistent long-term funding, as do the wide variety of technologies for increasing the efficiency of energy use. Improved nuclear reactor designs should be further explored on paper to see what contribution they can make toward solving nuclear energy's formidable economic, safety, waste, and proliferation problems. It goes without saying that basic research on the greenhouse phenomenon deserves the highest priority. And, finally further policy analysis of most of the issues raised in this discussion is badly needed.

If all this is begun in the next few years, prospects for averting an unacceptably large greenhouse warming might soon appear dramatically better than they do today. The task itself will be easier to accomplish now than later. And if, by some completely unexpected turn of events, the greenhouse warming disappears as a global concern, the world will still be better off.

Role of a Hegemon

There are those who subscribe to the theory that the presence of a hegemon in the international political system will contribute significantly to international cooperation. Labelled 'hegemonic stability theory', it 'predicts that the more one such power dominates the world political economy, the more cooperative will interstate relations be' (Keohane, 1984 : 34). Although taken from the field of IPE, it can be applied to our more general question about international cooperation. The theory starts from the premise that, given its superior hierarchical power position, the hegomonic state may be able to 'help to create shared actions by providing rewards for cooperation and punishments for defection' (Keohane, 1984: 78). The powerful state may undertake such action for one or both of two reasons. First, the hegemon may impose a regime upon the international political system as it is willing to bear an undue par of the short-run costs of these goods, either because it regards itself as gaining in the long run, because it is paid in a different coin such as prestige, glory, immortality, or some combination of the two (Kindleberger, 1986a: 8).

Or second, the hegemon may have some feelings of 'conscience, duty, obligation, or such old-fashioned notions as noblesse oblidge' (Kindleberger, 1986b: 845). For whatever reason, some argue that a hegemon (or, more commonly today, a hegemonic bloc or a condominium—i.e., a group of powerful states) may have the power resources with which to impose a cooperative order upon the international system. And regardless of whether the motivation is based upon feelings of domination or obligation, they argue that there is a congruity between the concentration of power and the chances for the realisation of cooperation. This is a much disputed claim and one that has yet to be adequately resolved. Yet even Keohane, who argues that hegemony is not a necessary precondition for cooperation, recognises that '[t]he dominance of a single great power may contribute to order in world politics, in certain circumstances' (Keohane, 1984: 46). Without explicitly endorsing hegemony, many others also recognise the ability of those possessing power to create specific arrangements (for example, Young, 1989: 350). In a related analysis, other scholars have taken Olson's conclusion about group theory and suggested that goals will be more easily achieved in small groups—thus arguing that a select condominium in the international political system might be able to reach an agreement and impose its preferences upon the entire system. In these ways, it is clear that a hegemon or a hegemonic block may be able to aid the goal of international cooperation. With regard to the case at hand, the possible role of a hegemon can be looked at in two ways.

First, one must wonder whether a single state could deploy the resources necessary to impose a regime in this issue-area upon the international system. Generally, since the world's power resources are becoming more and more widely distributed, the ability of any one state to impose its will is being curtailed. Accordingly, a hegemon, or even a hegemonic bloc, is not able to impose any collective good as easily as in earlier centuries, before the complexities of interdependence weakened the efficiency of military power. But does this generalisation apply to our case under investigation?

At the present time, the major producers of carbon emissions are the United States, the Soviet Union, the European Community and China (Grubb, 1990: 73). Given the fact that production is fairly concentrated (these four actors produce more than one-half of the total global emissions (Grubb, 1990: 73)) one might think that the prospects for environmental preservation rest upon the ability of these four actors to come to some sort of agreement and impose it upon the international political system. But the nature of this issue is such that this possibility is not realistic. Unlike traditional indices of power (which are usually used to measure the presence of a 'hegemon'), it is relatively easy to become a major producer of greenhouse gases and thus join the ranks of the 'powerful' on the issue. And therefore any such hegemonic bloc would be considerably less stable than is normally the case. For example, what is a hegemonic block of these four actors today may be a group of different players tomorrow. Not only does this possibility weaken the case for hegemony but it also carried another and more important implication: it will ensure that the developing world is involved in any future negotiations because these states, with their potential to produce significant amounts of greenhouse gases, hold a significant 'bargaining chip' in the process.

Thus, the use of 'hegemonic stability theory' in this first sense—though somewhat applicable in other areas of IR research—will not be the most telling catalyst for international cooperation. There is, however, another form that a hegemon could take. One could envisage a hegemon as a 'crusader' state—a state which acts as a mediator and acquires special status because other states act on its initiative. This possibility will be taken up in the next section.

The concept of the 'political entrepreneur' is usually applied to individuals in a traditional domestic political science context, where it refers to a person who for 'career reasons, find[s] it in (her/his) private interest to work to provide collective benefits to relevant groups' (Hardin, 1982: 35). Nevertheless, this concept can, and has, been extended to IR; Young has observed 'the crucial role of entrepreneurship at the international level' (1989: 373). He also commented that:

> leadership in connection with the formation of international regimes is a matter of entrepreneurship; it involves a combination of imagination in inventing institutional options and skill in brokering the interests of numerous actors to line up support for such options (1989: 355).

At the global level, an entrepreneur can be defined as an actor that accepts the task of facilitating international cooperation in return for private benefits. These benefits may include intangibles such as an increase in international prestige or the returns dérived from acting fors simple altimistie reasons. On the advantages may come in the form of tangibles such as benefits in matters outside of the issue-area of the potential regime—in either international or domestic arenas. The political entrepreneur is willing to accept the initial organisational costs because it is antipated that these burdens will be smaller than 'the net discounted value of the benefits that the entrepreneur expects to capture for itself' (Keohane, 1983: 155).

With regard to IR, the application of this concept has usually been undertaken in one of two closely related ways: representing an individual person as an entrepreneur or representing an individual state as an entrepreneur. Though they are based.upon the same theory, let us look at each separately. Keohane has noted that '[i]nstitutional development is affected by particular leaders' (Keohane, 1988: 389). The global warming issue has been a case in point since individual international states people have certainly played a role in the issue. Gro Harlem Brundtland (former Prime Minister of Norway), by means of her involvement with the World Commission on Environment and Development and her participation in many international conferences, is one obvious political entrepreneur in this issue-area. There are also other heads of government who, by making key statements on the issue, seem to be starting to emerge as political entrepreneurs—Canada's Brian Mulroney and Britain's Margaret Thatcher are two (rather unexpected) examples. While we are considering the role of individuals, we must look beyond simply the leaders of states because other powerful individual personalities can also act as catalysts for cooperation. Mostafa Tolba, the energetic executive director of the United Nations Environment Programme (UNEP), has contributed significantly toward fostering cooperation. This list could continue by citing various entertainment personalities, pressure group leaders and others.

While at the national level, a number of states have been active as political entrepreneurs. As a point of departure, one need only examine the large number of conferences that have been recently hosted by various countries in order to recognise that many states are attempting to become political entrepreneurs—Canada's June 1988 Conference on the Changing Atmosphere; the Netherlands' meetings of 1989; and the various meetings of the IPCC are the most important in this regard.

Thus, it is clear that, for whatever reason, a number of entrepreneurs, both individuals and states, are starting to emerge with regard to the global warming issue. The extent to which this trend continues will play a determining factor in dictating when an agreement will be realised. Each given level of warming corres-ponds to a requirement to reduce greenhouse gases relative to some starting point, or alternatively, to invest resources in adjustments to a warmer world. Thus various 'mixes' of preventive and adaptive measures can be employed. Adaptation will include government-induced measures and 'natural' adaptation-ie actions undertaken by individuals in response to warming. Such measures will typically involve private costs for those making the adjustments. The overall aim should be to minimise the sum of the policy and private costs for any given level of warming.

Analysing policy options to counter the greenhouse effect is not a simple task. The challenge is how to make decision in the face of irreversibility, uncertainty and risk over long periods of time. The irreversibility of climatic change arises because past emissions of greenhouse gases ensure that the world is inevitably committed to some future warming. If atmospheric concentration of greenhouse gases were stabilised now, the climate would continue to warm for decades to come as the oceans released heat that they previously absorbed. If present emission continue, by the year 2030 the world would experience an actual temperature change of 1.0^0-2.0^0C warmer than today, with the ocean lag effect committing us to an eventual increase of 1.5^0-3.1 C warmer than today.

These difficulties in projecting climate change scenarios accurately and in predicting the resulting impacts in turn add to the risk and uncertainty surrounding the greenhouse effect. Risk relates to broadly quantifiable probabilities, eg the likelihood of a drought. Uncertainty relates to contexts in which probabilities are not known. Clearly, there are great uncertainties involved in global warming—scientific uncertainties over the precise atmospheric, geographical and climatic responses, forecasting uncertainties in predicting future changes and the scale of their effects, and finally, time-lag uncertainties over the urgency in divising appropriate policies to contain global warming.

Estimating the magnitude of impacts is unaffected by the presence of risk and uncertainty, but translating these magnitudes into decision-criteria is affected. Risk may be treated through expected value approaches, but this is unlikely to be satisfactory if the pay-off is a catastrophic event (eg major flooding by sea water). Uncertainty can be treated by decision-theory in which mixes of strategies and pay-offs are addressed according to the degree of risk-aversion.

Uncertainty can be reduce through the acquisition of more information, and in some instances improved research techniques, over time. Indeed, the informational approach to uncertainty is frequently used to justify a 'wati and see' response. But climatic change is irreversible, so that there is likely to be an additional cost associated with waiting to respond until more information is available. By the time we are sufficienty certain of the risks associated with climatic change, we may be already committed to a future temperature rise that increases the probability of the impacts actually occurring.

Once again, the proper approach is a cost-benefit one. The costs of 'doing nothing' need to be compared to the benefits. The costs are those associated with the additional committed warming due to delayed response. The benefits are the avoided costs of prevention and adaptation. Discounting, if permitted, will affect the calculation since avoided cost of action occur now, whereas the costs of delay occur in the future.

We can assume that society is risk-averse; that is, society prefers obtaining a sure income of, say, US$100 million over an uncertain expected income (ie the average income over all future probable states) of the same amount. In the case of global warming, the source of the risk is on the supply-side: society is certain of demanding the services of the environment in the future, but the prospect of changing climate and seal-level rises means that we are uncertain about the state of the environment in the future. Bishop demonstrated that, under such conditions of supply-side risk, society would be willing to pay a sum to avoid the future risks of climatic change—an option price (OP), which is greater than the expected loss of social welfare from all these risks (ESL). This risk-aversion premium is the option value (OV) to society of reducing the risks associated with climatic change.

Thus,

$$OP = ESL + OV$$

and (1)

$$OV = OP - ESL > 0$$

The positive option value on the supply-side may indicate the need for investment in uncertainty-reducing activities. This is clearly the case in the presence of general uncertainty (as opposed to the estimable risks) over the effects of global warming, where the expected value of new information, or quasi-option value, to society must be extremely high. Nordhaus recommends that the appropriate policy response in the face of uncertainty is to use the 'best guess' and proceed as if the world is certain, while at the same time assuming that the high expected value of new information requires that 'the best investment to ameliorate the CO_2 problem today is probably to expand our CO_2 knowledge'.

But in analysing the stream of future benefits and costs and every policy option (and consequent global warming scenario), we can go further than that. As the impacts of climatic change involve pure public risk, this risk cannot be pooled across individuals in society because one individual's assumption of risk does not reduce the risk borne by others. The policy implication is that a risk premium should be assigned to the evaluation of the risks associated with global warming such that when these future environmental risks are costed, these costs attain greater importance to the future risks associated with climatic change.

One option would be to use a lower, risk-adjusted interest rate to discound future costs in terms of their present values. But as indicated in Equation (1), in the presence of supply-side risk, society is already demonstrating its willingness to pay to avoid these future costs, through which the risk aversion premium, or option value, can be measured. Given the long-term nature of climatic change, the main losers in the case of the greenhouse effect are most likely to be future generations. Thus the issue of climatic change also touches on question of intergenerational equity.

Concern over intergenerational fairness, and ultimately 'sustainable' development, may require compensation of future generations for the irreversible loss of natural capital today. In the context of the greenhouse effect, pollution by current generations will harm future generations, at best by imposing risk, and at worst by causing serious adaptation problems. Consequently, the choice is between climatic change with actual compensation of future generations and climatic change without compensation. Moreover, if the compensating investment is not made in the present, compensation is not an option in the future.

How would such compensation be calculated? If we knew the actual damage, Y, associated with climatic change, we could set aside a sum of money, X, today which would grow at some interest rate, r per cent to be equal to Y in the relevant future year. That is,

$$Y = X(1 + r)^T$$

However, as we do not know T or Y, the computation of the required compensation cannot be made. Some writers suggest a proxy in the form of the sum that current generations would be willing to pay to prevent future risks of the damage Y, ie the option price embodied in Equation (1). Estimating option price may be extremely difficult if, as in the case of climatic change, the future is expected to include low-probability, high-loss events such as natural disasters. However, Brookshire *et al* have measured expected damages from earthquake losses in California and individuals self-insurance through purchasing 'safe' site housing, which suggest an implicit option price. Similar approaches could be extended to the risks associated with climatic change.

This discussion suggests that compensation of future generations can take two forms: direct compensation by the present generation in the form of final goods, capital and technology, or indirect compensation in the form of willingness to pay now to reduce future risks and to accumulate knowledge to reduce general uncertainty. Direct requests of final goods are problematic, because the preferences of future generations may be different from current preferences. In terms of capital, there is a strong argument that the minimum criterion for any compensation for the future is that they inherit a total stock of welath no less than that enjoyed by the current generation. This means that, if some environmental assets are to be depleted as a result of climatic change, they must be compensated for by current generations building up other forms of wealth. Of particular concern would be the excessive depletion of essential life-supporting natural capital-tropical forests, coastal zones and biological diversity.

If global warming results in an irreversible and non-substitutable loss of capital, then indirect compensation is the best means of ensuring intergenerational equity. Cost-effective preventive and adaptive measures today, coupled by increased research and knowledge of climatic change, are the best compensatory investments for the future. Moreover, many of the actions might generate other benefits, increasing the benefit-cost ratio of taking action now. Energy conservation is a conspicuous example-the benefits include energy savings and acid rain reduction.

6. Towards International Greenhouse Agreement

ACTION BY single countries to combat global warming will be ineffective except through a 'demonstratioin effect' to other countries. International agreement is necessary and will need to come to grips with a major concern: the design of incentives to secure cooperation.

There are several reasons for thinking that compensatory financing will be necessary as an incentive for the developing world to cooperate. First, the greenhouse effect may have its most dramatic impacts on many developing countries. Sea-level rise could be potentially catastrophic for low-lying and deltaic countries such as the Maldives, Guyana and Bangladesh. Even the agricultural and ecological impacts are heavily weighted against the developing world (as noted above). Second, developing countries are in no position to afford the measures needed to adapt to and mitigate the effects of climatic change. Third, tropically forested developing countries have a special role to play in conserving forests as carbon 'sinks'.

The implications for rethinking aid and monetary transfers and therefore formidable. The debate has centred on first, mechanisms to induce tropically forested countries not to deforest; second, the scale of foreign aid required from industrialised countries; and third, the creation of a 'carbon fund' through which international transfers might be made. All these approaches reflect the wholly correct idea that the greenhouse effect is the equivalent of a 'game with a large number of players. In order to secure a cooperative solution it is necessary to make 'side payments' to induce unwilling partners to enter the game. Unwillingness in this context may simply reflect an inability to do anything significant because of lack of resources. But it will also unquestionably reflect game strategies by poor countries to acquire the largest sums possible from the developed world in return for forest conservation or for foregoing some of the industrialised world's technologies. This game is already being played in international arena on the Montreal ozone protocol, with China and India particularly looking for major technology transfers and cash compensation.

An interesting presumption has been that a global warming convention must closely follow the Montreal ozone convention. But Montreal may be a poor precursons for a wider warming convention. For one, the number of key countries involved in the global consumption of CFCs as well as the overall economic costs of reducing their use is relatively small, thus facilitating the negotiation of a cooperative solution. Moreover, there is evidence that, prior to the Montreal agreement, some of the more important producers and consumers of CFCs (eg the USA, the EC) already intended to take substantial unilateral action.

Consequently, the conditions existed for an international agreement on CFCs limitation, but the protocol may have been less significant than originally thought for achieving reductions in the global levels of CFCs.

In contrast, in the case of the greenhouse effect, the lack of incentive for any one country to take unilateral action, coupled with the limited effect of such unilateral reduction on global emissions, means that full international cooperation is necessary for major gains in global welfare. Unfortunately, the number of countries required to cooperate is large, which reduces the likelihood of agreement. In addition, for each country, the costs relative to the benefits it receives will be high: the additional costs of abating greenhous gas emissions will rise very steeply as fossil fuels must be substituted for and energy is conserved. Additional complications are that a global agreement on reducing greenhouse gases would have to take into account not just the limitations on fossil fuel use but also on global deforestation and other human activities (eg agriculture) contributing to emissions of these gases. This will require cooperation by both industrialised and developing countries. Thus, unlike the ozone case, the costs for abating all greenhouse gases will vary significantly across countries and emitting countries, making international compensation necessary, especially for the developing world.

The complexity and scale of the constraints on international action to reduce greenhouse gases are gradually being appreciated. A Ministerial Conference on Atmospheric Pollution and Climate Change in the Netherlands has called for a two-phase approach. The first phase requires primarily industrialised countries to abate their emission of greenhouse gases so that world levels are reduced by the year 2000 by 30 per cent, relative to 1986. However, reduction by OECD countries alone are not considered sufficient to deal with the global warming problem. A second phase, emphasising global cooperation on joint intenational action, is required. The objectives would be to continue reductions in emissions from fossil fuel use in industrialised countries, phase out CFCs world-wide and reduce deforestation while stimulating reforestation globally. The establishment of an international funding mechanism (the 'carbon fund' idea) with initial funds of US$500 million a year, is considered essential to these objectives.

Lagging behind these movements toward a global warming convention has been the discussion of appropriate policy instruments to achieve reductions in greenhouse gas emissions. Part of the problem is that environmental policy as it has been practised to date has been mainly in the 'classical' mould of standard-setting and emissions monitoring. Where charges and subsidies have been employed, they have been largely used for 'administrative' purposes or to raise revenues rather than as incentives to reduce pollution.

However, the standard setting approach is inadequate for dealing with the uncertainties of the greenhouse effect. These require a policy regime based on a enforceable upper limit on carbon emissions, imposed through an incentive system that is flexible enough to adapt to changes in that limit as new information becomes available. Hence greater reliance on variable, or market-based, incentives is called for. Both the use of 'carbon taxes' and marketable permits may be applicable. A carbon tax on energy would be levied proportionally to the potential carbon content of fuels, thus providing a direct incentive to all users to reduce their carbon emissions.

Even more flexibility may be built in to a marketable permit system: overall emission limits are set and permits to emit carbon up to these limits are allocated to polluters, who in turn are allowed to engage in permit trading. Carbon polluters with high abatement costs might prefer to buy permit, whereas the low abatement cost polluters might prefer to sell. The overall emission limits could be increased by the authorities issuing new permits, or alternatively, reduced by their intervening to 'buy back' permits. Consequently, new innovative approaches to economic instruments are needed if the world is serious about the need to reduce significantly its carbon emissions.

The cost-benefit mode of thinking offers a framework for analysing global warming policy issues. This remains true even in the context of both the impossibility of monetizing all costs and benefits and the extreme uncertainty about the timing and scale of impacts. The prime task is to determine an acceptable level and rate of warming, which in turn means having a much clearer understanding of the costs of control measures and the damage costs of impacts. The appropriate criterion for determining the policy mix is cost minimisation, which must encompass the costs and benefits of delayed action. Traditional approaches to uncertainty may not be adequate for evaluation of risks. Above all, the analysis suggests a premium on action now rather than later. This reflects the results of risk-averse approaches to uncertainty in which outcomes can be catastrophic, and the multiple benefits from some policy options such as energy conservation.

Finally, global warming can be dealt with only through international agreement. The context is one of game theory, and the pressing need is to design incentive systems for global cooperation. The Montreal protocol on ozone may be a poor guide to the prospects for a greenhouse agreement. The most urgent need is to develop appropriate policy instruments and compensatory mechanisms. The construction of a resilient, global greenhouse gas regime requires that state elites share a consensus as to common values and norms of behaviour; submit their respective states to observe the rules and procedures of the regime; and participate in the institutional arrangements established under the regime.

The signing of the Climate Change Convention is *prima facie* evidence that most national leaders agree that the global climate system must be conserved and restored, and that it must be used rationally in ways that are compatible with ecological imperatives. The rest of the interlocking elements of the regime have been left largely unspecified. Some analysts believe that states so hedged their commitments under the Convention that it merely maintains 'polluter sovereignty' and is so vague as to be almost meaningless.

One is less pessimistic, however. One is convinced that a fully fledged regime can be constructed that will achieve the goals of the Climate Change Convention. Achieving this goal will first and foremost mean activating the norms implicit in the Convention as to mutual reciprocity between, and differential responsibilities of, the rich and poor nations. Specifically, the articles on financial assistance and technology transfer must be elaborated in protocols and implemented.

Donors may place conditions on financial transfers to the South to abate greenhouse gases. One outlines the contrary positions held by potential parties to a Climate Change Convention on conditionality and additionality of resources for greenhouse projects in developing countries. The biggest likely demand on resources provided by the North to the South will be to fund technology transfer. The desirability and even the meaning of technology transfer are contentious, however. One may outline two important positions that have emerged in debates on this topic before turning to three aspects of multi-pronged technology transfer strategy for greenhouse gas reduction. These are cost reduction; technical assistance and training; and information dissemination and transnational research collaboration. For these commitments to be implemented, parties will require confidences that other beneficiaries under the Convention are not cheating. Procedures must be established to monitor and to verify compliance with convention commitments along with mechanisms to resolve disputes and to enforces compliance.

Transferral of financial resources and technologies on a scale envisaged in this study would transform North-South political-economic relations. One may conclude by asking whether 'ecological' issues such as climate change may strengthen the South's bargaining position in its geo-political and geo-economic relationships with the North. My answer to this question is only a qualified 'possibly'. Nonetheless global greenhouse regime may succeed for three reasons:

1. the likelihood of continued technological innovation and associated reduction in the cost of emission abatement;
2. the contribution of science on comprehension of the greenhouse effect;
3. the influence of social movements on governmental policy.

The language of the Convention implies- although it is no where specifically stated that the commitment of parties of the Convention should encompass the costs of actually reducing greenhouse gas emissions in developing countries. (As developing countries carefully avoided committing themselves to abatement at this stage, they did not obtain a precise matching commitment from the developed world to fund their abatement costs but only vague statements of intention.

A variety of conditions on providing or accepting financial assistance to abate greenhouse emissions or to emeliorate the impacts of climate change have been advanced. These include the possibility that donors will tie the aid to greenhouse abatement activities and/or require that recipients reform their energy prices and institutional structures; and on the part of possible recipients, that acceptance of such aid in no way infringes on the exercise of national sovereignty in determining the best use of development assistance and that its provision in no way reduces existing flows of official development assistance.

Some analysts have argued that transfers to the South ought not to be for any purpose. Michael Grubb, for example, suggests that the transfer should only be convertible into provision of technical assistance and equipment needed to abate greenhouse gas emissions, and not cash. Conversely, the Group of 77 have argued that funds provided for incremental abatement costs 'will be to a great extent of a compensatory nature.'

There are strong arguments that compensatory payments should be linked closely with greenhouse gas reduction activities. Most important, the resource transfer calculated was based on an obligation to-pay index applied to marginal carbon abatement cost argues. This estimate did not define marginal benefits of avoiding greenhouse gas induced climate change. Thus, it does not represent an estimate of compensation that the North might owe the South for hang pre-empted atmospheric space or for climate change induced damages.

In practical terms, it is also impossible to calculating monetary values for compensation. There are other advantages to providing compensation in the same dimension as the images imposed by climate change. An in-kind approach makes explicit the nature of the international transaction. It thereby avoids any connotations of buy-off of recipient elites.

From a pragmatic perspective, technology transfers financed by funds placed at the disposal of developing countries would increase economics of scale in supplier countries, thereby reducing the cost of supplying the aid in the first place. Linking compensation to specific uses may also help to circumvent elite corruption in recipient countries. It could also increase political support in donor countries.

However, linking resource usage reaches its limits when dealing with the most vulnerable states where urgent local development priorities require that resources be applied across a variety of economic and welfare projects, not limited to greenhouse projects. The most advanced expression of this non-linkage is the proposal for an insurance fund for the most-affected island and coastal states. Some mix of compensatory and linked-compensatory resource transfer therefore seems inevitable.

Other conditions that might be placed on the transfers studied above include sectoral policies aimed at reforming energy prices or institutional arrangements in a recipient country. In China, for example, coal prices bear little relationship to supply and demand. In the Soviet Union, subsidised natural gas prices foster a high rate of leakage of methane out of transmission pipes. Conditions related to project self-financing by the recipient state, to environmental performance, and to expanding the role the private sector (including permitting foreign investment in abatement activities) might also be included by donors. If financial and resource efficiencies are to improve in developing country energy utilities, many deep-seated causes of poor institutional performance must be rectified, including: overstaffing; inappropriate skill mixes; shortages of middle level and technical staff; low wages; rigid and politicised hiring and firing practices; political interference, graft and corruption in procurement and billing activities; and inadequate training facilities.

As international energy expert Russell de Lucia puts it, The primary problems are associated with institutional matters and market structure not technology or know-how. In short, the energy sector in developing countries is often so irrational and inefficient that donors will be very reluctant to provide substantial technical or financial assistance unless prices are revised upward to reflect cost, and energy utilities are privatised. Developing countries assert strongly that they will not accept greenhouse-abatement financial assistance unless such transfers add to rather than substitute for current official development assistance (ODA) flows. The ODA recipients are concerned that existing ODA will be diverted from local development priorities to global environmental concerns, to redress problems created by the donor countries in then first place. They also hold that it is impossible upto distinguish between development projects that benefit only their own country, and those that damage or restore only the global environmental protection.

'Additionality', however, is a deceptively simple word. ODA flows fluctuate from year to year in most donor budgets; some donor countries have announced a gradual increase in ODA up to about 1 per cent of annual GNP; others are allowing their ODA to slide to even lower levels. There is no simple (or even complicated) way to ascertain what would have happen to ODA if fanners do not fund greenhouse projects in recipient countries. Furthermore, many of these projects are justified in traditional developmental terms with or without consideration of global environmental concerns of the donors. The issue is largely symbolic, therefore, with the United States declaring in 1991 that its contribution to one important mechanism for greenhouse funding, the Global Environment Facility, comes from existing ODA flows on the one hand; and Norway creating a new budget line separate from existing ODA to fund the Facility on the other.

Defusing this issue requires two things to happen. First, overall ODA levels (minus constant or increase. Second, greenhouse-related projects in China, India and elsewhere (especially those funded by the GEF) should deliver enough developmental benefits to allay fears that greenhouse related projects are an environmental diversion that benefit only the donor states. Countries like China have also insisted that they will only participate in a Climate Change Convention that does not impinge on their national sovereignty. This stipulation preseumbly includes retaining control over setting of priorities for the use of external assistance in domestic development projects.

Finally, donor countries may insist on using bilateral rather than multilateral aid mechanisms to transfer greenhouse abatement assistance. In part, this preference would arise from the sheer scale of the transfers discussed above that would exceed the total current United nations budget and fears that a bloated, inefficient international bureaucracy could not hope to meet the challenge in a timely fashion. Again in part, it would follow from the pursuit of narrower national interests in tying the aid to their own suppliers of greenhouse abatement goods and services and in traditional bilateral aid relationships. In so far as greenhouse projects are funded multilaterally, donor countries prefer the World Bank's Global Environment Facility rather than setting up a new mechanism for climate change (as suggested by China). Developing countries have objected strongly to the non-representative and World Bank dominated decision-making system at the Facility. Given the strong statement in the Conventionon reforming the GEF, it seems inevitable that either the South must be allowed to participate in the decision-making at GEF on an equal basis, or bilateral funding will dominate the global climate change arena.

The promise of energy efficiency

Policy discussions concerning global warming have been based on the assumption that CO_2 emissions from developing countries cannot be limited without slowing their economic growth. The promise of energy efficiency is clear. However, only a correct shift in development funding and investment to support energy efficiency would allow this promise to be fulfilled. This paper shows that such an assumption is wrong. In fact, these nations can limit fossil fuel use and CO_2 emissions while at the same time expanding economic growth. The path to this is energy efficiency, which allows resources to be shifted from wasteful expenditure on energy to more productive sectors. The result is less CO_2 is emitted while Third World nations benefit from more rapid and sustainable economic growth.

Towards the close of the 1980s global warming became generally recognised as the most serious environmental problem facing the earth. The spectre of a rapidly warming planet has prompted us to begin searching for ways to cut back emissions of greenhouse gases, including carbon dioxide (CO_2), methane (CH_4), nitrous oxide (N_2O) and chlorofluorocarbons (CFCs). Reducing emissions of CO_2, responsible for roughly half of global warming, will be far more difficult because CO_2 emissions come largely from the burning of fossil fuels; coal, oil and natural gas, which are so central to our economies and our lifestyles. OECD nations are just beginning to face the challenge of curtailing our dependence on fossil fuels. To date, the most cost-effective and promising path to limiting the use of fossil has proven to be reducing the amount of energy we waste.

The free market response to higher energy prices in the wake of the 1973 oil embargo was unexpectedly positive—thousands of innovations were devised for using energy more effectively. Driven by higher oil prices, a huge range of more energy efficient appliances, vehicles, and industrial processes have allowed OECD nations to continue robust economic growth with little overall growth in consumption of fossil fuels. These energy efficiency gains now save the USA some $160 billion/year. Large, as yet untapped potential for energy efficiency suggests that efficiency gains could outpace economic growth in the coming decades, allowing OECD nations to cut CO_2 emissions while achieving continued economic expansion.

But regardless of the steps taken by Western nations to reduce CO_2 emissions, these efforts may well be overshadowed by increased in CO_2 emissions from developing nations. The portion of CO_2 emitted from fossil fuel burning by developing nations has quadrupled within 40 years, growing from just 7 per cent in 1950 to 14 per cent in 1965, and 28 per cent in 1987. Already, just three developing countries—China, India and Brazil—emit more CO_2 from burning fossil fuels than all of Western Europe. Between 1980 and 1987 global CO_2 emissions from developed nations were unchanged, while developing country CO_2 emissions grew at 5 per cent year. At these rates of emissions growth, developing nations will be responsible for more than half of global CO_2 emissions within 20 years. This would more than negate any conceivable efforts by industrialised nations to cut CO_2 emissions.

Meanwhile, despite very large investment in expanding energy production, more and more developing nations face power shortages that limit economic growth. About one quarter of public spending in developing countries and about one fifth of development assistance goes to the energy sector, yet funding is increasingly inadequate to meet the fast growth in energy demand in the Third World.

This is in large measure because developing nations have not shared in the energy efficiency gains (and huge financial savings) made in Western nations. This study seeks to show that a range of energy efficiency options in developing nations are not only far less expensive than expanding energy supply, but also offer the most promising path to limiting CO_2 emissions. The fact that OECD gains in energy efficiency have not been transferred to developing nations is, in part, a measure of failure of development assistance to date. But concern over global warming is prompting a new look at energy efficiency as a strategy to make energy affordable for developing nations while limiting CO_2 emissions. The potential is enormous and the need is urgent.

Those who are concerned about the consequences of the "greenhouse effect" overlook the benefits that would follow from the general adoption of a vegetarian diet. If there were a halt to the consumption of grain-fed livestock products (while retaining the present pasture lands to support animals for meat and milk), at a *per capita* grain consumption of 200 kg per year for a healthy vegetarian diet, only about 60 per cent of the present land under cultivation would be needed to feed the world population. With the further modernisation of techniques in developing countries, that 60 per cent would be more than sufficient to feed the projected world population in the next century. This would not only stop the deforestation that contributes substantially to the increase in atmospheric CO_2, but would enable us to refoest 40 per cent of the present agricultural land, leading to large-scale absorption of atmospheric CO_2. Such an extensive mopping-up of carbon dioxide by young growing forests may be the only way to prevent the worrying consequences of the greenhouse effect, at least until alternative energy sources can effectively replace fossil fuels.

Legislation to effect a change in our dietary habits may be difficult to achieve. But fortunately, it is step that we could fruitfully take at the individual level without having to wait for government action, the laws of supply and demand would go a long way towards making sure that reduced meat demands would decrease the pressure on land and pace the way for rapid afforestation.

It is disturbing that you should publish such a prejudiced review (*Nature* 320, 318; 1986) of Sonny Kleinfield's book, *A Machine Called Indomitable,* containing derogatory remarks about Raymond Damadian, whose struggles have given the world a powerful tool for medical diagnosis. The reference to Damadian as a "misfit" does harm to a journal with a reputation for even-handedness and fair play. In his review Peter Newmark seems to take popular acceptance as the sole criterion for scientific excellence. If this were true, Copernicus, Lavoisier, Darwin, Planck and Einstein were all "misfits", rejected by their contemporary peers.

Newmarks's comment that I, though successful in casting a spell over Damadian, nevertheless "remain a voice in the wilderness", is also uncalled for. I suggest he reads the review (*Nature* 311, 682; 1984) of my book, *In Search of the Physical Basis of Life,* where the reviewer, Dick, though a long-time scientific opponent of mine, stated with fairness and general accuracy: Since the 1950s majority views of the cell have moved significantly nearer to those of Dr. Ling.....".

At first, Charles Owen admits, the goal was simply to win a design competition. But developing a concept to combat global warming came to mean more to him and his students at the Illinois Institute of Technology (ITT) in Chicago.

"It became a mission to raise public awareness," says Owen, a professor at ITT's Institute of Design. "We could do something about global warming, and if we do it right, we could help the world economy as well as take steps as an insurance policy in case it's as bad as some scientists predict." To fulfill that mission, ITT graduate and fourth-year students developed Project Phoenix, named in the hope that Earth could arise from the ashes of fossil-fuel burning as renewed as the mythical bird. The massive two-pronged plan proposes to reduce fossil-fuel use to decrease the generation of carbon dioxide—one of the chief gases contributing to the greenhouse effect. ["Global Warming," Aug., Sept., and Oct. '89]. Then, to remove carbon dioxide from the atmosphere, it posits regreening deserts and adding vast green areas to oceans.

The students began work in 1988, using research data from published scientific papers. By that time, entries from Owen's class had won two out of three of the prestigious International Design Competition awards in Osaka, Japan. ["The House That Computers Built", April '84; "Water Works", Nov. '88]. While Project Phoenix hasn't won a competition, Owen hopes it will help resolve the greenhouse war. Global warming is caused by atmospheric gases—the greenhouse gases—that trap heat. Many experts believe an increased greenhouse effect could lead to climate shifts and rising coastal water levels. Scientists estimate that 225 billions tonnes of carbon are removed annually from the atmosphere by forests and oceans. However, 228 billion tonnes of carbon area added, of which 5.5 billion are from the burning of fossil fuels and 2.2 billion are released by deforestation. Compounding the problem is the rising need for electrical power, now generated mainly by burning goal, oil, and gas.

Replacing fossil fuels with other energy sources, such as solar, seems a likely part of the solution. But because solar energy generation is confined to daytime and further limited by cloudy days, Project Phoenix suggests transferring the bulk of the power production to solar-power-generating satellites. In geosynchronous orbit 22,245 miles above Earth, satelites 5.6 miles across would beam 10 gigawatts of energy to collectors on Earth in the form of microwaves. The huge satellites would be constructed in space. "Your wouldn't build them on Earth and then take them up," says Owen, "That's a logistics nightmare.

"So we looked to the moon as a source for materials," he continues. "If you do that, you need to get people back and forth to the moon for maintenance, repair, and assembly. And that makes a sensible place for the long-proposed space station—a way station between the moon and Earth." A reusable shuttle would to the station; another would travel from the station to the moon or satellites.

To build the satellites, lunar materials such as silica and alumina would be fired into an orbiting receiver. A lunar factory in orbit would then process the materials, growing silicon crystals in microgravity for photovoltaic cells, for example, Robots would do much of the satellite assembly. On earth, a receiving antenna would convert microwave energy to DC. The electricity would be distributed through a grid or use do electrolyse water to make hydrogen. To avoid interfering with radio waves and exposing people to the low microwave levels, the receiving antennas could be located in sparsely populated areas.

Solar-power satellites, such as these suggested by Project Phoenix, are a serious option for the future, say experts. "These ideas are not pie in the sky," explains Gregg Maryniak, executive vice president of the Space Studies Institute in Rocky Hill, N.J "We in the space science community are delighted that the students have found solar-power satellites, because it's the best option," he adds.

"The science is well understood; now it's an engineering challenge," agrees Peter Glaser, who introduced the idea in 1968 and is now vice president, space operations, at Arthur D. Little Inc., in Cambridge, Mass. "Developing solar-power satellites is like designing an airplane. Today, we're at the *Spirit of St. Louis* stage," says Glaser, referring to the plane Charles Lindbergh flew nonstop from New York to Paris in 1927. "If somebody at that time said, 'Can you build a 747' we'd probably say, 'Yes, but it will take a little while'.

The second half of the plan calls for adding green areas to absorb more carbon dioxide. Some 3.3 million square miles of once forested land is now barren for various reasons; 1.9 million square miles of this total are now unused. "We'd put two million square kilometers [772,000 square miles] under forestation, which would absorb a billion metric tonnes [1.1 billion tonnes] of carbon annually," says Owen.

Just as once lush areas can become deserts, reasons Owen, the process can reverse. "By protecting plants until they can live on their own, you can get the system to work for you," he says. Hexagonal air-supported plastic domes, about 820 feet on a side, would shelter developing ecosystems. "When the domes are removed, the plants will absorb heat, aspire and produce water in the air column above," says Owen. "This will get the temperature down below the dew point so you'll get rain again". Under the domes, native desertlike plants will grow first. "Ten years down the line, when you begin changing the area's climate, you could introduce plants that live in a more temperate environment—plants that would naturally live on the periphery of that particular ecosystem," Owen says. Eventually, economically valuable crops could be grown. Power generated by solar arrays could help to irrigate the plants and crops by moving air and pumping water from underground aquifers or from nearby sources.

"In terms of harebrained schemes, greening the ocean is probably the wildest," admits Owen as he introduces the other half of the carbon-absorption scheme. In deep ocean water, huge areas are relatively empty of sea life. Project Phoenix envisions adding systems of large floating "wetlands"—covering as much as 386,000 square miles—

to absorb 0.6 billion tonne of carbon annually.

Mangroves (land-forming trees that grow in salt water), sargassum (floating algae), and mollusks will form one type of floating wetland. Mangroves would grow on an anchored net system; their interlocking root systems, clinging oyster clumps, and detritus would strengthen the structure. Like the mangroves, beds of kelp—a harvestable food—would be seeded in grid structures below the surface to make up the other type of floating island. Propeller pumps would bring up cool, nitrogen-rich bottom water.

The islands could provide another benefit. "Coastal wetlands are gradually becoming cement and harbors," Owen says. "Thses new areas could become a habitat for displaced fish and shellfish. "This idea isn't too farfetched, says Jack Rudloe, president of the Gulf Specimen Marine Laboratory in Panacea, Fla. "There have been lots of studies on artificial reefs," he says, adding that even crab buoys and oil rigs can become rich mini-ecosystems by giving marine animals a place to live. To draw awareness to global warming, Owen is sending copies of Project Phoenix to research centres, universities, and heads of state. "To avoid the kind of Middle Eastern conflict we've just gone through, we must no longer be limited by who controls an oil well or a coal mine," says Glaser. "The stakes are enormous." Toward this end, Glaser places hope in students such as those at the Illinois Institute of Technology. "What's most important is they become familiar with these ideas so later they can have an impact on the companies that will be involved in these projects."

There is such evidence for linking economic growth to pupulation growth. On the population growth side it has a long tradition of both growth-optimism and growth-pessimism. Karl Polyanyi stated, *"Natural rates of increase of population may actually be an index of either cultural vvitality or cultural degradation. The original meaning of the word 'proletarian', linking fertility and mendicity, is a striking expression of this ambivalence."*

Polanyi went on to state that in a market economy, *"poverty is nature surviving in society."* This cleverly exposes the self-fulfilling optimism of growth-economists and the technique of creating (but then ignoring or denying) externalities. This economic rationale even asserted that disbenefit is caused by nature, rather than by society in its pursuit of wealth and its creation. Nature could be called to account when some embarassing phenomenon such as 'poverty' or even 'population growth' (when linked to poverty) reared its ugly head. Just as 'conquest' of nature has been 'needed', 'conquest' of poverty became the inevitable consequential need as well. This has continued through to the present with conquest of poverty veering between alleviation and eradication.

What we find is no inclination to discuss poverty as a consequence of wealth creation. There is an insistence with 'describing' it (from a 'distance') and perhaps dealing with it' as 'an externality'. The following by Herman van der Wee is typical in this way of much of the literature. It seems to be 'objective', but 'subjectively lacking analysis', projects fallacy, *"Economic development in the Third World during the twentieth century and especially after the Second World War was accompanied by a big drop in the death rate while the birth rate fell much less. Even more than in the West, the demographic revolution led to a complete breakdown of the traditional balancing mechanisms, and the unprecedented population blast followed. He continues, "Western techniques to reduce deaths are more easily absorbed by the Third World than techniques to reduce births. This phenomenon is at the heart of the whole problem... The introduction of techniques to control marital fertility is a much more difficult matter. An efficient spreading and general application of control techniques demands the active participation of the population.* This has all the authority of a bill board. The fallacy projected comes from sustaining the implict denial of disbeneficiaries. It implies justification for growth optimism and the application of economics to externalities. In fact nature can take 'any form' for as long as it is 'externalies' and made the object of the 'attention' of a disembodied economic process. [An extreme and increasingly frequent example of the darker side is the effect of radiation on foetuses and the horrifying extent of biological deformity possible consequent on exposure to this. As Merleau-Ponty put it, *"Analytical though interrupts the perceptual transition from one moment to another, and then seeks in the mind the guarantee of a unity which is already there when we perceive. Anaytical thought also interrupts the unity of culture and then tries to reconstitute if from the outside."* From the very beginnings of the market economy Adam Smith was growth-optimistic and very specific about the linkage between economic growth and population growth. *"The liberal reward of labour. . . as it is the effect of increasing wealth, so it is the cause of increasing population. To complai of it is to lament over the necessary effect and cause of the greatest public prosperity.*

And a Andrew Tylecote shows, this growth-optimist theme has continued up to the present. *Two hundred years ago it was taken for granted—by Adma Smith, for example—that an upswing in economic growth would lead to a corresponding upswing in the growth would lead to a corresponding upswing in the growth of population. Birth-rates would rise, death-rates fall, and there would be an inflow rather than an outflow of migrants. But since that time we have seen, in the countries which have grown richer, that the birth-rate has fallen to an extraordinarily low level—so low that in spite of the similar fall in the death-rate, and at least some net imigration, these richer countries population have been virtually stagnant, while poorer peoples multiplied at an unprecedented rate. Does this not cast doubt on the relation between income and population? Certainly the long-term effect of economic growth on population is a complex one. Once it has had time to affect women's status in society, and people's attitudes to children, then the way is open for the 'perverse' [growth-depressing] effect of prosperity on fertility. But such shifts are not the immediate and automatic response to a few years' fast growth. When we look at the statistics, the old implicities turn out to be not a bad guide to population growth over the long wave. It is, and has been, generally true that in the short run an increase in the rate of economic growth makes for an increase in the birth rate and a decline in the death rate while a fall in the growth rate has the opposite tendency."*

This conclusion underwrites Adam Smith's point about the deterministic link between the two with economic growth as the forcing factor. However, it doesn't isolate population increase consequent on deprivation. Richard Douthwaite in "The Growth Illusion" (Resurgence 1992) however points to the linkage between economic growth and population growth via poverty. In the context of nineteenth century British industrial society Douthwaite cites a Dr J P Kay commissioned by Cambridge University to report on low living standards in Europe of the 1850s. *"A low standard of living always tends to stimulate improvident marriages, to unduly increase the numbers of the population and to engender pauperism, vice, degradation and misery."* This is the proletarian truism identified by Polanyi, embracing the emiseration defined by Marx.

This truism is now being revived by an increasing number of commentators but this time with the element of advanced pessimism to the fore over population growth. The Malthusian extent of this pessimism is to be found in innumerable 'stark' statements over the past decade, particularly in the run-up to the United Nations Conference on Environment and Development. The World Commission on Environment and Development made the following statement six weeks prior to the UNCED. It links poverty, environment and population growth and it is an advance to realise that out of self-interest, poverty needs alleviating on a global scale. *"The time has come for a massive attack on mass poverty, poverty, environment and population can no longer the dealt with-or even thought of—as separate issues; they are interlinked in practice and must be linked in policy formulation. Unless poverty is alleviated, there is little chance that we will be able to stabilise world population, which has grown by 500 million since the Commission last met.*

Human resource development, which is essential for sustainable development, itself requires effective population management through programmes that recognise the linkages between poverty and population growth.

This—albeit starkly honest assessment of the symptoms of the problems—does not attempt to explain what the cause of the poverty/population growth-loop is either. If anything it would imply that only a deficit of economic growth is the problem. In a conventional sense this is undoubtedly true. But it is a flawed observation because here again the causal links to the dominant paradigm of wealth creation remain unidentified. What is not acknowledged is what creates poverty. In other words it too doesn't ask Akio Morishima's embarrassing question.

Forty five heads of states (including many of the most populous, such as China, India, Indonesia, Bangladesh, Nigeria, and Egypt) signed a *"Statement on Population Stabilisation by World Leaders"* and presented it to the Congress of the United States on April 24, 1987. It included the following: *"At present there are 76 million more births than deaths on our planet each years. If present rates continue, by the year 2,000, there will be 100 million more births than deaths. A billion people have been added in the last 13 years and the next billion will be added in 12 years. Degradation of the world environment, income inequality, and the potential for conflict exist today because of over-consumption and over-population. If this unprecedented population growth continues, future generations of children will not have adequate food, housing, medical care, education, earth resources, and employment opportunities. We believe that the time has come now to recognise the world-wide necessity to stop population growth within the near future and for each country to adopt the necessary policies and programmes to do so consistent with its own culture and aspirations."*

The World Bank's "World Development Report 1992" also makes the link between population growth, poverty and environmental degradation, but doesn't task question. *"Rapid population growth, agricultural stagnation and environmental degradation have been common to most Sub-Saharan countries in recent decades. These three factors have been mutually reinforcing...Stagnant incomes and the absence of improvements in human welfare have impeded the demographic transition...There is strong evidence that economic stagnation is delaying declines in fertility."*

These statements draw attention to the population growth linkage to poverty and its feedback at least into environmental degradation. Remember the earlier statement:—"*environmental arguments for slowing human reproduction rates in the South pale before the truth that most environmental offences with global impacts are committed by rich minorities with hazardous production patterns, energy use, and consumption styles.*" The central question behind all these observations has to be;—what is the relationship between economic growth and the creation of wealth on the one hand, and population growth and the creation of poverty on the other?...but it doesn't get asked.

That's to say, it doesn't get asked by and within the dominant paradigm. The reason is obvious—you don't ask embarrassing questions. Economic-growth-optimists would say 'there is no linkage'. The dialectical position would be that one set is the opposite to the other, but 'in practice' still amenable to 'cure' through continued economic growth.

The forty five heads show an understandable (if regrettable) disinclination to bite the invisible hand which might smack them.

The paradoxical and deconstructionist position would simply be that since everything has its opposite, they are inseparably two sides of the same coin-that one is the inevitable consequence and companion of the other-that Caliban must inevitably recognise his own image (and the sooner the better). The paradoxical position is irreducible and therefore impossible to 'prove'. Proof requires externalities. The living proof of 'disbeneficiaries' will have to suffice. However, beyond all this deconstruction, we shall present some charts which attempt to show loss of quality and equilibrium. These are the beginnings of an approach to assessing quality and enfranchising disbeneficiaries.

It was predictable that the discussion of the so-called population question should have settled into uneasy consensus around the economic growth-optimist rationale of demographic transition. The politics of the Right and of the Left has been a contest between the "me-first" instincts of the right and the "me-too" reactions of the left since at least the French revolution, culminating in the near-miss fallacy of "me-only politics" at the height of the cold war. Demographic transition theory extended to developing countries as part of the benefits of the emerging global market, presumes either that developing countries too will attain comparable and unsustainable per capita rates of resource consumption (which is impossible) or that the von Weizsacker recommendation of reduced per capita resource consumption in the North will have to be achieved in the name of setting a sustainable example. The only other possibilities are the catastrophes of continuing conflict with each other and with the biosphere.

Industrial development consolidated the practice of gains in wealth through growth for individuals, corporations and particularly men, in a trajectory of increasing exclusivity. The creation of poverty, female poverty and the growth of poverty and 'externalities' followed. With the links between (a) poverty and population growth and (b) wealth and economic growth and (c) wealth and poverty growth and (d) all of this linked to environmental damage configured as well, the paradoxical relationship between wealth and poverty ontologically linking economic growth and population growth seems *consistent with the anecdotal comment cited which makes these linkages too.*

Economic growth aggravates population growth in a deterministic way. In a global paradigm, demographic transition theory cannot rescue us from this bind. The wealthy end of population growth can perhaps be 'cured' through economic growth to demographic transition. But continued economic growth in the wealthy sector to cure population growth at the poverty end of the scale-with the same ends in view-is futile. If this proposition is accepted, it creates an intellectual dilemma of enormous magnitude for the proponents of economic growth who are also opposed to the continuation of population growth and who pose demographic transition theory as a solution. It may seem obdurate to reason so intensely, but if this proposition is accepted, the anti-population growth proponents of continued economic growth must accept the impossibility of decoupling population growth from economic growth and work therefore either for the rise or fall of both together, or have laissez-faire do the same job instead.

Since the time of the formalisation of the political economy of growth, economic solutions to poverty had to be sought. Poverty, as Karl Polanyi describes, only arrived into political consciousness with the creation of the market economy. Various positions were adopted on the poverty issue. But the ad hoc consensus, in the face of egalitarian pressures, was that the growth of the creation of wealth cold outpace any growth of poverty. This would resolve the social casualty dimension through the increased availability of resources for laissez-faire distribution through the markets. This was the acceptable face of the rationale of the political liberal right. The political solutions of the left were to compensate for at least the extent of the failure of this rationale by promoting early modern forms of social welfare arrangements based on various forms of so-called intervention in the economy.

All the above follows the crossing of a significant threshold in the history of the relationship between society and population growth in the second half of the eighteenth century and is reflected in economic theory. The mercantilist trading ethos of the zero sum game as practised benignly between those who bartered, and malignly between those who bullied, was supplanted by the growth optimist political economy of trading absolute advantages, comparative advantages and observing opportunity cost theory. Among other things, this discipline of political economy attempted to dissolve the idea the exchanges in the economy between trading partners had to be a zero sum game. What was asserted instead was the idea that net benefits could accrue to both sides in a trading relationship. The application of advantage and opportunity cost theory could maximise these benefits through international trade. The same shift from zero sum game (in the form of the notion of a just price) to profit maximisation occurred within society. With the simultaneous arrival of the industrial revolution, transnational capitalism and technocratic progress were married at a stroke. All these changes embraced this hidden determinism of poverty and population growth and the creation of disbeneficiaries.

Winner/looser scenarios supposedly gave way to greater winners and lesser winners, and net benefit were achievable by the increasing number of poeple. This was formalised by Pareto with his theory of optimum efficiency. This theory depended on no more than being able to assert that within the totality of transactions in the economy, whilst there had been costs, there had been increased benefits with no net dis-benefits to any first, second or third parties. Because this analysis was also linked to the expectation and actuality of increased choice, notions of economic and social limits receded and disbeneficiaries 'vanished'. The ethos of unlimited growth and improvement became the norm right up until the catastrophe warmings of the present day.

The presence of a scientific consensus is another key precondition. International cooperation will remain elusive as long as there is disagreement about the issue being discussed. Consensus will need to include an agreement upon the nature and identification of the problem under investigation; an agreement upon the data collection process; and an agreement upon the method of data interpretation.

With regard to the interpretation of the problem under investigation, it has been recognised that '[t]he degree of ideological consensus and agreement over causal relationships, regardless of the nature of the issue, is an important variable in explaining cooperation' (Haggard and Simmons, 1987: 511). Thus, the extent to which the actors hold a common perception, interpretation and understanding of the problem will help to facilitate cooperation.

Also, in order to achieve a cooperative agreement, there must be an acknowledged credible process whereby information is gathered on the issue (Brown *et al*, 1977: 235). Whatever the means, it must be agreed that the raw information being gathered is accurate and that the analysis is acceptable to all parties involved. Stein noted that '[t]he nature of knowledge also determine[s] actor preferences and thus the prospects for regimes' (1983: 136). For without a consensus on the credibility of the information, the actors will not be working from the same information base and will therefore hold different understandings of the issue.

In 1990, the scientific consensus upon the global warming issue is by no means concrete. The theory upon which the argument is based is accepted by a large portion of the scientific community. It is also agreed that there has not only been an increase in the carbon dioxide levels in the atmosphere, but that there has also been an increase in the global average temperature since the beginning of this century. What remains to be resolved is whether or not the observed warming is causally linked to the greenhouse phenomena. Despite all of the attention that has been paid to the case for global warming, there are still a number of questions that need to be answered.

There are those who suggest what the temperature change could well be just a result of the variability of solar output. Furthermore, there has been a suggestion that the effects of alterations upon the Earth's orbital pattern may be having an effect upon the globe's temperatures. Finally, it may simply be that the fluctuations are a reflection of the high noise to signal ratio. Since the current warming trend can only be traced back uninterrupted for twelve years (average global temperatures were stable or decreased from 1940 to the late 1970s), it may be that people are reading something into the data that they should not be. Thus there remain some questions as to whether or not we are experiencing global warming. Scientists have noted that it will be at least ten years before anyone will be uncategorically certain that we are experiencing global warming.

Even among those who agree that the greenhouse effect is upon us, there is no consensus upon what the future may hold as some important questions about the natural processes involved still remain unanswered. At this point, scientists do not know if increased cloudiness will mean that more heat will be reflected back to space before it enters the troposphere or whether more heat will be trapped in the 'greenhouse'. Obviously, either process would be an important feedback in the process—though one would be negative and the other one positive. Furthermore, it is unclear how the physical, chemical and biological activities of the oceans will be affected by global warming. Finally, the role of plant life in a carbon dioxide-enriched atmosphere is not clearly understood. These are just a few examples of the uncertainties which are hampering present efforts at understanding the greenhouse effect.

It is clear that in order to achieve some sort of cooperative agreement, scientific consensus will be needed. In order to build this consensus, better theory and more data will be required. However, such a requirement once again raises the consideration about data sovereignty. For the reasons outlined previously, some states may be reluctant to fully participate in the efforts to achieve scientific harmony and therefore the fulfilment of this precondition may be hindered.

Therefore, the goal of scientific consensus may be hampered by a number of factors. And until such a unity is realised, scientific disagreement may well continue to be used as a reason for inaction; either because the dissenting player honestly does not accept the majority argument or because the player has other reasons for avoiding a cooperative arrangement, but realises that ostensible disagreement with the science and its findings is the most palatable rationale to cite in support of the political case.

The answers begin with where we stand today. As noted earlier, the correct measure is not how much warming has already occurred but rather what warming the atmosphere is committed to, because the time greenhouse gases are emitted is the point of irreverersible (so far as we know) commitment. Veerabhadran Ramanathan and his colleagues at the National Centre for Atmospheric Research have estimated the effect of emissions from 1860 to 1980 to be approximately 0.8^{0}C—that is, a commitment to this much warming has already occurred. Allowing for the uncertainty caused by various feedback effects, this translates to an eventual warming in the range of 0.5 to 1.5^{0}C above the preindustrial global average temperature.

Irving Mintzer at the World Resources Institute has constructed a tool that can illuminate future warming trends. His computer model combines the climate-altering effects of all the major greenhouse gases. Individual components of the model an be separately manipulated, reflecting hypothetical policy and economic changes. Mintzer's "base-case" scenario reflects conventional wisdom regarding population, economic, and energy growth from 1980 to 2075. No policies are introduced to slow the rate of a greenhouse gas buildup: no major effort is made to arrest tropical deforestation or to improve the efficiency with which energy is used. In this scenario, environmental concerns have little effect on energy policy. The model is quite conservative in projecting substantially lower rates of growth in CFC emissions than are occurring today, even though it envisions no policy steps to limit these emissions.

For this base-case scenario, the model predicts a commitment to an average global warming of 0.9 to 2.6^{0}C above preindustrial temperatures by the year 2000. A warming well within this range—2^{0}C—would put the global average temperature beyond that experienced in recorded human history. Yet politically speaking, 13 yeas is the blink of an eye for reaching and implementing a global agreement. The Nuclear Non-Proliferation Treaty took only seven years to negotiate and secure sufficient ratifications to enter into force, but it is a notable exception. The Law of the Sea Treaty, more the norm, took 14 years. Perhaps the best comparison is to the Vienna Convention to Protect the Ozone Layer, it took 10 years to reach an agreement on a framework, and the convention's regulatory teach are still being negotiated. The convention is a notable precedent in that it has been negotiated in the context of considerable scientific uncertainty and it, like a possible greenhouse agreement, is an effort to anticipate and avert—rather than clean up—an environmental problem.

Looking a little further ahead, Mintzer's model predicts that by 2030 the commitment to global average warming will be 1.6 to 4.7^{0}C, or about 3 to 8^{0}F, above preindustrial temperatures—that is, the equivalent of a doubled preindustrial carbon dioxide level. The warming commitment will be substantially greater at the poles. This is a very different climate regime than today's and it is only as far ahead as World War II is behind us. While immense changes can occur in 45 years, it is not a long time for certain transformations. For example, it has generally taken 50 years for new energy supply technologies to capture half of the global market.

Mintzer has also used his model to investigate two scenarios that are significant diversions from current trends. In his "high-emissions" scenario, the use of coal and synfuels is encouraged, the deforestation of tropical forests continues unabated, and minimal environmental costs are reflected in the price energy. CFC emission rise with income growth at the high historic rate at which they grew in the United States over the past few decades. By contrast, in the "slow-buildup" scenario, strong policies are adopted to implement solar energy, improve energy efficiency, discourage solid fuel use, and arrest deforestation. CFC production is frozen by international agreement at the 1985 level. In the high-emissions scenario the date of commitment to the equivalent of a doubled preindustrial carbon dioxide level advances from 2030 to 2015, whereas in the slow-buildup case it is delayed until about 2080. Thus, if the model is correct, this particular set of policies, hypothetically implemented in 1980, would mean a difference of six decades before the planet is committed to this benchmark warming level.

A different way to approach this question is to ask what are the consequences of delaying action until a significantly higher degree of scientific certainty about greenhouse mechanisms and impacts becomes available. Two decades is a reasonable guess for how long this might take. If an additional decade is then needed to reach international agreement on a set of policies, 2015 might be the earliest date for more confident action. According to Mintzer's projections, the planet would then be committed to a 0.25 to 0.8^{0}C greater warming than would be the case had the policies envisioned in the slow-buildup scenario been adopted in 1985. Thus, if no action is taken to alter current trends, over the next 30 years the planet will become committed to an additional warming about half again as great as that believed to have been committed to during the entire 125 years of the industrial age.

In sum the greenhouse clock is ticking. A substantial global warming (0.5 to 1.5^{0}C) is already inevitable. Since the rate of greenhouse gas emissions is accelerating, additional warming increments will be committed to much faster than in the past, meaning faster climate change and less time for societal and ecological adaptation. A reasonable estimate of the path we are now travelling shows a warming commitment of 0.9 to 2.6^{0}C above preindustrial temperatures by 2000, and 1.6 to 4.7^{0}C by 2030. On the other hand, history is not necessarily destiny. Politically optimistic but economically realistic policy changes undertaken now could significantly slow the anticipated warming. Waiting for greater scientific certainty would probably mean a 0.25 to 0.8^{0}C additional average warming to cope with—a not insurmoutable, but by no means insignificant, difference.

The rough outlines of a greenhouse policy debate are becoming clear. Their are three dominant views. One holds that it is too soon to begin to think about doing anything other than intensive research. This is the position, for example, of the US Department of Energy and of many scientists. Without question, the current scientific uncertainties are very large. On the other hand, there is highly suggestive evidence that a greenhouse warming is already under way; there is even stronger evidence that the rate of greenhouse emissions is accelerating; and there is ample evidence to indicate than the costs of a rapidly changing climate are very high. Moreover, scientific confidence on the greenhouse issue is an elusive goal. My estimate that scientific uncertainties may be resolve within two decades might be optimistic. And Mintzer's analysis suggests that there could be high cost associated with waiting. Thus, a strong case can be made in favour of acting now, *if* economically sensible and politically feasible stems can be identified.

Among those favouring some kind of action, one group, who might be termed "adaptionists," emphasiss learning to live with the greenhouse warming. Another group emphasises the need to slow and eventually halt the warming. This latter group might be called "preventionists," although this is obviously a misnomer because, as we have seen, the warming is apparently under way and would continue for decades even if all greenhouse emissions were halted tomorrow.

Some adaptionists proceed from the belief that because unilateral action to avert a warming is infeasible, and international cooperatioin impossible, adapting is the only option. Other adaptionists believe that climate warming will liekly yield net positive benefits, and that because some nations will gain and other lose from the effects of climate change, strategies based on independent adaptation make the most sense. Heavily represented in this group are climatologists who have argued for years in favour of policies to build greater societal adaptability to natural climate fluctuations.

There are five principal problems with an adaptionist approach. First, adaptionists usually underestimate the likely negative effects of a warming or at least the risks of such. Most such calculations give short shrift to the human costs that would accompany, for example, flooded coastal zones or shifts of hundreds of miles in areas suitable for farming. Although roads and dikes might be build and new crops bred, the human costs of disruption, dislocation, and retraining are unavoidable. The ecological risks are also generally not considered. Unmanaged ecosystems are less able to adapt, especially to unnaturally rapid changes, than human society. Abundant geologic evidence links periodic mass extinctions in the oceans to decreasing temperatures. Populations within parks and conservation preserves that could not migrate or adapt would be lost, and biologically vital coastal communities would be severely at risk. Similarly, adaptionists generally do not factor in the likelihood of unpredictable, unpleasant surprise. The stunning discovery in 1985 of a "hole" in the Antarctic ozone layer underscores the potential for sudden threshold effects instead of smooth incremental change. In recent Senate testimony, Wallace Broecker, a professor of geochemistry at Columbia University, raised the possibility, based on new evidence, that "future climate changes may come in sharp steps, and these steps may have complicated geographic patterns. If so, then we must begin to think in terms of climate surprises that may, without warning, cause shifts in temperature and rainfall patterns. In light of these and the other risks described earlier, it is a poor bet for any government to count on emerging from a rapidly changing climate as a "net winner".

A second difficulty with the adaptionist approach is that it either depends on being able to predict the specific regional (or even local) impacts of changing climate or else requires that steps he taken to adapt to change without knowing, for example, whether the new climate would be wetter or drier. Current ability to model and predict geographically specific impacts is very limited, and the ability to do so with confidence is generally thought to lie far in the future. Hence, the lead time for action under this approach may simply be too short (and certainly shorter than for a preventionist strategy, which relies on the ability to predict the broad outlines—rather than the local or regional impacts—of global climate change.) Adaptation without knowing the likely regional impact is possible. For example, a region could at the same time conserve water supplies and develop reservoir and flood control systems. However, building margins of safety in both directions is expensive and would certainly be politically hard to swallow.

Third, adapting may widen the already yawning gap between the developed and developing countries. Adaptation requires strong research capabilities, heavy capital investment, and a government that can mobilise and direct resources. All of these are extremely scarce in the developing world. Thus, unless the developing countries as a group are net winners from a changed climate, a "go-it-alone" adaptionist approach is likely to leave these nations worse off relative to the developed countries than they now are.

Fourth, adaptionists must at some time in the future become preventionists. No one believes that society as we know it can adapt to, say a 15^{0}C global average warming. Thus, the planet's warming trend must eventually be slowed and then stopped. The longer society waits, the more rapid and disruptive the slowdown might have to be. It could also be much more expensive than it would be today it, in the interim, a heavy commitment had been made to coal, tar sands, oil shales, and other carbon-rich fuels. This would be especially true—and tragic—for developing countries, which have the choice, now, of choosing a more energy-efficient path than that travelled by the now-industrialised nations. Later actions to halt the warming might also be far more internationally divisive than cooperative action would be today. Assume for the sake of arguement that in a warming world some countries experience net gains. There would also be—perhaps next door—net losers. International cooperatioin is difficult to imagine today in the presence of so much uncertainty about which nations, if any, stand to gain, but it would be infinitely more difficult in a world split into countries with demonstrably opposed interests in continuing a climate-altering regime. Suppose, as some have suggested, that the Soviet Union would be a net winner and the United States a net loser. (Soviet scientists, it should be noted, do not on the whole share this view.) The prospects for international cooperation would be far dimmer than in today's shared uncertainty.

Finally, even today, the adaptionist strategy poses political challenges at least as formidable as the preventionist approach. As one adaptionist report puts it: "A country with stable government, balanced budgets and low unemployment is in a strong position to adapt to a warming; a political entity that harbors autocratic or corrupt leadership or one with excessive numbers of people living in hunger and poverty has small reserve to meet the further stresses of a changing climate." Moreover, this strategy "will require fundamental changes—in the economic and political goals and procedures" of nations, and "a shift of priorities and incentives from short term performance towards the identification and pursuit of long term goals for society. By these criteria few governments would qualify, and the prospects look dim indeed.

Several offset deals appear to have been undertaken quite independently of legislation or anticipation of regulation. Applied energy Services (AES) of Virginia has also undertaken sequestration investments in Guatemala (agro-forestry) and Paraguay and is in the process of setting up another project in the Amazon basin. The Guatemale project is designed to offset emissions from a 100 MW coal fired power plant being built in Uncasville, Connecticut. The intermediary for the project is the World Resources Institute and in Guatemala the implementing agency is CARE, The project involves tree planting by some 40,000 farm families. Carbon sequestration is estimated to be 15.5 Mt of carbon. The US$14 million cost includes US$2 million contribution from AES; US$ 1.2 million from the government of Guatemala; US $1.8 million from CARE, with the balance coming in kind from USAID and the Peace Corps. Note that the motivations for involvement vary. AES's involvement relates to its concern to offset CO_2 emissions, whereas other partners are concerned with the local development and environmental benefits the deal brings.

Dixon et al. report the sequestration cost as US$9 tC overall, but inspection of the data suggests it may be less than this, US$9 tC would be expensive for carbon sequestration alone, but there are other benefits from the scheme, including local economic benefits. In the Paraguay deal, AES is planning to advance money to the (US) Nature Conservancy for investment in some 57,000 ha of endangered tropical forest. AES expects to sequester some 13 million tC at around US$1.5 per toone of carbon. Local benefits include eco-tourism, scientific research, recreation, agroforestry and watershed protection. The private sector carbon offset deals to mid 1993.

Financial transfers may take place without any regulatory push. The consumer demand for green products has already resulted in companies deciding to invest in conservation either for direct profit or because of a mix of profit and conservation notices. The Body Shop is an illustration of the mixed motive, as is Merck's royalty deal with Costa Rica for pharmaceutical plants and Pro-natural's expanding venture in marketing indivenous tropical forest products. There is, in other words, an incentive to purchase or lease exotic capital in the same way as a company would buy or lease any other form of capital.

The deal between Merck and CO, the world's largest pharmaceutical company, and INBio (the National Biodiversity Institute of Costa Rica) is already well documented and studied (gamez *et al.*, Sittenfield and Gamez, and Blum). Under the agreement, INBio collects and processes plants, insect and soil samples in Costa Rica and supplies them to Merck for assessment. In return, Merck pays Costa Rica US$1 million plus a share of any royalties should any successful drug be developed from the supplied material. The royalty agreement is reputed to be of the order of 1 per cent to 3 per cent and to be shared between INBio and the Costa Rican government. Patent rights to any successful drug would remain with Merck. Biodiversity is protected in two ways—by conferering commercial value of the biodiversity, and through the earmarking of some of the payments for the Ministry and Natural Resources.

How far is the Merck-INBio deal likely to be repeated? Several caveats are in order to offset some of the enthusiasm over this single deal. First, Costa Rica is in the vanguard of biodiversity conservation, as its strong record in debt for nature swaps shows. Second, Costa Rica has a strong scientific base and a considerable degree of political stability. Both of these characteristics need to be present and their combination is not typical of very many developing countries. Third, the economic value of such deals is minimal unless the royalties are actually paid and that will mean success in developing drugs from the relevant genetic material. The chances of such developments are small—perhaps one in one to 10,000 of plant species screened. INBio has undertaken to supply 10,000 samples under the initial agreement. There is therefore a chance of one such drug being developed. But successful drugs could result in many hundreds of millions of dollars in revenues. Finally, there are two views on the extent to which deals of this kind could be given added impetus by the Biodiversity Convention. The convention stresses the role of intellectual property rights in securing conservation and is sufficiently vaguely worded for there to be wide interpretation of its provisions. But is also appears to threaten stringent conditions concerning those rights and technology transfer and it remains to be seen how the relevant protocals are worded. If so, parties to the convention may find private deals being turned into overtly more political affairs with major constraints on what can be negotiated.

Other examples of direct deals on biodiversity prospecting include California's Shaman Pharmaceuticals (Brazil and Argentina) and the UK's Biotics Ltd (general purchase and royalty deals), while Mexico and Indonesia are looking closely at the commercialisation biodiversity resources.

The demand for direct investment in conservation is not confined to the private sector. The demand for conservation by NGOs is revealed through debt for nature swaps, which are further examples of these exotic capital trades (for an overview see Pearce and Moran, and Deacon and Murphy).

Newcombe and de Lucia have drawn attention to another potentially very large private trade which has global environmental benefits. Investment by the private sector in the developing world is invariably constrained by risk factors such as exchange rate risks, repayment risks, political risks and so on. In so far as this investment benefits the global environment, as with, say, the development of natural gas to displace coal, the existence of the risks reduces the flow of investment and hence the global environmental benefits. But these risks might be shared (bought down) by having an international agency, such as the GEF, provide some funds or services which help reduce the risk. Given the scale of private investment flows, the potential here is enormous. Nor is there any reason why it should not benefit biodiversity, either indirectly as a joint benefit of other investments in for example, raising agricultural productivity and hence in reducing the pressure for land degradation, or directly through afforestation schemes.

With population set to double in the next 80 years or so, it is unrealistic to expect to save the world's current biodiversity completely. Much of it will be lost. But much of it can be saved. There is a major role to be played to regulations and agreements, but the prospects for real action must lie in (1) leveraging the flow of funds that will occur under international agreement; and (2) exploiting every possibility for mutual economic gain that also generates biodiversity conservation benefits. The idea of exploiting mutual economic gain will strike many as morally unacceptable, but that viewpoint—while heartfelt—partly explains why we have the problem in the first place. Taking the moral high ground may be satisfying for the moralist, but it has not yet produced the wave of resources needed to contain the problem nor is likely to. The answer is to pioneer on all fronts. The economic approach looks especially promising. This paper has argued that there is substantial potential for mutual gain through global trades. This should not divert attention from the continuing need to correct domestic policies in the developing and developed world alike. All forms of economic failure need to be addressed.

Protecting the Atmosphere

Each morning, as people begin the day, they take part in a process that is altering life on the planet. As 500 million cars and other vehicles are started, millions of small pollutant clouds are released into the air. As work gets under way in factories and power plants, on farms and in forests burnt to clear land, the pollution cloud grows.

Poor air quality is a global health hazard—from among alerts in Los Angeles and Sydney to worsening traffic pollution in Lagos and Bangkok and the coal-fired blackness of Eastern Europe. More than 1 billion people breathe air that does not meet minimum health standards. Respiratory diseases, cardiac problems and cancer are the result.

Global problems demand international solutions. Pollution of the atmosphere will be one of the primary concerns of government leaders when they meet for the United Nations Conference on Environment and Development in Rio de Janeiro, Brazil, from 1 to 12 June, 1992. On the agenda of the Conference, known as the Earth Summit, will be proposals to prevent acid rain, the warming of the planet and the destruction of the ozone layer.

Global Warming

Air pollution has been viewed as a local problem: on a bad day, it hangs in a yellow-grey cloud overhead; on a good day, the winds will take with them the visible reminder that the air around us is being used as a rubbish dump.

Out of sight but still potent, the pollutant gases accumulate in the atmosphere. Evidence suggests that here the gases are trapping the heat of the sun, causing the planet to warm as if it were in a greenhouse.

Carbon dioxide (CO_2)—the main cause of the "greenhouse effect"—is entering the atmosphere at the rate of 8.5 billion tons a year. Mostly the result of the burning of fossil fuels (coal, oil and natural gas), CO_2 is being spewed from exhaust pipes of vehicles and industrial smokestacks. The second largest cause is the burning of forests to clear land.

Other gases are also suspected of contributing to climate change, notably chlorofluorocarbons (CFCs, comprising 20 per cent of "greenhouse gases"), methane (16 per cent), ozone (8 per cent) and nitrous oxide (6 per cent). However, CO_2 from energy-related activities was responsible for about half of the global warming effect between 1980 and 1990, according to the Inter-Governmental Panel on Climate Change. The Panel, made up of several hundred scientists from 35 countries, was set up by the UN in 1988 to assess changes in global climate patterns.

The best hope for avoiding the life-changing impact of global warming will be to reduce the amount of CO_2 released into the atmosphere. Getting agreement from the main producers of CO_2 is central to negotiations for a convention on climate change, now under way as part of the Earth Summit process.

"Accelerated climate change must now be seen as the primary risk to the human future," notes Maurice Strong, Secretary-General of the Conference. "Like a cancer spreading through the body of our global society, the effects are likely to be irreversible by the time the symptoms have become painful."

Unless the international community can come to an agreement on reducing the global output of CO_2, scientists believe it is likely that life on Earth will soon be irreversibly altered. Sea levels will rise, and small island nations and low-lying coastal lands in many countries will be engulfed. Weather patterns will change, affecting agriculture: where there were rains, there may be none; where temperatures were worm, they may be cooler. Fresh water will become scarcer as groundwater evaporates, run-off is reduced and salt water leaks in and contaminates underground sources. The impact on the lives of millions of people will be devastating.

Developing countries fully share the risks of global climate change. In fact, because their environments are often more ecologically fragile and their economies more vulnerable, they are likely to be less able to deal with the projected consequences of climate change than more industrialised countries.

Who Produces CO_2?

At present most CO_2 is produced by the industrialised countries, a quarter of it by the United States. Of the remainder, according to 1986 figures, 19 per cent was produced by the former Soviet Union, 14 per cent by the countries of the European Community, 10 per cent by China, 3 per cent by India and 1 per cent by Brazil.

This pattern will change. As developing countries grow economically, they can be expected to contribute significantly to global atmospheric problems unless assisted to use cleaner, more efficient technologies and environmentally sustainable sources of energy. These are available but either are more expensive than conventional technology or require research and development. Generally, a country consumes energy at a faster rate at the beginning of its industrial growth than at later stages.

In some developing countries, energy consumption is so low that an increase will be essential just to reach a basic standard of living. In others, people are beginning to obtain the comforts of modern living—cars, appliances and refrigeration. They will all, unless given assistance to develop differently, adopt the model used by industrialised countries.

Most developing countries plan to meet their energy needs with massive investments in fossil-fuel technology. China aims to ensure that every household has a refrigerator. Even if the units are small and energy-efficient, this will require the energy equivalent of the output of 20 large nuclear reactors.

Forests Keep the Balance

In the past, the balance between nature and the activities of humans was maintained, in the case of CO_2, primarily because much of the planet was covered with forests. Trees convert CO_2 into life-giving oxygen; thus forests are viewed as "sinks" for absorbing carbon.

The increased output of CO_2 in the developed world after the industrial revolution was accompanied by widespread deforestation: more than two-thirds of the forests of Western Europe have disappeared since Roman times; 95 per cent of the virgin forests in the United States have been felled in the last 150 years.

The focus of global preservation efforts is now on tropical rain forests, which are mostly located in developing countries. As much as 17 million hectares of tropical forest are levelled each year, to be sold as timber for export and to clear land for agriculture, grazing and settlements. The clearing of old forests in North America also continues at a devastating rate: in Canada, 7 million hectares of forests were burnt in 1989; 2 million hectares are levelled in an average year.

Member States taking part in United Nations talks for a legally binding convention on climate change are attemtping to reach agreement on reducing the output of CO_2 so that by the year 2000 the amount released into the atmosphere annually will be the same as it was in 1990, 5.5 billion tons. European countries and Japan appear willing to adopt such measures; the United States, so far, is opposed to setting definite targets for CO_2 emissions. It is likely that such an agreement will also emphasise the need to preserve global forest cover.

The Transition to Sustainable Energy

Reducing CO_2 emissions will require the use of environmentally cleaner fuels, particularly for industry—which uses between one-third and one half of all energy produced globally—and for transportation, the second largest consumer of fossil fuels. Massive use of fossil fuels is considered unsustainable not only because it generates pollution but also because supplies are being depleted.

Nuclear power also cannot be considered a sustainable energy source because of the need to dispose of mountains of hazardous radioactive waste from nuclear reactors and the dangers from accidental emissions of toxic gases into the air.

Renewable energy sources—including hydroelectric, geothermal, solar, photovaltaic, wind, tidal and biogas—have not been widely used. They provide only 7 per cent of global energy, excluding traditional biomass (vegetation-based fuel, primarily wood-burning), which contributes 14 per cent. Almost all renewable energy currently used is from hydropower, which on a large scale has severe impacts on the environment, caused by the destruction of land flooded by dams, and on the people and animals who must be removed from that land.

Wood and other biomass fuels provide the only source of energy for more than 2 billion people in developing countries. Hall of the forests cleared in those countries are cut to provide fuel for cooking and heating. Here too, environmentally sound alternatives must be found. Severe fuelwood shortages are expected early in the next century. People responsible for feeding families, usually women, already walk miles each day looking for wood that is increasingly scarce. This is not sustainable in its impact on either women's lives or the environment.

Improving Energy Efficiency

Until renewable energy sources are developed, interim steps can be taken to burn fossil fuels more efficiently and as cleanly as possible. Improving energy efficiency to get the maximum amount of production from the least amount of fuel reduces not only fuel consumption but also the volume of pollution. Fuels can be burnt more cleanly, both at the point of energy conversion and when wastes are emitted, using techniques such as filters on smokestacks.

World-wide, energy savings of 40 per cent or more could be achieved by making existing technologies more energy-efficient. Studies in industrialised countries have concluded that emissions of CO_2 could be reduced by 15 to 30 per cent in 10 to 15 years if efficiency were strongly promoted. The potential for improvement is at least as great in developing countries, since many key sectors such as manufacturing and transportation use energy less efficiently in developing than in industrialised countries.

In many developing countries, the only option at present may be to build plants using fossil fuels. Improved operations and maintenance techniques can reduce environmental degradation, but the extra investment needed to reduce emissions is difficult for Governments to finance. In China, the United Nations is working to modernise a coal-fired power plant to reduce both the consumption of coal and the emission of pollutants. It is a pilot project for 112 similar plants. If more countries were given assistance, they could bypass the pollution-intensive phase that the industrial world has gone through, and instead move directly to more efficient technologies. In doing so, the full consequences of global warming might be avoided.

Many government policies and business practices can promote energy efficiency. Changes in manufacturing processes, such as the making of steel, can sharply reduce the amount of fuel required. Buildings can be better designed and managed. Cars can be designed for improved fuel economy (some Japanese and Swedish cars get over 80 miles to a gallon). Changes in products made can also lead to lower energy intensity (the amount of energy used per dollar of output). It is increasingly evident that public transportation systems in both industrialised and developing countries must be given priority if commuters are to be persuaded to leave private vehicles at home.

Energy Policy is Environmental Policy

A variety of powerful interests will have to be confronted before the availability and use of renewable sources of energy will be significantly increased. However, Governments can protect the environment by using market and pricing mechanisms—such as taxes, tradable entitlements, incentives and disincentives—to encourage greater use of environmentally sound energy and the development of cleaner technologies. In 1983, Sweden raised the tax on oil to discourage its use. When oil prices were falling in 1986, Denmark raised fuel taxes and Portugal removed subsidies on some oil products to ensure long-term efficiency.

Introducing Renewable Energy

Technologies for using solar, geothermal and tidal power have been developed but vary in sophistication and economic competitiveness. Renewable energy provides 5 per cent of the total energy requirements in Australia, Austria, Canada, Denmark, Sweden and Switzerland. Four solar energy plants are under construction in the United States to provide the residential power needs of 800,000 people.

In developing countries, the United Nations is helping to design solar systems for industry, agriculture and domestic power supply. The development of a highly efficient, affordable storage battery would be a key advance. Solar and wind energy could then be stored for use after dark and on calm days.

Research on alternative fuels for cars was given a boost in the United States by a law recently passed in California requiring that by 1992, 2 per cent of all cars must have zero emissions, and 10 per cent by 2003. Yet even existing battery-powered cars—which have a driving range of up to 40 miles before they must be recharged—could present a viable alternative for many people worldwide. Drivers in the United States average only 28 miles a day.

Wind projects operating in China and India with United Nations support are providing electrical energy for local needs. The UN is also testing an integrated energy system—using solar, photovoltaic and wind sources—in four Pakistani villages to evaluate the use of such systems in remote areas. It has also played a pioneering role in the development of geothermal energy. In Yemen, a UN biogas project is providing 70 to 80 per cent of household energy requirements in small villages.

Although there is a clear need for more research in renewable energy technologies, spending in that field by the countries that are members of the International Energy Agency fell from $1.5 billion in 1981 to about $500 million in 1989. Some renewable technologies are already competitive; more research will lead to lower costs and greater reliability, especially in photovoltaic and biomass energy. It is estimated that using a mix of energy sources could cut emissions of carbon dioxide by as much as 68 per cent.

Today, the world depends on fossil fuels for 85 per cent of its energy needs. The risks of continuing business as usual are the pervasive health problems and deteriorating quality of life caused by air pollution, as well as the catastrophic flooding and loss of secure food and water supplies projected to arise from global climate change.

Leaders attending the Earth Summit will attempt to devise a blueprint for a brighter future. In the area of energy policy, this will involve rethinking daily practices for both individuals and institutions, and will require changes in national and global policies and spending priorities.

Key Issues for Climate Change Negotiations

The major greenhouse gas emitters have not yet committed themselves to major emissions reductions. Nor have the international donors backed up their words with money and action. The authors of this chapter explore a way that the greenhouse management contract between rich and poor states can be constructed in an open, and equitable manner.

These issues are not solely political and economic, however. They are also technical and scientific in nature, relating as they do to complex and poorly understood issues of climate and ecology. In part 1, four authors explain these issues and produce a technical and scientific foundation for determining who is responsible for climate change. Kirk smith explains the science that underlies comparisons of different greenhouse gas emissions from different nations and time periods, that is, the indices that can be used. Implicitly or explicitly, an index of some sort must be used so that choices among options can be made. He concludes that there are some hidden value judgements in choosing indices and that the choice of index depends strongly on the particular policy question being asked.

Sussan Subak compares the results of applying five different indices to the question of relative national contribution to global greenhouse gas emissions. These are:

1. cumulative carbon dioxide emissions from fossil fuel combustion only;

2. Cumulative carbon dioxide form fossil fuel and land use changes;
3. current annual carbon dioxide emissions;
4. an expanded list of current emissions including methane from landfills and fussil fuel production;
5. a comprehensive range of current greenhouse gas emissions from energy, land use change, and agricultural sources.

When allocated on a per capita basis, each measure produces a different distribution of national responsibility for past and present contributions to climate change. Subak shows how the scientific and technical dimensions of buildingindices of responsibility have significant political and economic implications for different polluters. In short, depending on which index is chosen for the protocols that allocate responsibility, the Convention will impose differential burdens on states of widely varying characteristics. The outcome will be widely varying incentives and disincentives for future action under the Convention . It is incumbent on policy makers, therefore, to pay careful attention to these technical issues.

Drawing on these technical and scientific foundations, Dilip Ahuja, Kirk Smith and Jole Swisher present a simple, transparent method to determine who should pay the cost of creating a global greenhouse regime. A composite induce is proposed that includes both ability to pay on the one hand, and historical contribution to climate change on the other. The former index confronts the issues of equity and economic realism that will affect participation rates of the poor. The latter index embodies the polluter pays principle and reflects the practical politics that the poor, small polluters are not likely to constrain their behaviour unless the wealthy, big polluters recognise that they have occupied the available 'ecological space' and must compensate latecomers for this pre-emption.

Smith, Swisher and Ahuja's approach provides a powerful philosophical and partial underpinning for discussions of the distribution of cost associated with managing climate change. If accepted, it would influence the outlook of key parties even if they are unable to accept specific number based upon it in actual negotiations. Such indices can be recommitted to investigate alternative yardsticks of ability to pay and historic responsibility for climate change. Whatever the final numbers, what is crucial to creating an effective global green house regime is that it rests on these twin principles of equity and polluter pays responsibility. Peter Hayes confronts directly the deceptively simple question: who should pay? He introduces a method to calculate the likely costs to developing countries of complying with a global convention of climates change. He calculates the incremental cost of abating carbon dioxide emissions from the use of fossil fuels by the following procedure.

He begins by estimating projected emissions and required reductions of carbon dioxide that meet stringent IPCC emission targets which would restrain the growth of realised temperature and sea level to 0.1^0 C and 3 cm per decade respectively. To achieve this goal, the global permitted emission in 2025 is about 2.7 gigatonnes of carbon as carbon dioxide. This target is about 60 per cent of projected global emissions in that year, or about 50 per cent of emissions in 1990. Each country is required to reduce its emissions from projected 1995 levels so that it eventually reaches its fraction of this global permitted emission in 2025. This fraction is set to equal current national sink rights distributed to nations on the basis of current population and land area (although some other allocational criteria could and probably should be used to avoid problems associated with defining sinks. High, medium and low marginal abatement cost curves are applied to these profiles of carbon abatement over time. In this way, the method generates a stream of annual incremental abatement costs for each country.

Hayes applies the quantitative allocational rules developed into the range of numerical estimates of the cost of carbon emission abatement and coastal protection from. He presents two rules that abatement and coastal protection. He presents two rules that have been proposed to allocate the cost to various parties of meeting emission targets imposed by a climate change agreement. These rules are' obligation to pay,' based on each nation's historic emissions and ability to pay; and the UN scale of payments.

Under the UN scale, the OECD (North) pays about 77 per cent, the former Soviet Union and Eastern Europe (East) about 14 per cent, and the developing world (South) about 9 per cent of total UN cost. In the obligation to pay (OTP) index, the North's OTP is about 73 per cent; the East's about 20 per cent; and the South's about 7 per cent. Hayes argues that negotiations are likely to proceed therefore by countries making bids to vary their contribution relative to the UN scale until consensus is reached. The obligation to pay index provides a sound, transparent baseline against which to measure the fairness of departures from the UN scale.

Hayes treats the South's incremental cost, minus the South's obligation to pay, as the responsibility of the wealthy countries of the North and transfers it to the North's account. The annual transfer from North to South is estimated at $29-34 billion in the medium and high marginal abatement cost cases respectively. (There is no economic case based on incremental cost for transfer from the North to the South in the low cost of abatement measures and to increase scientific and technical capability in developing countries as reasons for providing funds in any case.) He concludes by examining how the substantial funds required might be collected and transferred by carbon taxes, or earned by the sale of tradeable permits or abatement services that would also push countries toward equalising their marginal abatement costs at a global level.

An alternative approach would have been to construct a global marginal abatement cost curve from national marginal abatement cost curves; and to apportion reduction activity to each coutnry up to the marginal cost that delivers the desired total emission each year. The reduction activity would be paid for according to the relative obligation to pay for each year. The reduction activity would be paid for a according to the relative obligation to pay for each country (as updated periodically). On this basis, no country would be asked to reduce emissions more than the then-current marginal cost level, and no nation would remain unpenalised for failing to undertake reductions found to be chapter than the current marginal cost criterion.

While attractive in principle, it remains difficult to put this approach into computational practice due to the lack of meaningful global and national marginal abatement cost curves as well as difficulties associated with allocating and quantifying emission rights over time. The small island states most vulnerable to the impacts of climate change have been among the most vocal proponents of a strong Convention. In Chapter 7, Michael Wilford enumerates a proposal emanating from the alliance of Small Island States that in insurance fund be established to cover the costs of adaptation to sea-level rise. Precedents exist for this approach in the oil and nuclear industries, but neither apporaches the scale or scope of a fund that would cover losses implied by the greenhouse effect. The moral influence of these small states whose very existence is at stake is evident in the Convention article cited above which declares that vulnerable states are eligible for extra assistance.

A major difficulty that hampers calculations of the required funding by the rich countries of the incremental costs of developing countries is out ignorance as to the shape of the latters' emission abetment curves at various levels of required reduction. Part III contains the work of eleven authors who levels of required reduction. Part III contains the work of eleven authors who are deeply immersed in the empirical calculation of incremental costs associated with climate change and greenhouse gas abatement. These authors present abatement cost curves at the national or regional level in Asia, Africa, Australia, and Eastern Europe/Russia. All conclude that significant cost saving will likely accrue at the outset of carbon reduction programmes, although the absolute cost levels vary widely.

In India, Jayant Singh and Amulya Ready show that while emissions are likely to grow, there are substantial opportunities to abate emissions or to fix additional carbon at low costs or a net saving. They demonstrate that India could largely offset its carbon emissions by fixing carbon inforestry reservoirs, thereby emphasising the importance of the sink issue in determining responsibility for cost. However, they also identify public and private institutional and informational obstacles to the realisation of this potential. They argue that a basic needs economic strategy will itself enhance development and reduce emissions. They conclude optimistically that for the first time ever, fundamental interests of the rich and poor countries are aligened in the climate change area- provided that these multiple barriers to abatement can be overcome.

In West Africa, Ogunlade Davidson notes that although Africa contributes only a small fraction (about 3 per cent) of total global carbon emissions, its energy usage and greenhouse gas emissions will grow substantially. He estimates that emissions can be reduced in this region by between 13 and 36 per cent (depending on the country) by the year 2025, simply by introducing economically justified carbon conservation measures. As in other developing countries, he finds that a significant number of financial, institutional and technical obstacles exist which block the region from implementing these abatement options.

Because the generally poor management of energy and related institutions also hinders the effective implementation of these measures, he concludes that institutional reform is an essential ingredient of a carbon abatement strategy. He also finds that lack of investment finance is a major obstacle to energy development in the region. Weak capital markets and heavy indebtness require major economic reforms. These steps along, however, will be inadequate to the task unless supplemented by external financing.

In Brazil, Jose Moreira and Alan Poole present an aggregate cost curve for abating carbon emissions from Brazil's fossil fuel and biomass energy use which incorporates eighteen categories of abatement technology Brazil is unusual in that a large fraction of its electricity is generated by hydropower and a large amount of alcohol from biomass as used in the transport sector. Moreover, the authors did not include steps related to re-or de-forestation in Amazonia in determining the potential for Brazil to abate or to offset its energy-related emissions. Nonetheless, they identified abatement potential that amounts to about 16 per cent of projected energy-related emissions, much of which can be obtained at negative cost (that is, at a savings).

In Thailand, Peter du Pont, Somthawin Patanavanich, Mark Cherniack, and Michael Philips demonstrate that the most rapidly growing source of carbon emissions, the electric power sector, can be curbed significantly at a low or negative cost. They project that Thailand's carbon dioxide emissions from fuel combustion will double over the next decade, from 24 to nearly 50 million tonnes annually. They estimate that an aggressive demand side management effort in the power sector could reduce emissions by 2.5 million tonnes annually by the year 2001 at an average cost of conserved carbon of about US$190/tonne. While still nascent, Thailand's electric utility has aggressive programme to tap this potential saving and may offer a good model for other countries to emulate.

In Central and Eastern Europe, Stanislav Kolar produces aggregate cost curves based on detailed local research to demonstrate that carbon abatement can be achieved with major economic savings until high levels of carbon abatement are reached. Kolar points out that all the countries of the former Soviet Union are grossly energy inefficient. Equally, they also offer massive and relatively cheap carbon abatement. He concludes that energy efficiency is their most effective means of reducing carbon dioxide emissions and can achieve the twin goals of economic development and environmental protection. He concludes that these states have considerable flexibility as to which combination of price reforms and regulations would serve best to realise this potential.

Finally, in Australia, Hugh saddler examines estimates of the cost and scope of emission abatement measures and reviews estimates of the impact on the Autstralian economy of achieving various levels of abatement. Relative to business-as-usual scenario, he reports that about 20 per cent of prejected emissions can be abated with economically justified carbon abatement steps. He also notes that the local manufacturing-versus-import content of equipment needed to implement this strategy is a crucial determinant of the macro-economic impact of carbon abatement—a—variable that many developing countries may do well to examine carefully.

Generally, these studies indicate that abatement is possible at negative cost or a savings at the outset of the abatement strategy, but that costs will become positive fairly quickly. The studies also point to the obvious and urgent need, for demonstration carbon abatement programmes and additional research into costs to obtain much better cost estimates necessary for the formulation of sound policy.

Politics will not end once robust analysis and widely accepted estimates of the cost are available. Part IV returns to these *realpolitik* considerations which will intrude into future negotiations over protocols to the convention. For example, big wealthy countries anticipate a new wave of technological innovation associated with greenhouse gas abatement. As the major donors, they will seek to tie resource transfers to exports of their own equipment and services. For their part, recipient states will seek minimal ties on these funds. Aid flows justified on the grounds of greenhouse abatement will be no more or less susceptible to mismanagement, waste and corruption than existing development assistance.

Pater Hayes outlines some of the practical political issues that will arise in negotiations over resource transfers from the North to the South on the scale justified by the earlier chapters. Simply moving money across the North-South divide may only worsen existing development difficulties by creating an ongoing technological dependency. Studies of energy efficiency and related carbon abatement show that a wide range of scientific, technological and managerial capabilities must be created in developing countries to obstacle to the emergence of an endogenous technological capacity needed to reduce greenhouse gases—but they are only one of a range of issues that must be resolved before such programmes can be realised.

Hayes also analyses the potential for regional greenhouse initiatives in Pacific Asia as a precursor to a global climate change convention. He concludes that demonstration abatement projects in developing countries of Asia and the pacific are needed urgently to demonstrate the viability of schemes such as tradeable permits and trade in abatement services. He emphasises that 'first in, first served' will dominate the emerging markets for greenhouse gas regime. It remains to be seen whether the parties to the Convention can muster the domestic political will needed to meet the commitments contained or implicit in the Convention.

In this respect, the 'review and pledge' procedure implied by Articles 4 and 10 of the Convention is particularly important . In Article 4, countries undertook to prepare and to communicate national greenhouse gas inventories using common methodologies and to implement mitigation measures on sources and sinks of emissions. In Article 10, they created an international body that will 'assess the overall aggregated effect of the steps taken by the Parties in the light of the latest scientific assessments concerning climate change.' Implementing these two commitments will create an iterative dynamic that will lead to stronger action under the Convention in the future.

Action, however, will require that resources be allocated to match the rhetoric of the Convention. The parties to the Convention must adopt a transparent method of calculating obligations and cost if the commitments on funding and technology transfer are to be fulfilled. The studies reported in this book illustrate the complex and difficult issues that must be addressed in such a method.

At this stage, however, it is not the specific method nor its results that are important. Rather, what is vital is that negotiators of protocols to the Convention develop parallel ways of thinking that faciliate communication and agreement on these concerns. We hope that this book contributes to this task.

With respect to the global warming problem, the GEF should be part of a precautionary policy. Its aim should be to support those activities and investments that would leave the international community better placed to address the global warming problem, should the need arise. Hence, the investments and activities it supports should be based on the *premise* that global warming will take place.

In a global warming scenario, the achievement of energy efficiency must be an important element of the policy of reducing CO_2 emissions. But, however economically desirable energy efficiency may be, it will not by itself prevent carbon from accumulating in the atmosphere. Global warming could be delayed by energy efficiency, but it can be *prevented* only by widespread recourse to the non-net-carbon-emitting or backstop technologies, the most promising of which are renewables.

Thus, *the best strategy of addressing the problem of global warming is a two-pronged strategy of (1) "buying time" with technologies that delay global warming, and (2) preventing global warming with non-fossil fuel or backstop technologies.*

Buying time is best done through energy efficiency, environmentally more benign fossil fuel technologies (such as natural gas) and emissions reduction and GHG sequestration. And, the most promising non-fossil-fuel or backstop technologies are those using renewable energy, primarily solar and biomass, the costs of which have declined appreciably over the past two decades.

In a scenario of global warming, it would be necessary to turn to them increasingly in order to stabilise GHG accumulations at some 'safe' level.

With respect to its energy-related investments, the primary function of the Global Environmental Facility (GEF) is to develop cost-effective approaches towrds addressing the global warming problem. Investments should be directed towards that mix of technologies which can achieve the maximum reduction in emissions for a given investment. The identification of such a mix requires information on the unit cost of emissions reduction for the various technologies and the magnitude of the emissions reduction achievable with these technologies.

If this information were available, one could adopt a least-cost emissins-reduction strategy based on cost-emissions-reduction-technology (CERT) curves. These curves are constructed by choosing technologies in the order of increasing unit costs of emissions reduction as successive elements of the mix so that their cumulative emissions reduction potential achieves the desired goal. Thus, one can either estimate the total investment that is required for the mix of technologies to achieve a given magnitude of emissions reduction or one can identify which mix of technologies is likely to achieve the maximum emissions reduction for a given total investment.

Least-cost-emissions planning is a worthwhile approach to move towards because (1) it takes into account both the cost-effectiveness of a technology in reducing greenhouse gas (GHG) emissions as well its emissions-reduction potential; (2) it treats the supply-side and demand-side options for reducing GHG emissions on equal terms and does not discriminate against either of them; (3) it ensures that different technologies are compared and prioritised on the basis of their cost-effectiveness (unit costs and potential emissions-reduction); (4) it provides some idea of how much reduction in emissions is achievable and what cost; and (5) it constitutes a powerful heuristic for developing an investment strategy and portfolio for reducing GHG emissions.

Unfortunately, least-cost emissions-reduction planning is easier recommended than implemented—there are many conceptual and methodological problems in computing the costs of interventions and in estimating the benefits or effectiveness.

In particular, there are a number of specific questions: (1) how to allow for the prospective costs of GHG emissions (or of reducing them) when appraising GEF projects; (2) what discount rate to use in comparisons of costs, bearing in mind inter-generational concerns and the importance of cost-effectiveness; (3) how to estimate the benefits of innovation, in particular the contribution of an investment to reductions in the costs of future investments; (4) how to reduce the large transactions costs in projects having good economic potential.

Assigning a Cost (or Shadow Price) to GHG Emissions

A vital step in assessing the cost-effectiveness of technologies is estimating the cost (or shadow price) of GHG emissions. The shadow price needs to be based on an accumulations constraint rather than on an annual emissions limit.

Though it is not known with certainty and consensus what is the safe level at which carbon and other GHG accumulations should be stabilised, it turns out that, unless a 'runaway' feedback occurs, use of the backstop technologies is consistent with stabilising accumulations at any one of several levels over the long term. Thus, the ground rules derived are not too sensitive to the magnitude of the safe level.

It is propsoed that, for GEF projects, the cost (or shadow price) of GHG emissions should be the extra marginal cost of turning to non-GHG-emitting or backstop technologies, i.e., the marginal costs of renewable energy technologies should set upper bounds to the shadow price (or imputed tax) to be placed on GHG emissions. This means that the shadow price grows at a rate determined by the discount rate until the long-term costs of using the backstop technologies is reached.

The resulting formula for estimating the marginal cost is similar to that for computing the optimal price of a depletable resource. According to the depletable resource formula, the price of a resource (such as fossil fuels) equals the "extraction cost" plus a "user cost" term which rises over time until the backstop or replacement technology (renewable energy) becomes economical. In the context of global warming, the resource in question is not the *availability* of fossil fuels but the *safety margin*. This safety margin (defined by the limit to the use of fossil fuels set by concerns about climate change) can be deemed to be a resource because it is used up or depleted with increase in GHG accumulations.

It is vital that the upper bounds of the shadow price should be set by the costs of the *marginal* technologies likely to be brought into use—not by the lowest costs of the most promising technologies. With this approach, technologies at the margin, and of course those below it, automatically qualify; only outliers do not. Further, extra weight is given under this criterion to those applications (such as the use of photovoltaics for rural electrification in favourable locations) which reduce any added cost of turning to renewables.

The preliminary estimates suggest a figure of about \$35/tonne of CO_2 rising at the discount rate of 10 per cent per year up to \$600/tonne. Alternatively, an undiscounted figure of \$35/tonne migth be used in NPV calculations of costs. Such estimates should be raised or lowered periodically as further evidence on costs is gathered. They would also eventually be raised or lowered according to the emerging evidence on global warming.

Human induced global climate change is now in an unnecessary condition of gathering fossil GDP driven risk. This contravenes the 'prevautionary principle' which says that "risks which do not have to be taken, should not be taken." The dangers of fossil-dependence are great and these are alternatives to fossil fuels. In order to understand the political and the ecological nature of this risk, the global relationship between fossil-GDP-economics and global ecopolitics must be assessed.

This assessment should recognise that the climate change threatened global economy is a single system within which equity is now inextricably linked to survival. 'Equity" is the trans-temporal equal sharing of rights, risks, irresponsibilities, rewards and restraints. Our global survival prospects depend on understanding this and preventing powerful minority fossil fuel interests from continuing to raise the levels of risk and obliging the majority to face risks which violate universal survival interests. The UN Climate Convention may yet be the legal instrument which embraces these fundamentals. Preventing dangerous anthropogenic interference with the climate system is its stated objective.

Vested interest in a 90 per cent fossil global economy

To this end, the case for initiating immediately the displacement of fossil fuels in favour of renewables, is now more compelling than is the case for refining their use through'efficiency gains' so as to justify their retention. Efficiency arguments are relevant, but alone are incapable of solving the climate problem. They should not be used to postpone fossil fuel displacement (This will be examined in greater detail later on in this paper). The only arguments which can be (and are) put up for efficiency alone are those inequitably based on unnecessary and dangerous risk-*enhancement* in violation of the precautionary principle. The fundamental judgement against the continuation of fossil energy investment and dependence is really only opposed by vested fossil fuel interests.

These interest groups have resorted to the so-called "no-regrets" arguments. These energy-efficiency and cost-effectiveness arguments are divorced from the precautionary principle. The no-regrets idea is, "if it makes economic sense to do it anyway, do it anyway". The arguments try and minimise the public's perception of the risks of climate change whilst trying to maximise the image of their proponents as caring and competent in the management of risk. The proponents' track record however, has been in the management of financial risk. And this has been in a context of unsustainable resource deployment and unaccountability over the creation of social and environmental disbeneficiaries. In reality these vested interests are a monopoly of self-interested primary fuel and energy producers in a global fossil economy which is still 90 per cent dependent on the combustion of fossil fuels for energy.

The IPCC and CO_2 emissions cuts

If further deterioration in the global climate is to be avoided, the Intergovernmental Panel on Climate change (IPCC) atmosphere stabilising CO_2 emissions cuts must be initiated promptly. The IPCC stated in 1990 that in order to stabilise the rising atmospheric concentrations of CO_2 at 1990 levels, immediate minimum 60 per cent cuts in the emissions of CO_2 were required it was not clear whether that represented an optimistic or a pessimistic view of where the levels of atmospheric CO_2 should be stabilised to avoid interference with the climate system. The IPCC did however, exclude from that estimate the effect of what they described as the probably positive feedbacks which would further accelerating the warming trend.

Where must the emissions cuts commence?

It is in the lead economies of OECD from where these emissions cuts must be led. The Convention text already acknowledges this requirement. Historically, the industrialised countries are responsible for at least 83 per cent of the CO_2 output (the principal greenhouse gas) from industry. Currently, OECD alone is still responsible for nearly 50 per cent with only 16 per cent of global population, and the US is responsible for 50 per cent of this amount with only 3 per cent of global population.

In order to solve the problems of human-induced climate change, a per capital fossil GDP indicator needs to be established. Three years after the IPCC warned of the need to reduce CO_2 emissions, emissions are still rising, dangerously provoking global climate change. The difficulty is the global economy and its Gross Domestic Product (GDP) are currently 90 per cent linked to the combustion of fossil fuels for energy. Inevitably fossil CO_2 emissions and GDP trajectories are very closely inter-related. We can therefore measure what GCI calls this 'fossil GDP'.

However, despite this context, rising per capita GDP is still taken as a valid monetary measure of economic benefit. It is the World Bank's principal so-called 'development indiactor'. But attention is not given to its fossil fuel component, and rising per capital GDP is assumed to be a universally achievable phenomenon giving rise to a universally accurate indicator of benefit. The truth is in a fossil GDP economy, universally rising per capita GDP is only possible (if at all) in the context of rising CO_2 emissions, rising environmental impact and rising risk of climate change adversity.

Consequently we believe that: "per capital fossil GDP is—and should be made—the principal indicator of environmental impact globally." It is for this reason and in the understanding of equity as defined above, we assert that 'responsibilities for' rather than 'rights to' environmentally finite and vulnerable resources-like stable climate—should be allocated by income.

We will return later to the detailed consequences of using this indicator to calculate the "sustainable per capita fossil GDP threshold" and what this reveals about consumption globally.

It Pays to Pollute-The view from the World Bank

The World Bank and others advocate the reverse saying that environmental rights should be allocated by income. The 1992 development report contains a description of why they feel CO_2 emissions entitlements should be allocated by income. Allocating rewards to income is normal in the bank's normal line of business. It is the business-as-usual incentive of positive feedback for capital. However, in the management of climate change this approach is clearly problematic. Rising global temperature indicates net positive feedback (or self-reinforcing reactions) coming into play in the global climate system. For e.g. heat trapping gases accumulating in the atmosphere raise global temperature and this in turn causes other temperature raising phenomena to manifest-examples of this are polar ice melting as a consequence of temperature rise which causes the incoming heat from the sun to be less efficiently reflected back into space with more heat retained-as more heat begets more heat, the oceans warm up releasing more CO_2 and so on—this is a sampling of positive feedback.

On the other hand, the global economic system seeks growth operating on a series of deliberately applied positive feedback mechanisms such as interest and profit. Here the practice of seeking financial gain is facilitated by rewarding capital and capital transactions with the incentives of interest, growth and/or commissions. In fact, 'growth' is an expression of net positive feedback within any frame chosen for observation. In the economic process, investors and traders naturally try to take out more than they put in for personal corporate reasons. But the global climate crisis is really about this very point-if you take out more than you put in, the system will eventually compensate for the imbalance. The neo-classic economic law of diminishing returns should really be viewed in this context: the current net fossil GDP growth phenomenon (ie in spite of the diminishing returns), and the growth of risk (ie the diminishing prospect of climate stability).

By proposing as the Bank does, that rights to emit CO_2 should be allocated on the extent of fossil GDP income which a country derives from the global fossil economy, the Bank and others are effectively formalising the marriage of the deliberate positive feedback of the financial system on the one hand, to the already positively fossil fuel sensitised climate feedbacks on the other. De facto, this approach exists already. Formalised, it would simply consummate folly. But the Bank describes it as "more feasible" than their Aunt Sally alternative. The Aunt Sally is where rights are allocated by population and then regarded as implicitly infeasible because the high income countries "have exhausted their right to emit."

What they should be recognising is that by allocating responsibilities by income, causers of climate change would be targeted. This would enable responsibility for remedial actions to be specifically allocated to those who have been and continue to be responsible for creating and unnecessarily sustaining this crisis. Contrary to the assertion made in the report by the Bank, this approach seems to have more prospect of a political—not—to-mention an ecological—future. It is also more consistent with the "polluter pays principle" than the Bank's adriot shift to the **"it pays to pollute principle"**.

At the heart of this crisis there is this great conundrum. Individual well-being can now only really manifest in the context of seeking the well-being of all. This is not so much a moral question as an economic question. Self interest needs to be redefined to embrace the universal interest. The indicators of progress, development, impact and change which we choose, need therefore to be universally valid as well. The narrow self-interest of profit and the business of seeking this in the short term at the expense of second and third parties, will continue to ensure the irresponsible investment decisions which still dominate the global fossil economy. In consequence, and because the threat of global climate change is so comprehensive, we must re-educate personally narrow assumptions about well-being to assimilate this fundamental sense of per capita income accountability and the universal interest.

What is the appropriate context in which to position energy efficiency arguments?

We now look more closely at energy efficiency and explain why we believe it is subsidiary to the primary issue, namely-the early start to fossil fuels being displaced in favour of renewables.

First the meaning of energy efficiency. Energy efficiency is expressed as a ratio. It has a technology component, ie how efficiently does a device or a system match the energy it uses to produce the usable energy it puts out. In economics, energy efficiency is taken as the ratio between the money or GDP generated by the energy/economic system on the one hand, and the energy units required by this system to generate the GDP on the other. In the context of the global fossil economy, it is the ratio of dollars of GDP generated per unit of carbon (from CO_2 emissions from the fossil fuels) used to generate the dollars of GDP.

For example:

Global GDP in 1990 totalled around twenty and a half trillion dollars ($20,500,000,000,000);

Global CO_2 emissions from industry in 1990 (according to US energy department statistics) contained around six billion tonnes (that is metric tonnes) of carbon (6 gigatonnes, or 6 GTC, or 6,000,000,000 metric tonnes of carbon);

Fossil fuels supplied 90 per cent of the energy required to generate the global GDP, so the twenty and a half trillion dollars should be reduced by 10 per cent to make the calculation of the dollar/carbon ratio more precise; this gives a figure of $18.45 trillion;

The 1990 global average energy efficiency ratio was therefore $18.45 trillion/ 6 GTC or $3,075 per tonne of carbon from fossil fuel burning.

For comparison the:

Japanese energy efficiency ratio for 1990 was $10,839 per tonne of carbon,

US energy efficiency ratio for 1990 was $3,852 per tonne of carbon,

Chinese energy efficiency ratio for 1990 was $596 per tonne of carbon.

Of the group, Japan was clearly the most energy efficient economy, China very inefficient by comparison (there are grounds, for regarding the available Chinese GDP figure as an under-estimate) and the US is about average. However, if every one of the 5.2 billion people on the planet in 1990 had emitted industrial CO_2 even at the 1990 efficiency rate of the average Japanese person (ie 2.35 tonnes of carbon per person), the actual global CO_2 output of 6 billion tonnes would have been more than doubled to over 12 billion tonnes of carbon. If every one of the 5.2 billion people on the planet in 1990 had emitted industrial CO_2 at the 1990 rate of the average US person (ie 5.23 tonnes of carbon per person), the actual global CO_2 output of 6 billion tonnes would have been more than quadrupled to over 27 billion tonnes of carbon.

The point of this observation is to show that in the context of rising energy demand and the need for universally valid indicators as already argued, even by the standard of the conventionally most energy efficient economy on the planet (Japan), five-fold universal average gains in energy efficiency (or an 80 per cent reduction in the amounts of carbon per \$) would be required to meet a 2.4 GTC maximum output level—or the IPCC atmospheric CO_2 concentrations stabilisation requirement of a 60 per cent cut in the 1990 levels of CO_2 emissions. Even assuming such a technological miracle could happen, it would take years to achieve by which time the cut requirement would have increased because of further atmospheric CO_2 build-up.

So the real point of this is to back up the main assertion at the outset-namely, that the case for the immediate displacement of fossil fuels in favour of renewables, is more compelling than is the case for retaining them whilst refining their use in the name of efficiency gains. The efficiency gains will never outpace the positive growth in emissions unless this displacement is initiated immediately. The leadtime for such a technology transition is considerable and even with an urgent action plan under way, it is not a foregone conclusion that we will beat the clock of fossil GDP-induced climate adversity, especially if the second hand of corporate efficiency arguments continues its sweep unchallenged.

A sustainable per capita fossil GDP threshold

We already know that the global average energy efficiency ratio for 1990 was \$3,075 per tonne of carbon. We can also calculate the 1990 global average per capita carbon emission from industrial CO_2. It was 6 GTC divided amongst 5.2 billion people or 1.15 metric tonnes of carbon per person. This figure reduced by 60 per cent-consistent with the IPCC stabilisation cut-gives an "allowable" 0.46 tonnes per person per annum.

The sustainable per capita fossil GDP threshold for 1990 is calculated by seeing what globally averaged level of dollars could be generated, given the \$3,075 per tonne efficiency average and the allowable 0.46 of a tonne per capita. The answer is \$1,415 per person (or 0.46 tonnes of fossil carbon) and this is the sustainable per capita fossil GDP threshold for 1990. The results of applying this figure to the global economy are striking.

GDP *debitors* and *creditors* were assessed using the following method. The 'sustainable per capita fossil GDP threshold' figure of \$1,415 is allocated to each country by population. These *sustainable fossil GDP quotas* are then compared with the actual 1990 GDP values for each country. When a country exceeds its quota, we call it s *fossil GDP debitor*. The amount by which a country shortfalls its quota, we call *'production dumping'*. When a country shortfalls its quota, we call it a *'fossil GDP debitor'*. The amount by which a country shortfalls its quota, we call *'subsidy*. Also *'Transfers'* are calculated in the following way. The total amount of the transfer required, equals the total amount of the subsidy. The debitors settle this collectively with amounts calculated in proportion to the size of their debt. For example the US debt represents 33 per cent of the total value of the debt, therefore they put up 33 per cent of the value of the transfer. After deductions in this way to liquidate the subsidy, all countries commit the balance of their economies unconditionally to the pursuit of sustainable development and displacement of fossil fuels. The actual fossil GDP quotas and related population imbalances on 1990 values.

The are approximately proportional to the actual imbalances described. The proportions, expressed as percentages, are summarised in the table alongside. The actual full country breakdown appears on the full page chart overleaf entitled. *All countries as 1990 fossil dollar debitors and creditors with proportional transfers and sustainable development commitments."*

In total, the energy subsidy received at the present by the debitor countries from the creditor countries, is worth $3.1 trillion annually on 1990 values. It is a serious distortion of the global political economy and shows clearly where the unsustainability of the present economic order is to be found and in what proportion. The energy subsidy actually assists the unsustainable production-dumping of $14.6 trillion worth of fossil-fuel products on the global market by the debitor countries. At least the current subsidy should be dealt with immediately.

The fact this runs annually in the order of magnitude of trillions of dollars, makes risible the OECD countries' response to the issues outlined in this paper. They have yet to grasp the policy implications of their own enormous and unrestructured fossil debt overhang. What is required is structural adjustment of the fossil debitor economies and an end to the energy subsidy and the fossil production dumping. The diversionary response monies theoretically proposed so far by debitor countries are incapable of dealing with the structural issues. $6 billion are being offered conditionally through the World Bank's new *"Global environmental facility" (GEF)* for global warming and biodiversity projects. Monies are even projected net of the value of the 'local' benefit of any project. The World Bank's professed global outlook will founder if it continues to identify the universal interest based on the rationale of *"it pays to pollute"*.

Equity and survival are linked, because it is clearly dangerous, anti-democratic and in violation of the precautionary principle to allow a subsidised 36 per cent of the world's population, controlling 93 per cent of the fossil derived income, to continue to raise the level of climate change risk which the global population is obliged to take in consequence. Income differentials of this kind are clearly the appropriation of the survival rights of the many by the inequitably and unsustainably subsidised few.

Adopting procedures like those outlined here, will allow us to phase out fossil fuel led GDP growth consistent with the precautionary principle. We can then concentrate on the strong and sustainable economic growth and development of democratic, decentralised, renewable, energy systems. This is the prerequisite of global ecological recovery.

For interest we include these final graphics, US and Japanese CO_2:GDP curves 1960-90, floating, fixed and as shares of global. It is clearly precipitous to allow 5 per cent of world population to control 50 per cent of the unsustainably generated wealth. We also comparing the 1991 UNDP assessment of skdwed distribution, with GCI's. UNDP's income split is 6:1; GCI's income split at the sustainable fossil GDP threshold, is 16:1.

To settle the scientific debate about how mankind is changing the earth's climate, UNEP and the World Meteorological Organisations jointly set up the Intergovernmental Panel on Climate Change (IPCC) in 1988. The panel made a presentation of its findings at the Second World Climate Conference in Geneva late last year. IPCC chairperson Bert Bolin warned that there was no longer time to use the remaining uncertainties spout global warming and climate change as an excuse for inaction.

"What we are doing is irreversible", he said. "Climate change is on. We don't know how much or when, and we don't know what's tolerable. We do know there's delay between increasing greenhouse gas concentrations and climate change, and we know climate change will be with us for decades, if not centuries".

The planet is already committed to regional drought, storms, shortages in water supplies, disruption of fisheries and extinction of many plant and animal species, as global warming changes weather patterns and disrupts natural ecosystems. But the current predictions go global waning cannot take account of the reactions and interactions of the planet's land, ocean and ice masses in response to the rise in temperatures which we have already initiated.

The conference's final statement was a particular warning about 'feed-back' process. By which the greenhouse effect triggers off reactions that, in turn, exacerbate the greenhouse effect. These are likely to increase greenhouse gas concentrations in the air leading, in most cases, to greater global warming than predictions have so far indicated.

A problem arises in including feed-back reactions in computer models of a future climate because these processes do not behave in a predicatble enough way to give reliable results. It is impossible to quantify the effect of feed-backs when they are all feeding of each other. It is rather like the feedback shriek from a sound system—it is impossible to separate individual sounds or notes from the noise. Similarly, when the responses of plant communities, land masses, oceans and ice begin feeding back on each other, the uncertainty of each divided effect is magnified by all the others. It there fore becomes impossible to produce reliable predictions of their overall effect on the climate system. What we can predict, however, is that if they do interact synergistically in this way, their combined effect will be much greater than the sum of the individual effects taken separately.

In addition, none of the calculations of the concentrations of greenhouse gases and their likely warming over the next hundred or so years take into account the feed back that will come from the biosphere, and the planet's plant and microbial committees in particular, as temperatures rise and the climate changes. These biospheric reactions are only now being recognised, and it will be 10 to 15 years before they are understood well enough to incorporate them into computer models.

Alarmingly, however, nearly all the feed-backs that have been identified so far promise to increase greenhouse gas concetrations in the atmosphere, and the overall extent of global warming and climate change which we can expect in the future.

Many of the biospheric feed-backs hinge on expected changes to the carbon cycle during which carbon is stored by land masses and oceans, then released to the atmosphere, and finally absobed into land masses and oceans once more. The scientific consensus is that we can expect significant amounts of extra carbon dioxide to be released into the atmosphere in future, as plants and micro-organisms change their behaviour in reaction to warmer temperatures. Plants and soils store almost three time-share carbon than is found in the atmosphere. It has been estimated that the planet's ecosystems on land store around two million tonnes of carbon, with solids strong twice as much as plants. Consequently, even small changes in carbon storage by ecosystems could have large effects on the amounts of carbon dioxide released into the atmosphere.

A reduction in the storage capacity of plants and soils is likely to arise as a result of changes in plant respiration rates—the process by which plants absorb oxygen from the air and release carbon dioxide in its place. In a warmer world, plants will respire more rapidly, releasing more carbon dioxide into the atmosphere. Rough calculations made by IPCC scientists suggest that these increases in perspiration rates are likely to add ' one or a few' thousand million tonnes of carbon to the atmosphere each year. In 1987, total manmade emissions of carbon totalled 5.7 gigatonnes (one gigatonnes equals 1,000 million tonnes). Another potentially significant biospheric reaction will result from the expected death of any forests which do not adapt rapidly enough to rising temperatures. As these forests rot, they will release large quantities of carbon dioxide and methane.

Research suggests that the area covered by the northern boreal forests will diminish dramatically, from a present twenty three per cent of the world's total forested area, to less than one per cent. Boreal forests store substantially more carbon in their soils than do the deciduous forests that will be an additional boost to atmospheric concentrations of carbon dioxide as the northern boreal forests disappear. As temperatures continue to rise, greater quantities of both carbon dioxide and methane will be released from the Tundra and permafrost in high northern latitudes. These regions contain large and relatively undecomposed stocks of carbon in their soils. Warming will accelerate decomposition rates, allowing the strored carbon to be released to the atmosphere.

And as the permafrost melts, large quantities of methane trapped in the permafrost either as gas, or in frozen organic matter, or as methane hydrates will find their way into the atmosphere, further adding to global farming and its consequences. In another related feedback process, warmer temperatures in higher latitudes will speed up the activity of methane-producing bacteria which live in the feasible, because higher latitudes are expected to warm two to three times more than the global average which, so far, is predicted to burned 3^0 c by 2100. Methane, on molecule-for-molecule basic, traps 63 times more heat than carbon dioxide over a 20 year period.

Then there is nitrous oxide, a greenhouses gas which lasts in the atmosphere for centuries, and is an even more efficient greenhouse gas than methane. The IPCC scientists have reported that in regions where rainfall increases as temperatures rise, we can expect an acceleration of the nitrogen cycle, and a consequent increase in the amounts of nitrous oxide produced by solid bacteria and released to the atmosphere.

Apart from these large reactions from plant and microbe communities which will enhance global warming, there are other important feed-backs which will also swing into action as temperatures rise. In a warmer world there will be more evaporation, and so more water vapour is a major greenhouse gas, and increased amounts of water vapour will add significantly to the warming. Warmer temperatures will also substantially reduce the extent of the planet's snow and ice cover, bringing another large feedback into play. There has already been a dramatic decrease in the planet's snow cover since 1980. As snow and ice continue to melt, there will be fewer white areas on earth that reflect the sun's heat directly back to space. Instead, larger areas of the planet will be darker, and therefore able to absorb more of the sun's heat and increase global farming.

And then there are a number of reactions within the oceans, virtually all of which will increase the carbon dioxide in the atmosphere. For a start, the means will not conveniently absorb the extra carbon dioxide in the atmosphere at the same rate as it is emitted. If atmospheric carbon dioxide increases by 10 per cent, the oceans will only absorb an extra one per cent. In addition, as the surface waters of the ocean warm, they will not be able to absorb as much carbon dioxide as they do at present. IPCC scientists say this could add' a few per cent' to concentrations of carbon dioxide in the atmosphere.

It is also thought that, as the oceans warm, dissolved organic carbon will decompose more rapidly, again releasing increasing quantities of carbon dioxide into the atmosphere. Moreover, global warming is expected to adversely affect plankton, the tiny marine organisms which play a major role in oceanic absorption of carbon dioxide. As sea temperatures rise, the ability of these plankton to absorb carbon dioxide will be reduced. This process is likely to be attenuated by ozone depletion. Marine biologists have now established that the increasing amounts of ultraviolet radiation allowed through a diminishing ozone layer will significantly reduce the ability of plankton to soak up carbon dioxide.

The combination of knowledge about unavoidable warming and climate change, and the large uncetainties about feed-backs, has sparked urgent calls for action from the world's scientists. They insist the immediate reductions of 60-80 per cent in carbon dioxide emissions, and similarly drastic cuts in emissions of other greenhouse gases are essential. Underlying the strength of their calls is the fear, expressed repeatedly throughout the Second World Climate Conference that , without action, Earth's systems could eventually pass a certain critical threshold beyond which life forms might not be able to survive. Whatever this threshold might bẹ is unknown, but the scientific consensus is that anything more than a doubling of greenhouse gas concentrations poses and unacceptable risk of escalating, and potentially uncontrollable, feedback sections. At present, these gas concentrations are set to double by about 2025. Action to achieve the scale of reductions being talked about by the scientists presents possibly the greatest challenge the human race has ever faced, UNEP executive director, Dr. Mostafa K. Tolba told the conference that nothing short of action which affects every individual on this planed can forestall global catastrophe. "Nothing less than a complete change in attitudes and lifestyles will succeed", he said.

The consensus is that is up to the industrialised nations to take the action first. They started the problem in the first place; they are responsible for three-quarters of all man-made carbon dioxide emissions; they have the technological knowledge already available which would allow immediate and substantial reductions in greenhouse gas emissions, and could enable developing nations to do the same. Some 20 nations have already announced plans to stabilise or reduce their emissions but only global actions will forestall global catastrophe. It remains to be seen whether the industrialised nations will act in time. Contemporary humanity faces á fundamental challenge created by future unprecedented changes in the global climate—the challenge of securing sustained development of people around the globe. All countries, regardless of their present levels of development or types of political systems are affected by the adverse effects of the climate variability.

The Hamburg Congress took place at a unique juncture in time when governments have become aware more than ever before about the seriousness of the environmental and societal implications of climate change. There is now a higher degree of receptivity to the warnings about possible costs of inaction. These warnings have been underscored by the series of calamities demonstrably associated with climate variability. World attention has been further mobilised by the gravity of the message in the report of the Brundtland Commission "Our Common Future" and of the 1988 Toronto Conference on the Changing Atmosphere. The Hamburg Congress served as yet another stepping stone in the process of increasing the awareness of the impact of climate change on the future options available to the international community.

Many conferences during the last fifteen years issued alarming warnings about climate change but were able to capture only scant attention. Little action was generated to reduce or reverse human activities that contribute to the change in the global climate. Now that world attention is focusing on this problem, what is the next step that could generate serious actions? This has been the principal concern of the Hamburg Congress.

The Hamburg Congress

1. reviewed the current understanding of the process of the climate change;
2. outlined the known as well as the potential impacts of variability and climate change on the socio-economic development of different countries and regions;
3. discussed the nature of policies to respond to the climate change and variability and
4. considered strategic options indicating the possible operational role for different actors concerned with this problem.

The Hamburg Congress was more than just an expert meeting convened to deliberate on the scientific findings and uncertainties. It was also an occasion to identify reliable evidence that causes current concerns and to define possible courses of action that could mitigate the potentially deleterious future situation. In seeking viable options, the Congress was aware of the need to identify all possible constraints that could impede concrete actions and to include the special difficulties that might be faced by many developing countries in pursuing these options.

The Hamburg Congress attempted to treat these options not as the responsibility of any one group but as the joint responsibility of North and South as well as East and West, just as these are joint concerns of scientists, policy-makers, agricultural and industrial producers and the public-at-large. The unique combination of participants at the Hamburg Congress consisting of prominent social and physical scientists, policy and decision makers, representatives of international organisation, non-governmental organisations and industry facilitated the task of exploring a wide range of realistic solutions seeking to avoid either a sense of panic of a sense of procrastination.

It was the unanimous view of the participants of the Congress that noticeable increase in the global warming has taken place during this century, estimated around 0.7^0C and that the increase is consistent with the predictions about the emission of greenhouse gases due to human activities. At this rate, a further warming of the earth to the extent of 1.5 to 4.5^0C by the middle of next century is very likely. There was a high degree of consensus that the chief sources of the warming are Carbon Dioxide (CO_2), Methane (CH_4), Nitrous Oxide (NO_2), Chloroflourocarbons (CFCs) and Ozone (O_3).

There was also major agreement on the origins of these pollutants chiefly attributed to the present manners of energy utilisation; current land use patterns such as in agriculture and forestry; and some industrial practices. There were no major differences on the predictions of general consequences due to the global climate changes, such as sea-level rise and disruptions in the patterns of production systems. However, lack of sufficient understanding on the local and regional impacts of climate change was evident. Unprecedented levels of international cooperation will be needed to deal with the inputs of climate change.

The Congress reviewed a number of possible approaches to control, reduce or eliminate the sources of pollutants. While majority view prevailed on the need to immediately initiate concrete measures, while further research moved on, there were some differences as to whether some of these measures are implementable under current conditions. Some argued that warming trends may in fact be beneficial, at least to some parts of the world, and questioned the need for control measures.

The impact of the warming trends on developing countries; the effect of their development activities on future climatic changes, and the prerequisites to integrate the developing countries' special concerns in the global efforts received special attention during the deliberations. The Congress called for increased research efforts focused on clarifying outstanding issues and at the same time to elevate the tempo of political discussions to seek concrete directions to prevent potential disasters.

There was a positive recognition of the past and present role of the United Nations system, especially of the WMO and the UNEP, in supporting the research efforts to collect data, analyse and monitor the phenomenon of climate change and variability as well as to promote dialogues at the national and intergovernmental fora on all aspects of the issues of climate change.

SCIENTIFIC ASSESSMENTS

Warming Trends

The predicted climate change within the next century could be as large as between the last ice age and now. An estimated increase of 1.5^0—4.5^0C by the year 2050 in the global average temperature is considered very likely. Analysis of historical records of surface temperatures measurements suggest that the planet has already warmed by about 0.5—0.7^0C.

These predictions are based on a variety of modelling efforts to simulate the highly complex climate systems to include the influence of human activities on the climate. Presentations on the state-of-the art on climate modelling showed reasonable agreements in the large-scale predictions but differed significantly on regional features. Some of the other inherent difficulties in modelling, such as interactions and feedbacks among subsystem models, between atmosphere and the rest of climate system as well as influence of changes in ocean circulation etc. were pointed out. With the availability of more powerful computers, it will be possible to synthesise the subsystem model for different applications.

The discussion on modelling highlight the complications in separating the influence of human actions on the climate from considerations of natural climate processes.

Greenhouse Gases

Despite the current uncertainties of the models, there was a clear scientific agreement that the increase in temperature is taking place due to the combined effects of the greenhouse gases in atmospheric concentrations of, mainly, carbon dioxide (CO_2), Methane (CH_4), Nitrous Oxide (N_2O) and Chlorofluorocarbons (CFCs).

Nearly half of the increase in global warming is attributed to the increase CO_2 concentration. The rest of the greenhouse gases will contribute to the other half of warming. The measurement of these gases have shown steady increases during the last decade. The world will continue to warm, even if greenhouse gas emissions are reduced to zero in the immediate future. Emissions of these gases between 1880 and 1985 have committed the planet to a substantial additional warming. The total realisation of this warming is delayed by the thermal interia of the world ocean.

Sources of Greenhouse Gases

The main mad-made sources of CO_2 are fossil-fuel burning (coal, oil and natural gas) and land use (deforestation and soil degradation). One estimate puts this source of CO_2 in the order of 16,000 to 29,000 million tonnes per year. The ocean and land vegetation absorb part of this CO_2. Deforestation reduces this absorption. A large number of sources of other trace gases have been identified.

The spatial sources of emission of CO_2 also varies considerably. One estimate states that in 1985, the United States produced about 25 per cent of fossil-fuel-derived CO_2, while the contributions of the Soviet Union, Western Europe, China and Japan were 20, 15, 10 and 6 per cent respectively. It is expected that in future the developing countries and the centrally planned economies will contribute increased share of CO_2 due to a rise in population and growth in industrial activities requiring higher levels of fossil fuel consumption.

IMPACT OF WARMING TRENDS

Global Impacts

One of the direct consequences of the global warming will be world-wide rise in the sea-level, due to thermal expansion of oceans and melting of the ice. One assessment concludes that a warming of 1.4 to 4.5^0C would cause a rise in global sea-level of 40 to 120 cms. Such a sea-level rise would cause flooding in coastal areas, increase salt water intrusion into acquifers, inundate wetlands and destroy fishering resources.

Since the same activities responsible for greenhouse gases also emit other pollutants that cause stratospheric ozone depletion and tropospheric acid deposition, additional impacts such as reduced agriculture and forest yield, increased exposure to ultra-violet radiation and greater stress on natural ecosystems are probable. If greenhouse gas emissions grow rapidly and climate changes abruptly, the economic damages for societies will be large and unavoidable. If, on the other hand, the rate of change is moderated by deliberate policy interventions, sufficient time may be available to develop adaptive responses.

Regional Impacts

While there were general agreement on the global effects, considerable uncertainties exist on the regional impact of the climate change. One of the accepts predictions of the climate change is that the greenhouse effect will not be uniform everywhere. Warming at the poles is likely to be 200 to 300 per cent the global average while it may be only 50 to 100 per cent of the average near the tropics. As the natural thermal gradients on the planet's surface is reduced, global patterns of winds, ocean currents as well as the timing and distribution of rainfall. Ocean currents that now moderate the climates of certain high-latitude countries may shift, causing cooling effects in come countries, even as the rest of the world experiences warming.

The impact of these regional variations are perceived differently by different analysts. Some expressed the view that the net effect of the regional variations would have adverse consequences to most countries of the world in terms of food, health, trade and security. Some others, however, felt that the increased warming may in fact be beneficial to many countries. As for instance, the increased melting of some permafrost regions may expose valuable land and mineral resources. Prospects of increased CO_2 may contribute to enhance agricultural and forest yields. Altered patterns of rainfall and temperature at the tropical latitudes in favour of increased precipitation may help to increase agricultural production.

As global average temperature changes, more clear understanding on regional impact, especially on the timing, spatial distribution, duration and intensity of rainfall and runoff as well as on the frequency of extreme meteorological events such as droughts in the tropics will be necessary. Current scientific knowledge is insufficient to assess the impact global warming on regional rainfall regimes and on the development process. If the amount and intensity of precipitation were to change, it may affect soil moisture conditions as well as woil erosion and desertification processes.

These differences in perceptions show the importance of focusing greater attention on regional climate variability than has been the case so far. The proposed research programmes such as International Geosphere-Biosphere Programme (IGBP) of the ICSU and Human Dimensions of Global Change (HDGC) programme could help to clarify some of these perceptions.

Arising out of the general consensus that the present ways of energy use, land use and some industrial practices are largely responsible for the warming trends a number of suggestions were made to seek alternatives. The special situation in developing countries and their capacity to undertake necessary policy measures received special attention.

The main focus of the discussions of energy options was towards conservation measures in domestic, industrial and transport uses; and on concerted efforts to develop renewable resources of energy such as solar, wind and biomass energies. The nuclear option as a means of reducing CO_2 emission was raised but the lack of means to deal with the triple problems of safety, waste disposal and weapons potential inhibited serious consideration. There is all the possibility of increasing CO_2 emissions due to the explosive growth of cities and towns in developing countries and their economic development programmes resulting in a shift toward greater use of fossil-fuels. The policy choices to design a long-term energy strategy, which slows down or reduces emission of greenhouse gases without penalising human welfare was not clear. Some felt that it would not be realistic to achieve reduction in the present levels of CO_2 emission rates by the developing countries but what may be possible is to minimise future growth rates.

In light of the known impact of deforestation on the atmospheric carbon budget, there was a strong support for policies encouraging afforestation and discouraging deforestation. Possible technological and policy options in farming systems; fertiliser use, irrigation and drainage were proposed.

Concern of Developing Countries

The constraints of developing countries to undertake policy measures to reduce the greenhouse effects was pointed out. These include their preoccupation to address problems of abject poverty, inequitable distribution and draining of resources for debt servicing. The ability of the developing countries to promote national policies towards sustainable development and their readiness to participate in global efforts will, to a large extent, depend upon the measures of international development cooperation to redress these basic concerns.

Despite the seeming enormity of the problem posed by the complexities of climate change and variability, there was a general optimism that progress could be achieved in solving them. The prominent suggestions for future efforts centred around:

1. Support for continued and comprehensive research efforts to monitor and analyse patterns of global and regional changes in climate and their socio-economic impacts in specific contexts;
2. Encouragement for ongoing national and international discussions within and outside the United Nations system and to seek the highest level political support to launch world-wide initiatives; and
3. Emphasis on providing the developing countries the necessary means to be able to respond to the challenges of climate change.

The seriousness of the threat posed by the creeping influence of many of the human actions on the global atmosphere is recognised more widely now, than ever before. Scientific evidence tends to confirm the predictions that the unprecendented rate of change of global atmosphere will create adverse climate changes and accentuate the problems of regional and seasonal climatic variability. Wide-spread consensus has developed on the probability of a rise in the mean surface temperature of the earth by 1.5^{0}C to 4.5^{0}C and the possibility of sea-level rise of the order of 30 to 150 cms before the middle of the next century, if, the present rate of increase in the concentration of greenhouse gases in the atmosphere are not curtailed; and on the extent of the resultant catastrophies to people and nations. All countries, regardless of their present levels of development or types of political system, will have to face this emerging threat. Therefore the issue is not whether the threat will materialise but how soon and under what conditions.

Very little differences exist among the experts on the findings that the chief atmospheric pollutants are Carbon Dioxide (CO_2), Chlorofluorocarbons (CFCs), Methane (CH_4), Tropospheric Ozone (O_3) and Nitrous Oxide (N_2O). The sources of these emissions are clearly attributed to a number of human activities related to inefficient use of fossil-fuels, many land use methods and some industrial practices requiring synthetic chemical used as coolants, solvents, propellants and foaming agents.

The issue is not anymore whether these pollutants contribute to the predicted climatic changes but how effectively their sources could be controlled, or eliminated.

The vulnerability of socio-economic systems to prolonged variations from the normal climatic conditions, has been amply domonstrated in recent decades. The dire predictions of the predominant consequences of atmospheric pollution relate not only to a global climatic change but also to the cumulative changes in the climatic variations, regionally and seasonally.

These changes portened enormously deleterious effects on a variety of social and economic systems that are essential for the survival of the humanity. The damages to the food systems, human health, animal and plant species, trading patterns and international peace and security have become entirely conceivable. Even if the predicted impacts are wide off the mark, prudence dictates that acting now would save greater tragedies than procrastination or no action at all. Inaction will be tantamount to a moral equivalent of ecological irresponsibility, creating a situation which might render future measures ineffective.

The issue, therefore, is not whether the predictions of socio-economic consequences are realistic but whether reasonable courses of action can be prescribed to achieve a balance between ecological and socio-economic concerns.

Given the realities of imbalances in the contemporary world, the distribution of the damages as well as the ability to cope with them will vary from country to country and from region to region. It is accepted that the activities in the industrialised countries, as of now, constitute the main source of greenhouse gases and hence bear the main responsibility to redress the consequences of climate change.

In the developing countries population growth and rising living standards are expected to contribute to the greenhouse effect at a rate as yet uncertain. The current social and economic difficulties including the debt burdens, if continued to persist will surely compound these effects. The issue, then, is not who contributes more to the climatic change but what could be done to promote an equitable system of international cooperation whereby the legitimate aspirations of the peoples of the developing countries for social and economic development are not unduly hindered and penalised. In striving towards the reduction or elimination of the human actions which influence the climatic change, there is a need to distinguish these actions amenable to immediate measures by industrial countries and those which would necessitate international initiatives.

Since 1979, the United Nations system has made efforts to articulate the concerns of global climate change and possible strategies and responses. The support for much of the present day analytical studies and conclusions were derived from the World Climate Programme (WCP) of the World Meteorological Organisation (WMO) acting in concert with other organisations and institutions within the outside the UN-system. These include the World Climate Research Programme (WCRP) with the International Council of Scientific Union (ICSU); the International Oceanic Commission (IOC) of UNESCO; the Climate Applications Programme (WCAP); World Climate Data Programme (WCDP); and the World Climate Impact Studies Programme (WCIP) implemented by UNEP.

The various assessments carried out by UNEP such as Villach (1985), Villach-Bellagio (1987) under the WCP; and the UNEP/IIASA Studies have served to develop the scientific basis for the recent deliberations in the national and world bodies. The WMO/UNEP Intergovernmental Panel on Climate Change, recently established, is intended to seek international consensus on problems of climate change and pave the way for a coordinated international response. The Advance Technology Alert System (ATAS) of the UNCSTD has helped of raise the technology assessment capacities directed towards major socio-economic concerns. The Toronto Conference on the Changing Atmosphere: Implications for Global Security (1988) was a landmark international initiative calling for national and international actions to achieve desirable targets to counter the ongoing degradation of the atmosphere. The Hamburg Congress amply illustrates an international initiative geared towards highlighting specific concerns of the international community and point out the need for attaining desirable goals. The International Geosphere-Biosphere Programme (IGBP) of the ICSU and the Complementary Programme on Human Dimensions of Global Change (HDGC) established by the International Federation of Institutes of Advanced Studies (IFIAS) together with the International Social Sciences Council (ISSC) and the United Nations University (UNU) are two major international initiatives that could help to clarify the scientific and human aspects of climate change. These initiatives and a host of other planned actions provide the needed practical means to sustain international attention and help translate international calls into specific national actions.

Successful implementation of measures to reduce the level of emissions of greenhouse gases would require the participation of every segment of national economic and social endeavours. Governments tend to balance their short and long-term economic and political interests. Industries tend to weigh the corporate interests; non-governmental organisations are concerned with the larger public interests and tend to confront policy and moral issues. The scientific community would have to continuously seek new or additional evidences to support their predictions and translate them into convencing policy terms.

The search for definitive courses of national action to address issues of climate change requires painstaking appraisal of all view points. An approach to accomplish such a task has recently been demonstrated by the 'Equete-Kommission' of the German Federal Parliament. It is not certain, as yet, as to how many more countries will engage in undertaking national policy dialogues in the near future; and how effectively will their conclusions be implemented—and what would be the combined influence of such national initiatives in arriving at international agreements. There is sufficient evidence, however, that the industrial countries should quickly move towards national measures, instead of waiting for international agreements to emerge.

The intensity and diversity of view points that emerged during the Hamburg Congress demonstrates the intractable complexities of the issues associated with climate change. Achievement of real progress will require approaches that will encourage analysis instead of arguments; persuasion instead of force; and communication rather than confrontation. It will help to promote the ability "to think globally but act locally" and the leadership to undertake actions now rather than await clarifications of all issues which may never happen.

Irreparable damages to the earth's atmosphere are being caused by the intervention of human actions on the natural processes. The forecast of consequent changes in the global climate are ominous. Fertile regions will become barren and food baskets will become empty unless timely actions are taken. These dangers will be further magnified by the three most devastating developments of the eighties: the arms race, the destruction of the environment and the international debt crisis.

We are not yet on the path of sustainable development, without which neither peace nor environment is safe. On the one hand, the struggle for basic survival forces people in poorer regions of the world to overexploit nature. The pressures of debt servicing are accelerating excessive exploitation of natural resources. On the other, the short-terms economic interests make many governments to go slow on ecological issues. The cynicism towards multilateralism has led to bilateral preferences. Not a single major joint action on any major global problem occurred during the eighties.

The issues of climatic change, having captured the world-wide attention, could help to breakddown the North-South and East-West barriers. There are hopeful signs that the nineties will provide the prospects for concerted action to prevent the most perilous consequences of climatic change, "next only to the dangers of a nuclear war".

- Drastic change will be needed in the industrial production methods, patterns of land use and energy use.
- Major modifications is habits will be required in life styles to enforce conservation of energy. The nuclear option to reduce CO_2 pollution seems too distant.
- Climatically benign technologies and ecologically sound agricultural production methods should be vigorously promoted.

Convinced of the urgency of the need to act now to address the complex issues of climatic changes and variability confronting the humanity; and

Being aware of the need to mobilise the support of all of the society for effective implementation of the goals and targets.

The Hamburg Action Plan Proposes that:

1. the issues of climate change be maintained in the global agenda to facilitate communication and seek early consensus on viable actions;
2. the research efforts to clarify contentious issues be supported on a priority basis;
3. the industrial societies redirect their traditional present day patterns of energy supply, chemical production, agricultural production and transportation habits towards the path of ecological modernisation;
4. the developing countries be assisted to undertake autonomous analysis and assessment of the implications of global climate change;
5. effective international action be taken to deal with the fundamental and persistent problems of developing countries to

reduce yet another potential source of climate change;

6. the bilateral and multilateral assistance programmes help to facilitate easy flow of environmentally sustainable technologies that could serve to reduce the emission of greenhouse gases;

7. the ongoing activities of the organisations of the UN-system such as WMO, UNEP, UNESCO and CSTD, be pursued in concert; and initiatives such as IGBP of ICSU and HDGC programmes of IFIAS-ISSC-UNU should be supported;

8. national policy studies in the nature of the 'Enquete-Kommission' be carried out in other countries, especially, the industrialised countries; and their recommendations implemented;

9. neither resolutions nor targets and not even the most convincing discoveries can propel the kind of urgent actions that are required, unless the wisdom and statesmanship of the world leadership at the highest level is sought and secured. In order to maintain the current momentum, the Hamburg Congress appeals for a summit meeting of leaders from all regions of the world, under the initiative of the United Nations Secretary-General to meet in 1991, to give the necessary political support to the fruits of deliberations on climate and development. In preparation for such a summit, a Second Congress of concerned experts, organisations and institutions should be convened in Hamburg in 1990, to focus on the issues distilled from the ongoing national and international deliberations.

Global climate change may present manking with its greatest environmental challenge to date. It appears more and more likely that increasing concentrations of carbon dioxide and other greenhouse gases in the atmosphere will cause average global temperatures to rise by 1.5 to 4.5^{0}Centigrade within the next half century or so.

Existing global circulation models are not capable of predicting the magnitude and timing of this global warming with any degree of precision. They are even less capable of predicting the regional distribution of temperature change and its associated effects. The models appear to agree, however, that temperature changes will be greater in northern latitudes than at the equator, that sea levels will rise, and that some areas will receive significantly less rainfall than they do today—and some significantly more.

If global warming occurs within the time frame and temperature range roughly predicted and if these general predictions about its distribution and effects hold true, there will be strategic impacts on all nations and on every sector of human activity.

To assess these strategic impacts, one must look at the capacity of individual societies and the international system as a whole for "anticipatory response". By capacity for anticipatory response I mean a nation's ability to defer current consumption in order to strengthen its ability to minimise future adverse impacts, even when the nature and timings of those impacts is uncertain. This capacity depends more on the political, economic, and social characteristics of a nation than on its physical characteristics.

It is already fashionable to predict winners and losers from global climate change with statements such as "the grain belt will move north" or "the tropics will get more rain." Not only are these statements difficult to prove, given the current state of scientific knowledge, but they disregard relative capacities for anticipatory response. For example, the Netherlands and Bangladesh each face a similar problem with sea-level rise, yet there is little doubt that the former has a far greater capacity for anticipatory response than the latter.

From a purely physical perspective, a small island nation such as Japan has far less room to maneuver in responding to global warming than a huge continental nation such as the Soviet Union. But Japan would appear to have the greater capacity for anticipatory response if one takes other factors into account. In general, one can say that low population density, high per-capita income and technological sophistication contribute to a nation's capacity for early, preventive response and that their opposites detract from it.

Another factor I see contributing to a nation's capacity for anticipatory response is the openness of its political system. At first blush it might appear that authoritarian states with command-and-control economies would have an easier time imposing changes in energy use, agricultural practices, and other resources uses to minimise and adapt to climate change. Experience indicates, however, that effective responses to long-term changes of this kind require a combination of market signals, development of new technologies, grassroot political support, and participation by non-governmental groups which is only possible in democratic societies.

The United States is an interesting case in point. As de Tocqueville pointed out.more than 150 years ago. American democracy seems to have a short time horizon and has not always been successful in developing and carrying out long-term policies. At times of crisis, however, the American people have been able to marshal their creativity and other resources with great success. One can only hope that a growing public understanding of global climate change, combined with strong leadership at the national level, will enable us to make political decision in the short terms that will preserve our options in the long term.

To assess the capacity for anticipatory responses of any one nation, however, is to miss the crucial point. The cummulative impacts of manmade greenhouse gas emissioins cannot be restricted to any one nation or region, but affect the atmosphere over the whole globe. No matter how well any one country does in minimising greenhouse gas emissions or adapting to climate changes, it is at the mercy of its neighbours. Therefore, we must look to the international system as a whole and in particular to organisations such as the United Nations Environment Programme to develop international strategies to deal with climate change.

It is in the interest of all of us that our joint capacity for participatory response be developed and strengthened. In the end, one of the greatest benefits that may emerge from meeting the challenge of global climate change is international cooperation on an unprecedented scale.

THE WHEELS ARE BEGINNING TO TURN

Global warming is an international problem that will require extensive and unprecedented cooperation among both developed and lesser developed countries. No single country, acting alone, will be able decisively to affect the global warming problem.

Moreover, within any one country no single sector is entirely responsible for the problem. The transportation industrial, commercial, and residential sectors all contribute to greenhouse gas emissions—just as many sectors of the economy may be affected by global warming. Thus no single policy initiative can effectively mitigate manmade climate change.

EPA is actively involved in both domestic and international activities related to global climate change. The Global Climate Protection Act of 1987 requires that the US President, through EPA, "shall be responsible for developing and proposing to Congress a coordinated national policy on global climate change." The Act also requires the Secretary of State of pursue international cooperation on limited global climate change. It requires the Secretary of State and the EPA. Administrator joint to submit, by the end of 1989, a report that will:

- Analyse current international scientific understanding of the Greenhouse Effect.
- Assess US efforts to gain international cooperation in limiting global climate change.
- Describe a US strategy for seeking further international cooperation to limit global climate change.

This is a very broad mandate that will require close cooperation with other federal agencies (including the Department of Energy, Agriculture, and Interior, NASA, the Army Corps of Engineers, the National Climate Programme office, and the Domestic Policy Council. The Act encourages coordination of domestic and international climate activities with research and impact analyses that are consistent with policy goals.

EPA and other federal agencies are attempting to develop a process for designing a national policy that will respond to the global warming problem. Such a national policy must include options for adapting to climate change, and options for stabilising the atmosphere and limiting emissions of greenhouse gases. It must reflect assessments of the feasibility of alternative technologies for stabilisation and mitigation, and assessments of the costs of these technologies. EPA has already identified, for example, improvements in "end-use" efficiencies—cost-effective ways of reducing emissions of greenhouse gases. Such end-use efficiency options include making appliances, buildings, and transportation more energy-efficient, enhancing industrial competitiveness with energy-efficiency research, promoting least-cost utility services, and promoting international cooperation to encourage energy efficiency on a global basis. All of these considerations should be included in a national policy framework.

At the international level, several organisations have recognised the need for multilateral cooperation and have become involved with the global climate change issue. The United Nations Environment Programme (UNEP) is responsible for conducting climate impact assessments. The World Meteorological Organisation (WMO) is supporting research and monitoring of atmospheric and physical sciences. The International Council of Scientific Unions (ICSU) is developing an international geosphere-biosphere programme.

EPA is actively involved in many of these efforts. On the multilateral level, the Agency is working with the Department of State and other federal agencies to support UNEP and WMO in establishing an Intergovernmental Panel on Climate Change (IPCC). The panel is developing a vigorous international process to address global climate change issues. It is channeling efforts into three tracks: first, to assess the state of scientific knowledge on climate change; second, to assess the political social and economic effects from a warming, and third, to assess potential response strategies.

The first meeting of the IPCC was held in Geneva in November 1988. At this initial meeting, the IPCC established three working groups. The first work group, chaired by the United Kingdom, is responsible for the timely production of reports on the assessment of available scientific knowledge on climate change. Brazil and Senegal are vice chairs of this group. The second group, chaired by the USSR, will assess the environmental and socio-economic impacts of climate change. Australia and Japan are vice chairs of this second group. The third working group will formulate response strategies. This group is chaired by the United States, with Canada, China, Malta, the Netherlands, and

Zimbabwe serving as vice chairs. Each working group is expected to deliver its final report to the IPCC by late 1990.

EPA also has bilateral relationships with the Soviet Union and China. Since 1972 the United States and the USSR have been actively cooperating in the field of environmental protection and have conducted joint research on a wide variety of environmental issues. In 1987, EPA also signed a bilateral agreement with the Peoples' Republic of China, which included specific provisions on climate change. In general, these bilateral agreements provide an excellent opportunity to facilitate international cooperation and mutual understanding of the climate change issue.

The Agency has studied the effects of global warming for several years, and by Congressional request is currently producing two reports. The first report examines the policy options that, if implemented, would stabilise current levels of atmospheric greenhouse gas concentrations. The second report studies the health and environmental effects of climate change in the United States, including such issues as agriculture, forests, water resources, as well as other ecosystems and societal impacts.

EPA, along with other federal agencies, will continue to be an active participant in the formulation of a policy-oriented federal research agenda and in future debates over strategies to adapt to-climate change or to limit emissions. The evolution of national policy in this area will be complex, and the formulation of a coordinated international response may take many years. Much work remains to be done by the research community to understand emissions, the interactions of emissioins with the biosphere, and the ultimate effect on man and the environment of increasing greenhouse gas emission into the atmosphere.

Short-term Measures

Air pollution and acidification can be death with in short term by countering their effects. The only possible long-term solution is to bring about a reduction of emissions.

The acidification of lakes and streams can be counteracted in the short term by liming. This raises the pH of the water, causes aluminium and other metals to precipitate, making them fall to the bottom, and leads to a decrease in the levels of mercury in fish. After liming many species of flora and fauna quickly return to their former habitat, and fish can again reproduce. Frogs, reptiles and fish-eating birds also benefit. On the other hand, problems may arise for bottom-dwellers, because of the accumulation of toxic metals on the bed.

Liming must continue, however, as long as there is acid rain. Otherwise metals will again separate from the sediment, and then do even more harm than before. In order to prevent a recurrence of acidification, liming will have to be repeated every three to five years. In running water, as well as in lakes with a rapid hydrological cycle, it must be carried out continuously.

In 1985 nearly 100 million kronor was spent on the liming of lakes and streams in Sweden, and 7 million kronor in Norway. But in these two countries alone it would cost several hundred million kronor a year to lime sufficiently to really compensate for the acid rain. Moreover, liming is of no avail against a continued leakage of aluminium and heavy metals from the soil into the water.

The harvesting of crops and the use of acidifying fertilizer always tends to sour arable land, making it necessary in most cases to apply lime, which makes it relatively easy to offset the effect of acidifying fertilizer simultaneously with that of acid rain. The extra amounts of lime need however to be substantial.

The use of acidifying fertilizers and whole-tree harvesting must be limited so as to correspond to the ability of forest land to resist acidification. Increasing the proportion of birch in conifer stands may cause the soil's pH value to rise somewhat. Generally careful forestry is also important—cleaning up and removing damaged trees so as not to favour insects and other parasites, and planting trees that are suited to the local environment. Liming is however not so entirely effective as it is on arable land. While it does raise the pH value of the soil, that is offset by certain nutrients coming into short supply (e.g. magnesium, boron, potsssium), and forest growth may be slowed down at least for 25 years. Damage to mosses and lichens has also been observed to result from liming. Moreover, the cost would be enormous.

Acid groundwater can be improved by placing a basic filter near the bottom of the well as a neutraliser. Alternatively the soil surrounding the well can be limed. But if the only object is to counteract corrosion, this can be done by substituting pipes of a material that is less susceptible than copper. While they are necessary in order to alleviate the worst damage, these measures do not of course offer any long-germ solution to the problem. They are also expensive, and the people who have to pay are not usually those who have caused the trouble in the first place.

Even in the sixties Scandinavian scientists were affirming that there was a connection between ecological damage to lakes and streams in their own countries and emissions of sulpur dioxide in Central Europe and Great Britain. At the United Nations Conference on the Human Environment, which was held in Stockholm in 1972, the Swedish government presented a report on transnational airborne pollution. Already then an eventual diminishment of the growth in Swedish forests was foreseen as a consequence of soil acidification. The report met with general scepticism. Nevertheless, the conference passed a resolution declaring in principle that nation-states were duty-bound to ensure that activities in their country did not cause environmental damage in others.

As nations then became increasingly engaged during the 1970s, several international research programmes were started at the initiative of the Scandinavians to study the transport of polluted air across national frontiers. One was the European Monitoring and Evaluation Programme, EMEP, with some 90 measurement stations distributed among twenty or so countries in western and eastern Europe. The first international conference on the effects of acidification took place in Norway in 1976, and in the following year Norway proposed, at a meeting of the ECE Environment Committee, the setting up of an international convention to restrict emissions of sulphur dioxide.

In 1979 an important political move was made at a meeting in Geneva of the ECE Council of Ministers. At this thirty-four countries of Europe and North America, and the European Community itself, signed a treaty, generally known as the ECE Convention, to control transboundary transports of polluted air. But it was not until in December 1982 that it had been ratified by a sufficient number of states, enabling it come into force in March of that year.

The Convention does not say in so many words what the signatory states must do in order to reduce their emissions. It simply contains a general statement of principles, with emphasis on the responsibility of the individual countries, which shall "endeavour to limit and, as far as possible, gradually reduce and prevent air pollution, including long range transboundary air pollution," to achieve which they shall "use the best available technology that is economically feasible."

The signatories were also to cooperate in the development of plans for the control of pollution and emissions. The current acidification of forest soil, and the damage to trees in many places, cannot be explained away, however, merely by reference to modern forestry practices.

In Central Europe, the high levels of air pollution and the very heavy fallout of sulphur and nitrogen compounds have seriously damaged both forests and soil. Local damage around major emission sources has been noted for a long time. What is new is a substantial increase in the extent and degree of the damage that has taken place over so short a period of time. In 1985 over half of the forest area in West Germany was found to have been damaged. The stricken area totalled about 3.8 million hectares. About half the forests of Holland were also affected, in Switzerland just over one-third, and in Austria one-fourth. Forest damage due to air pollution has been a serious problem in Czechoslovakia since the 1950s, and there are reports that about one-third of the country's forests are now injured and in 200—300,000 hectares of highlying forest the trees have died. In Poland the stricken forests cover an area of at least half a million hectares.

Ailing forests are to be found to a greater or lesser extent in most European countries. Reports of damage also come from France, East Germany, the Soviet Union, Hungary, Yugoslavia, Luxembourg and Belgium, as well as Scandinavia. All told, at least six million hectares of forested land were estimated to have been affected in 1985.

A damaged tree can exhibit a number of symptoms. It is often exteremely difficult to establish a connection between any one type of damage and its cause—whether it is due to climatic effect, insect attack, or air pollution. The damage may have arisen for many different reasons, and any one stress factor may sometimes trigger off a whole range of injuries. Distinctions must also be drawn between factors that create the necessary conditions for the symptoms to appear, factors that start damage, and those aggravating it. Air pollution is prominent in creating the conditions and in starting damage off.

Polluted air can affect trees both directly and indirectly. Direct effects harm needles or leaves, as when the protective layer of wax is corroded by the dry depositon of sulphur dioxide, by acid rain or ozone, or there is damage to the stomata, the minute openings which among other things regulate the evaporation of water. Membranes may also be injured inside the needle or leaf, causing loss of nutrients through leakage and upsetting the water balance. There will often be synergistic effects, too, from several pollutants.

Indirect damage occurs when the ground has become soured by acid rain, with a consequent reduction of the supply of nutrients and release of substances that are injurious to trees, such as aluminium. The combination of a more acid environment with the toxic effect of metals is damaging to roots. As a result of the diminished supply of nutrients and the toxic effect of aluminium, the root hairs may be unable to suck up sufficient water or any kind of nourishment. Possibly, too, the mycorrhiza symbiosis of root hairs with fungi may be disrupted so as to impede, if not entirely cut off, the exchange of nutrients between the tree and the fungus. All this impairs the vitality of the tree, and so its ability to cope for instance with disease and insects. The situation will become worse if the weather is unfavourable. One or two dry, hot summers together with a severe winter may make it difficult for a tree to survive when it is already stressed through lack of water and nourishment.

Most often however the ultimate cause of death is neither air pollution nor unfavourable weather, but parasites such as insects and fungi, which make an end of an already weakened tree.

Trees vary in sensitivity to polluted air. Deciduous trees are less sensitive then the conifers, partly because they present a smaller area of foliage, and partly because the leaves fall off every autumn and hence are not exposed to air pollution for such a long time as the conifer needles. In Scandinavia, spruce, pine, and beech have been to be the most damaged.

Damage to spruce usually takes the form of yellowing and browning of the needles, loss of needles, growth of adventitious shoots, loss of root hairs and reduced root growth, shortened needles, narrower annual rings, a higher incidence of ruptures in the trunk, and drooping lateral branches. Pine trees, too, suffer discoloration of the needles, and thinning of the crown through loss of needles.

Beech shed leaves and develop cracks and tumours in the bark. Shoot formation is disrupted, and the top branches, which normally ramify abundantly, turn instead into long sticks with small, claw-like shoots.

The forests of Scandinavia are important for the whole of western Europe. They are a major source of raw material for the wood-based industries not only of the Scandinavian countries, but of others as well. About 80 per cent of the Swedish and Finnish forest output is exported, the chief markets long having been the United Kingdom and what is now the Federal Republic of Germany. If the forests of Scandinavia should no longer be normally productive, logging will have to be cut back, jobs will be lost and the national economies deteriorate, at the same time as a key resource for European industry will have diminished.

The forest is of more than economic value. It is just as important for relaxation and recreation. We go jogging, camp, pick berries and mushrooms, hike, ski—and meditate—all in the forest. In short, the forest is a part of life—and not only of our own life. For thousands of insects and small animals, numberless plants, and many larger mammals such as elk and deer, the forest is a home. The forest is the most widespread natural feature of Fenno-Scandia, teeming with living creatures and organisms that are adapted to this environment and to each other.

The forest is also vital for maintaining the water economy and regulating climate, both locally and regionally. On steep mountainsides forests are prime safegurards against erosion. Switzerland and Czechoslovakia, for instance, are having problems of soil loss on account of dead or dying forests.

Not only trees suffer from acidification. Other plants do too, and again the effect may be either direct, through dry deposition, acid rain and/or ozone, or indirect, as a result of the soil turning acid.

The most sensitive flora are lichens and mosses, neither of which have a protective wax coating. They take up water directly through their leaves or thalli. Mosses and lichens both have a very active period of growth during the autumn, which makes them particularly vulnerable, emissions then being great.

Lichens can often serve as indicators of air pollution, especially as regards sulphur dioxide. Because the different species vary in sensitivity, zones are formed in cities and around areas of industry. In a typical downtown section, where there is the greatest concentration of pollutants, lichens are almost completely absent, but the cleaner the air gets as one moves away from the centre, the more species appear. In southern Scandinavia the overall load of pollution is so great that some species have vanished almost entirely.

Flowering plants and ferns have not suffered so extensively, at least so far. There is however a tendency for species that are resistant to acidification and thrive on nitrogen to increase, while a decline is noticeable among plants that require a high soil pH and an abundance of nutrients or those that find it difficult to compete in a nitrogen-rich environment. The nitrogen-fixing legumes are another vulnerable group. Increasing depositions of nitrogen in the form of nitrate are inimical to these bacteria which in symbiosis with the roots of legumes convert atmospheric nitrogen (N_2) into an available orm.

But the change may take place unnoticed, since an already established plant can withstand a great deal. On the other hand, reproduction and seeding are in danger. In the long run the flora will therefore be impoverished unless there is an abatement of the emissions. A handful of common species, such as nettle, heavy hairgrass, and raspberry, would be likely to proliferate at other plants' expense.

In Sweden, 150 plant species that for some reason are already under threat may be further endangered by acidification. Even with continued husbandry, it may be impossible to maintain the old cultivated plant communities in areas that have become severely acidified.

Mercury

Mercury in the lakes and streams of Sweden has long been a serious problem. Many lakes had to be blacklisted in the sixties because the fish contained so much mercury as to be dangerous to health.

Steps where then taken to reduce the discharge of mercury, after which it was felt that the problem had been more or less solved. It has turned out, however, that many lakes—especially those poor in nutrients and with alow pH—contained more mercury than is allowable. And many of them were situated far away from the points of discharge.

When a lake is blacklisted no fish from it may be sold or given away, and neturally none should be eaten. A lake will be blacklisted when the amount of mercury in a pike weighing one kilogram is higher than 1 milligram per kilo (the World Health Organization has set the threshold at 0.5 mg/kg). About 200 lakes and streams are now blacklisted in Sweden, but investigations have shown that 5,000 to 10,000 others are at risk. Even unacidified lakes may have mercury concentrations close to the blacklisting limit. Most of the pike with high levels of mercury have been found in southern and central Sweden, where acid rain has fallen most heavily, as well as along the coast in the north.

Mercury can affect brain development during the foetal stage. This means that pregnant women have to be extra careful, although Swedes in general eat little freshwater fish. On the other hand, the water itself is harmless; one can bathe in it, and even drink it.

Why then should the fish be poisonous, when the water is not? That is because the mercury is absorbed by various plants, phytoplankton, and small animals (such as insects and zooplankton). These organisms are eaten by larger animals (roach, perch, pike, etc.) and with each step in the food chain, the concentration of mercury increases. The danger comes when fish-eating birds, such as the osprey, and humans eat fish with a mercury content. The fish themselves seldom suffer harm.

Even though the discharges of mercury into the air have fallen sharply, the levels of mercury in fish have not declined everywhere. One reason is acidification. In lakes differing in acidity, but with identical levels of mercury, the fish in the acidified lake will usually have more mercury in them than those in the non-acidified one.

In acid soil mercury behaves differently from other heavy metals: instead of being released, it binds even harder. This explains why it used to be assumed that once mercury got into the soil, it would stay there. It is now suspected, however, that although mercury binds hard, especially to the humus in acid soil, acidification results in an ever increasing part of the himus—together with the mercury—being washed out into the lakes and streams. This means that effluents can affect lakes for centuries after they have taken place.

The elevated levels of mercury in the fish of acidified lakes can also be explained by the fact that acidification changes the composition of species. Predatory fish, such as pike, are forced to change their feeding habits when acidification eliminates their usual prey. So the predators shift over to other prey fishes, which, however, now contain more mercury. Thus when roach or dance have vanished because of the low pH value of the water, the pike have to eat perch instead—and perch contain more mercury.

Mercury levels in fish can be reduced by liming the water. By causing the pH value to rise, this permits many species to recolonise the lake, and the number of individuals per species will also increase. While there will be no less mercury in the lake itself, it will be spread among a greater number of individuals; thus any one pike will ingest less mercury.

Several acidifying processes take place naturally in the soil—one of the most prominent being connected with the uptake of nutrients by plants, mostly in the form of positively charged ions. The plants compensate for this by releasing positively charged hydrogen ions. Thus growth itself is acidifying, whereas the breakdown of dead plant material has the opposite effect. It follows that in an ecosystem where growth and decay are abcut equal, there will be no net acidification. But if the cycle is broken by harvesting, whether it be crops or trees, the acidifying process will dominate. This may however be offset by weathering—depending more or less on the characteristics of the soil. Nevertheless it is the chief cause of the acidification of farmland.

In a coniferous forest there is usually an accumulation of incompletely decayed plant residues, which may have a similar acidifying effect. But the problem of soil acidification does not really arise in forests until there is an oversupply of acid from outside.

Putting it simply, soil acidification may have biological effects in these three respects:

- Through lowering of the pH value.
- Through an increase in the levels of aluminium and other toxic metals in the soil water.
- Through a loss of plant nutrients such as potassium, calcium, and magnesium due to increased leaching.

Some ten years ago, there was a general belief among scientists that the buffering ability of the soil would suffice to counteract acid rain. That is now known to be untrue. Throughout much of southern Scandinavia the pH values of soils have fallen by 0.3 to 1 unit, and in some instances by as much as 1.5. Especially remarkable is the fact that pH values have not only gone down in the upper layers. The soil has become soured to a depth of one metre.

It is likely that the big decline took place in the 1950s and early 1960s, just when there was the greatest increase in emissions. A similiar development has been noted in other countries. In West Germany and Austria, for instance, the soil pH has fallen by 0.3—1.5 units in two to four decades.

Once the soil has become soured, the populations of earthworms and bacteria decline, and the function of decomposing dead material and animal matter is taken over more and more by fungi. Since these organisms do it more slowly, fewer nutrients are released, thus tending to aggravate the problem posed by their paucity in acidic soils.

Acidification takes place more slowly in soil than in water (soil having a greater buffering capacity), but it is also harder to halt, once the leaching of nutrients and the precipitation of metals is well under way. Even if emissions should be drastically reduced, the process will presumably continue for a long time to come. The pH value would not be likely to fall further, but the loss of nutrients and the release of metals will go on unabated in acidified areas. If emissions are not lessened, though, acidification will spread ever wider.

Crops thrive best with a soil pH higher than 6. The plants, water uptake is then most efficient, important nutrients are easily available, and most of the harmful metals are absorbed only to a very small extent.

Depositions from the atmosphere are not the primary cause of farmland becoming acidified. Presentday agricultural practices contribute in high degree to reducing the pH of the soil. A great many hydrogen ions are released in the soil when the ammonium of the fertilizer is converted to nitrate. Such kinds of fertilizer are believed to account for anywhere between 15 and 50 per cent of the total acidification. Since much of the biomass is removed in harvesting, only a small portion of the nutrients amassed by the plants is returned to the soil. Consequently there is a gradual deterioration of the soil's buffering capacity, and unless lime is applied the pH will fall.

Air pollution can however have marked direct effects on plants. Ozone in particular damages many crops and leads to reduced yields per acre. According to estimates made in the United States in the early eighties, each year ozone influcts a harvest loss of between two and four billion dollars. This is equivalent to 4—6 per cent of the total crop value. Similar damage in Sweden is estimated by the University of Agricultural Sciences to cost 60 million kronor per annum. Experimental studies have found that the yields of spinach, potatoes, and peas fall off by between 10 and 30 per cent, given the ozone concentrations that are common in South Sweden.

The Forests

There is hardly any disagreement among scientists as to the harmful effect of air pollution on forests. The amount of damage varies however from place to place, and as in extent, so what is true for the high-level forests of Central Europe, for instance, does not necessarily hold for Scandinavia.

In Scandinavia there are two main types of forests soil: podzols, the most common, and brown soils. Podzols are usually covered with conifers or a mixture of coniferous and deciduous trees. They are most common in areas of relatively heavy rainfall, and because they are coarse-grained (easily letting water pass through) and poor in lime, much of the readily available nutrients are washed away. Such soils are necessarily fairly acid. On account of their low trophic status they offer poor resistance to acid rain, and thus any increased leaching of nutrients can pose a grave threat.

Brown soils are rich in nutrients and have higher pH values than podzols. They are usually found where there is meadowlands and broad-leaved forest. Acidification can inflict equally sweeping changes in both soil types as well as in others that resemble brown soils. Thus pH levels can drop sharply. Moreover brown soils of ten abound in clayey minerals which can loose aluminium, resulting in high concentrations of this metal in the soil solution. If conifers are planted in place of broad-leaved trees, the brown soil gradually becomes transformed into podzol. That is because conifers tend to acidify more than decidous trees.

Forestry practices can cause the soil to become acidified in several ways. With whole-tree harvesting, involving the removal of the branches, twigs and othrs residue, only a small part of the nutrient stored in the tree is restored to the soil. The natural circulation is broken and the soil becomes acid. Extensive clear cutting can also hasten the acidification of surface water, since more of the rainwater runs off over the surface into the streams and lakes without having first been neutralized in the ground. Applying acidifying fertilizer also helps to acidify forest soil.

The pH Value In Soil

Measuring the pH is much more complicated in soil than in water. The structure and composition of soil itself are very complex, and a host of chemical and biological processes are going on all the time. The pH can be measured in several ways. One method will not provide the same information as another. It is usual to start off by dissolving a sample of soil in a liquid, e.g. distilled water. The suspension is allowed to stand for some hours and shaken from time to time before making any measurement. An alternative is to take a saline solution such as potassium chloride or calcium chloride. The potassium or calcium ions force out these hydrogen ions that are loosely bound to the soil particles. There will then be a greater concentration of hydrogen ions than when using distilled water. In other words, the pH value will be lowers,. The difference in values often turns out to vary from 0.5 to 1 pH units. What is often done is to measure the pH value in both of these ways so as to arrive at some idea of the pH of the soil water and the total amount of soluble hydrogen ions in the soil.

Measuring the soil pH is a rather uncertain test for acidification, and so additional measurements are made to determine the base saturation. This shows how readily available are the calcium, magnesium, sodium, and potassium in the soil. Acidification lessens the store of these substances, which are, all with the exception of sodium, important nutrients.

Forest Damage In Scandinavia

As early as the mid-seventies, after a period of dry summers, forest damage was observed in southern Sweden, chiefly affected being Norway spruce. But then for a while the damage abated. But in the autumn of 1983, again after a couple of dry summers, damage both to conifers and hardwoods was reported from large tracts of south Sweden, and in each of the next two years a count was made of needle loss in spruce and pine in an attempt to estimate the extent and spread of damage for the country as a whole.

The counts showed that the loss was greatest among spruce trees in southwestern and southern Sweden. These are precisely the areas where the acidification of the lakes is worst, and where most of the acid rain falls. Here one-fifth of the trees in middle-aged and mature stands of spruce had lost 20 per cent or more of their needles. These figures are not altogether comparable with those for Central Europe, since Sweden defines the words "damage" or "injury" more strictly.

In the northern Swedish inland, damage was just as extensive as in the worst affected parts of south Sweden, which may seem surprising, seeing the fallout there is much less than in the south. The reason may be that the northern forests are older, genetically different, and exposed to harsher climatic conditions. Damage is in any case usually greatest in the oldest forests. Trees on slopes and heights, on forest fringes as well as any standing out above the others, also suffer more than the average: in other words, the more exposed the tree, the greater the damage.

As yet too little is known to be able to say for sure how much of the damage has been caused by air pollution and acidification, and how much is due to other, natural causes such as drought, frost, wind and advanced age. But to talk about forest death in Sweden would be going too far; damage is the more accurate term.

The first survey of forest damage in Finland was made in 1985. This showed that 11 per cent was affected in the southern part of the country, with extensive damage, to conifers in particular. Close to major sources of emissions an estimated 120,000 hectares, of forest were said to be affected, and the most important course of damage is assumed to be sulphur dioxide.

In the past few years, forest damage, mainly to conifers, has also been reported from Norway and Denmark. The investigations so far carried out, however, serve neither to confirm nor deny the possibility that such damage may be due, at least in part, to air pollution. And no thorough surveys, similar to those made for instance in West Germany, have yet been carried out.

As regards tree growth, it is still difficult to draw any conclusions. No general diminution has yet been observed in Scandinavia. Several Swedish studies have however shown that damaged trees grow more slowly than healthy ones. But for the moment growth in general does not seem to have been adversely affected. Reduction may on the other hand not be immediately apparent, since a rich supply of nitrogen does, up to a certain point, tend to stimulate growth. Formerly the most important substance in short supply was nitrogen, but now the soil has been fertilized by increased deposition, and the trees' requirements of nitrogen have been so great as to permit the fertilizing to offset, at least for the time being, the adverse consequences of air pollution. Recent studies have shown, however, that the trees' requirements are beginning to be saturated in southern Scandinavia, and that the negative effects of soil acidificaiton may soon be expected to make an appearance.

It is not likely that acid groundwater will in itself be harmful to humans. But where the water is highly acidic (stage three), metals such as aluminium and cadmium may appear at elevated levels, since they are released from the soil when the pH is under 5. Samples of water from wells in the provinces of Dalarna and Bohuslan gave revealed high levels of lead, zinc, and cadmium even though the pH value of the water was between 5.2 and 6.4.

Cadmium is the most mobile of the ordinary heavy metals. Since it is one of the elements to which people in the industrialized world have already been exposed to in dangerously high concentrations, alertness is required to the danger of levels rising in the groundwater. Cadmium accumulates in the renal cortex, causing lesions. It is moreover an element that leaves the body very slowly.

In Scandinavia, the amount of cadmium in soil that is available to plants is now rising. This is due in part to acidification, but in the case of farmland more to its presence in fertilizer. The levels in may plants, including farm crops, have consequently increased. The uptake by plants increased still further when the groundwater becomes acidified. At the same time people who are already ingesting dangerous amounts of cadmium, for instance by smoking, will be at grave risk if more cadmium gets into the drinking water.

Copper is the metal of which water pipes are most commonly made in Scandinavia, as well as in many other European countries. When the groundwater becomes more corrosive, the copper is dissolved. Since this elementis generally in deficiency in Scandinavia, however, a slight increase is probably of no account, although according to some scientists, larger amounts may cause diarrhoea in small children.

Aluminium is the most common metal in the earth's crust. Normally it is tightly bound to minerals, but acidification makes it markedly more soluble. Aluminium in water is chiefly a problem for kidney patients. In dialysis it enters the blood stream directly without first having passed the body's normal protective barriers, which may cause skeleton and brain injuries. The problem can be avoided however by using water that has been chemically treated.

Lately it has come to light that even individuals who are not suffering from kidney trouble may be endangered by taking in aluminium. Increased doses of the metal are suspected of being able to cause Alzheimer's and Parkinson's diseases, both of which may lead to premature senility and death. Recent investigations by the Central Statistical Bureau in Norway have revealed a very probable connection between the concentrations in drinking water and lake water in southern Norway and the death rate from senile and presenile dementia.

Lead is also liberated by acidified water. In Great Britain, where it is commonly used for waterpipes, it is said that about five million people are using drinking water in which the levels of lead are in excess of those recommended by the EEC. Lead is liable to harm the nervous system, especially of children. Just how much human health is affected by elevated levels of metal in the groundwater will depend on two things. For anyone who has already ingested a great deal of a harmful substance, additional doses could be perilous. Also if much of a person's intake of a certain substance comes from drinking water, even small increases would be risky. In the long term, the increased circulation of metals poses a distinct threat to health. A long time may elapse, too, before the effects become apparent.

By far the biggest problem at present, however, is not so much the effect on health, as the ability of acidified groundwater to cause corrosion.

Nor do animals escape the consequences of acidification. Not only do fish vanish from acidified lakes. Frogs, too, which lay their spawn in the shallow waters along the shore, are affected by the disturbance of the natural balance.

Small birds may also have problems of reproduction when living by acidified waters. The eggs of pied flycatchers and willow warblers have sometimes been found for instance to be too thin for successful hatching. This is because the birds have ingested too much aluminium with the insects they eat, which have developed in the acidified water.

Fish-eating birds such as the osprey, loom, and merganser, find it harder to obtain food as more and more lakes become sour. When the soil turns sour, plants accumulate greater amounts of heavy metals. Then herbivorous animals ingest them with their food. Several species of game animals have been found to have elevated levels of cadmium in their livers and kidneys. These include elk, deer, capercailzie, black grouse, and hare. Terrestrial snails may be unable to form shells properly when lime has been leached from the soil. Some moths with softwood habitats, such as *Lymanthia monarcha*, have dwindled, with increasing numbers of dwarfed individuals. Here again there would appear to be some connection with acidification.

The fauna will has be affected by changes in the composition and structure of the vegetation. If the forests for instance are further damaged, there will be great changes in the animal species of the forest ecosystem.

As with the forests, so in the case of flora and fauna will it be necessary to reduce not only the direct impact of air pollution, but also its effects in acidifying soil and water.

Acidification was first noticed in the lakes, and its effects on lakes and streams are still what is best known about the process. The acidification of lakes and streams is closely bound up with that of soil, for the simple reason that 90 per cent of the water has previously passed through the ground, while a mere 10 per cent has fallen directly as snow or rain. The first victims of acidification are nutrient-deficient lakes in areas where the soil has a poor buffering ability.

In the south of Norway, fishkills were observed in the streams as far back as the second and third decades of the century. The most sensitive lakes of southern Sweden had already turned sour between 1950 and 1970, just when there was the greatest increase in emissions of sulphur. Subsequently the pH value in acidification-prone lakes has continued to fall, but at a rather slower rate.

The water of an acidified lake looks beautifully clear, but appearances are deceptive. Typical signs that it is severely acidified are the disappearance of fish and the reduced numbers of plant and animal species. The water is no longer turbid from floating Plankton. But the main reason for its becoming clearer is that the humus substances precipitate and sink to the bottom.

Many aquatic plants are also adversely affected. Rosette species along the water's edge such as lobelia, quilwort and shoreweeds will often be overgrown with bog moss, which thrives in acid water. Pondweeds may also suffer. Water-lilites, on the other hand, do not, since their deepgoing roots get nourishment from sediment layers that are beyond the reach of acidification. In a severely acidified lake, the fish will after a time have vanished entirely, bog moss will have spread out over the lake floor, and only a few plant and animal species will remain.

The first victims of acidification are crayfish, snails and mussels, certain types of zoo-and phytoplankton, and some species of mayfly. Salmon and trout, roach and minnow, and many sensitive insect species will be harmed as pH is further lowered, and once it drops to lowered, and once it drops to 4.5—5, all that usually remains are bog moss, the hardiest insects and certain species of plankton that are resistant to acidification. It is not the acidification of the water alone that takes a heavy tool of plant and animal life. In acid lakes there are increasing concentrations of aluminium in ion form, which is highly toxic to many organisms. The disappearance of fish is due to the combination of a lowered pH value and aluminium poisoning. Both of these factors are however a result of acidification.

The levels of other metals aiso rise, among them cadmium, zinc, and lead, and these are then taken up to a greater extent by animals and plants. Both aluminium and other metals flow into lakes from the acidified soils of the surroundings. Not all the biological changes arise from changes in the chemistry of the water. With the fish out of the way, certain insects on which they usually prey begin to thrive. Among those that especially benefit from the disappearance of predators are certain water beetles and dragonfly larvae. In the absence of fish, insects will become more and more the dominant fauna. This also happens in lakes that have been emptied of fish for reasons other than acidification. These insects do not really thrive better in acid water, but they multiply because the fish are gone. In other words, this is an indirect biological effect. It also shows that, contrary to the popular belief, acid lakes are not entirely dead lakes.

Acidification varies according to the season, especially in running water. The water of lakes and streams is most acidic in spring, at the time of the snowmelt when huge volumes of acidic water flow into them. On account of the rapidity of the flow, much of the water remains unneutralized by the soil's buffering system. Acid flushes may also occur in connection with heavy autumn rains, and can even happen in calcareous areas, for instance in parts of the Swedish and Norwegian mountains.

Polluted air can affect people in all sorts of ways, the effects ranging from slight feelings of discomfort to increased mortality. The causes may be single substances or several substances in combination. Large numbers of people breathe such air, especially in urban areas. Among those that are especially likely to suffer from the effects are asthmatics, children, elderly persons, those with heart or circulatory diseases, or reduced breathing capacity. Polluted air will also compound the dangers of an already unhealthy working environment.

The immediate effect may be anything from irritation of the respiratory tract to actuate disorders caused by breathing in air that has suddenly become highly polluted. In West Germany they have had to issue smog warnings several times in the last few years, especially during autumn and winter. The concentrations have sometimes been so high in the Ruhr and West Berlin, for instance, that people have been advised to stay indoors. Breathing in polluted air over a long period can result in chronic illness, with recurrent respiratory trouble and even damaged organs.

The concentrations of sulphur dioxide over urban areas in Europe will thus have to be reduced on account of their effects on health as well as the risks of atmospheric corrosion. As regards nitrogen oxides, the dioxide form is the most dangerous. In order to bring down the amounts of nitrogen oxides and hydrocarbons, it will be necessary to attack vehicle exhausts in the first place, both because they constitute the largest source of such pollutants, and also because they are emitted at such a low level. Moreover vehicle emissions contain many other dangerous substances, such as carcinogenic and mutagenic hydrocarbon compounds, soot particles, and so on.

Direct and indirect effects

Airborne pollution can affect the environment both directly and indirectly. Thus when sulphur dioxide and nitrogen oxide gases are present in high concentrations, they can cause damage to trees and lichens, affect people's health, and corrode structural materials, such as steel. These direct effects are often greatest in the vicinity of the emission sources.

Sulphur dioxide and nitrogen oxides can, however, also form sulphuric acid and nitric acid, which may be carried with the winds over long distances before descending in rain or snow. Thus soil and water can become acidified far away from the sources of emission.

Acidification is chiefly an indirect effect of the sulphur and nitrogen pollutants that our industrial society discharges in such quantities. The extent of acidification will depend (1) on how great the fallout is of such pollutants; and (2) on the resistance that soil and water can put up to acidification.

Sulphur dioxide and nitrogen oxides are the chief causes of soil and water becoming acidified. Sulphur compounds are responsible for about two-thirds of the acidification of rain, and nitrogen compounds for the rest. Part of the latter, however, do not acidify if they are absorbed by plants when they hit the ground—as in fact occurs to a great extent. The pollution from sulphur is thus actually greater than the figure of two-thirds would indicate.

Sulphur in gaseous form, sulphur dioxide (SO_2), is chiefly formed in the combustion of oil and coal. These fossil fuels came into existence several hundred million years ago. The plant material in the large tropical, humid fern-shaws that covered vast tracts of Europe hundreds of millions of years ago metamorphosed into oil, coal, and natural gas. Everything organic takes up sulphur, which explains its presence in fossil fuels. The concentrations vary, however. Crude oil usually contains anywhere from one-tenth of one per cent of sulphur to between two and three per cent by weight. In coal the levels vary even more, whereas natural gas contains considerably lesser quantities. Moreover the natural gas that is imported into Sweden is cleaned so as to reduce its sulphur content more or less to zero. World reserves of oil and coal are still huge. Enormous amounts of sulphur from these fuels will be released if we persist in current practices. So far less than one-twentieth of the sulphur that is stored in known coal and oil reserves has been set free.

Annual emissions of sulphur according to the estimates for 1978 of the European Monitoring and Evaluation Programme.

Country	*Area 1,000 sq km*	*Emissions 1,000 tons*
Albania	28.7	50
Austria	83.9	215
Belgium	30.5	405
Bulgaria	110.9	500
Czeckoslovakia	127.9	1500
Denmark	43.1	228
East Germany	108.2	2000
Finland	337.1	270
France	544.0	1800
Greece	132.0	350
Hungary	93.0	750
Ireland	68.9	88
Iceland	60.7	6
Italy	301.2	2200
Luxembourg	2.6	21
Netherlands	41.0	240
Norway	323.9	76
Poland	312.7	2150
Portugal	76.7	84
Romania	237.5	100
Soviet Union	3363.4	8100
Spain	500.0	1000
Sweden	450.0	275
Switzerland	41.3	58
Turkey	460.8	500
United Kingdom	244.0	2560
West Germany	249.6	1815
Yugoslavia	255.8	1475

The largest increase in the emissions of pollutants came after the war. Consumption of fuel, oil in particular, rose dramatically—by about 10 per cent per annum. By 1970, 15 times as much oil was being consumed as in 1945, when the war ended.

All told, around 30 million tons of sulphur are now discharged every year in Europe, including the Soviet Union. Most of this, about 80 per cent, comes from combustion of fossil fuels, the remainder from industrial processes. The largest single emitter of Pollutants in western Europe is Great Britain; taking the whole of Europe, the only country that emits more is the Soviet Union. The general trend today, however, is towards a slight decrease in emissions.

There is also a great deal of airborne pollution in North America, where some 16 million tons of sulphur are emitted each year, mostly in the United States. And for the world as a whole, the yearly discharge on account of human activities is estimated to be 50 to 75 million tons.

The atmosphere also receives natural doses of sulphur from the eruption of vulcanoes, from the seas and oceans, and from certain processes in the soil. For the world as a whole, the emissions due to human activities more or less equal the natural supply. In contrast, man accounts for about 90 per cent of the emissions of sulphur to the atmosphere in the industrialized parts of Europe and North America. That is ten times the level that can be considered natural. In Europe, the precipitation has been measured for its chemical composition for rather more than twenty-five years. These observations show how sulphur levels increased during the 1950s, but have been more or less constant or even falling since the mid-seventies.

What is acidification?

Emissions of sulphuric and nitrogenous pollutants are thus the main causes of acidification. What then actually happens to air, soil, and water?

On coming into contact with water, sulphur dioxide and nitrogen oxides form the powerful sulphuric and nitric acids. This will occur when the oxides are dissolved in drops of water in the atmosphere. The pollutants subsequently reach the ground and overlying vegetation in rain and snow. But acid oxides may also be deposited directly as gases, or be attached to dry airborne particles. This is known as "dry deposition" as opposed to the "wet" type that comes with the precipitation. The dry fallout also forms acids on contact with water.

A typical property of acids is their ability to release hydrogen ions (H^+). Acidification amounts to an increasing supply of hydrogen ions. The concentration of hydrogen ions in a solution is thus a measure of acidity, which is usually indicated by a pH value. A low pH value means a high level of hydrogen ions, and the substance is acidic. Water that is neutral, neither acidic or alkaline, has a pH of 7. The pH scale is logaritmic, which means that for every drop of one unit, the hydrogen-ion level increases by a factor of ten. A solution with a pH of 6 is ten times more acid than one with pH 7; at pH 5 the solution will be a hundred times more acid than one with pH 7, and so on. Chemical and biological processes, both in soil and water, are greatly affected by the hydrogen-ion levels.

There are however natural mechanisms opposing a reduction of pH value. These are termed buffer actions, in which a solution will use up hydrogen ions so as to neutralise the acid to a greater or lesser extent.

In a lake, the buffering function is mainly performed by hydrocarbonate (HCO^{3-}). Lakes in lime-rich areas continuously receive hydrocarbonate in large quantities, and therefore run no great risk of becoming soured. While relatively little of Scandinavia is highly calcareous, a large part of Denmark, as well as the Swedish provinces of Skane, Oland, and Gotland are so, as is the Oslo area in Norway, and certain parts of the Scandinavian mountain region. Much of Scandinavia's extensive primary-rock area receives relatively small amounts of hydrocarbonate ions, and consequently the lakes there are susceptible to acidification. When the supply of acids exceeds that of hydrocarbonates, the lakes will sooner or later be wholly deprived of the latter and the pH value will fall drastically.

Several buffering processes operate in soil, the most important being the weathering of various minerals from rock. As long as calcareous minerals abound in the soil, the pH value will stay fairly high. But should the pH decline because of a deposition of acid, a major buffering process will be set in, causing aluminium compounds dissolve. Free aluminium in the soil is suspected of damaging tree roots, but it can leach out to the groundwater and surface water, this being a major cause of harm to the fish populations in acidic lakes.

Back in the fifties, anglers and naturalists noticed that fish stocks in many lakes of southern Scandinavia were dwindling. In some places the fish had vanished completely. But it was not until the 1960s that scientists were able to link these effects to any specific cause. It was found to be atmospheric pollution, much of which had originated far away.

Tens of thousands of lakes and streams in Scandinavia are now acidified. We also know that the groundwater, which means millions of people's drinking water, is becoming acidic, with dangerously elevated concentrations of metals as one of the consequences. It has been known for some time that the forests of Central Europe were badly affected, and similiar damage has now appeared in Scandinavia. Again the cause is air pollution.

Air pollution affects vegetation as well as people's health directly. The numbers of damaged trees in towns and along our highways show that something is wrong. In industrial cities the world over, people are being exposed to overly high concentrations of pollution that is mainly traceable to the emissions from motorcars, factories and combustion plants.

Around 1980 attention began to concentrate on the effects that pollution was having on forests. The warning signals came mostly from West Germany, where damage to forests was becoming widespread. Acidification of the soil was advanced as the likeliest explanation. However, other untoward mechanisms were discussed, notably the effects of photochemical oxidants. These pollutants are formed from nitrogen oxides and hydrocarbons under the influence of sunlight. The most common oxidant is ozone, a substance that is known to be harmful to plants. Since photochemical oxidants are not acids, it is however incorrect to speak only of acidification as the cause of forest damage.

Matters are further complicated by the fact that pollutants may be carried with winds over long distances. Those that fall down and cause damage somewhere may have been released from points hundreds or thousands of miles away.

Winds are no respecters of frontiers, and pollutants are not evenly distributed. Some countries are net importers of pollution, others are exporters. That is why acidification is an international problem. No single state can disclaim responsibility. All must do their part in keeping our common atmosphere as clean as possible. And it can be done. The technology is already available. If we fail to take advantage of our opportunities, we shall be committing an outrage against ourselves as well as nature, and against coming generations. Our aim here is to provide answers, as seen from a Scandinavian viewpoint, to the following five questions:

- What is acidification, and what are the causes?
- What are its known and feared environmental effects?
- What can se do to solve the problems of acidification?
- Why is acidification an international concern and an international responsibility?
- What has been done so far, and what of the immediate future?

Over a great part of Scandinavia there is only a thin soil cover, and the bedrock is deficient in lime. Both soil and water are thus especially sensitive to acid rain. Moreover the westerly and sourthwesterly winds that prevail over northern Europe bring with them large amounts of air pollutants from the continent and British Isles. So it was not be chance the problems of acidification were first noticed in Scandinavia.

These problems are not confined however to this region. There are also acid lakes in Scotland, Holland, and West Germany. Damaged forests are to be found in most European countries, and contaminated air is eating away at historic monuments in Venice and Cracow, as well as in London, where even Westminster Abbey is being attacked.

It is hard to say for certain what parts of the world are most at risk, since there are susceptible areas almost everywhere. If industrialization and increased fuel consumption should cause the emissions of sulphur to increase in places that which have so far been unaffected, the same problems are likely to arise there as elsewhere. This may well happen in Brazil, southern Africa, India, and southeast Asia.

The quality of air in any one country is very much influenced by emissions in others, so it will avail little for any country alone to reduce emissions. Without international cooperation there can be no real solution.

Nitrogen

The other great cause of acidification are the emissions of nitrogen oxides. These consist mainly of nitrogen monoxide (NO) and nitrogen dioxide (NO_2), which are usually grouped under the collective term NO_x.

Nitrogen oxides are formed in all types of combustion. Unlike sulphur dioxide, coming entirely from the sulphur contained in the fuel or raw material, most of the nitrogen oxides are formed by the reaction of nitrogen gas in the combustion air with oxygen. Of the air we breathe, about 78 per cent consists of nitrogen gas and 21 per cent of oxygen. When this mixture is heated, the two substances interact to form nitrogen oxides. The higher the combustion temperature, the more nitrogen oxide will be formed. Also contributing to the process is the nitrogen in the fuel.

The largest single source of nitrogen-oxide emissions is road traffic. In Scandinavia about two-thirds of the nitrogen oxide entering the air comes from motor vehicles. Emission volumes are harder to estimate for nitrogen oxides than for sulphur. By and largre, these emissions are believed to have doubled during the 1960s. Since then they have continued to increase, although more slowly. In terms of nitrogen dioxide about 20 million tons are released every year in Europe, and mainly because of the increasing volume of traffic, discharges of nitrogen oxides seem to be still on the rise. Since the emissions of sulphur are now abating, nitrogen oxides are consequently becoming relatively more important as acidifiers of the environment. In the past five years, however, there has been no upward trend in the level of nitrates (dissolved nitrogen oxides) in rain and snow.

Certain types of fertilizers are another source of nitrogen pollution. Croplands treated with liquid commercial fertilizers and animal manure emit large amounts of ammonia (NH_3) to the atmosphere. Ammonia actually causes a rise in pH of the precipitation, but the effect is eliminated when ammonium ions (NH_4^+) in the rain or snow are converted by microorganisms in the soil or absorbed for instance by trees after descending to the ground.

The enormous emissions of nitrogen pollutants, ammonia, and nitrogen oxides also have the effect of everfertilization. Most plants are adapted to a nitrogen deficiency, but now they are getting a surfeit of nitrogen via the atmosphere, as has been found for instance in southern Sweden. This is feared to have negative effects on vegetation, and to cause nitrogen to start leaking out of the forest floor, unless the plants have enough capacity to absorb it all. Nitrogen leakage used to be exclusively confined to farmland. But the risk of leakage from forest land has now sharply increased owing to nitrogen saturation. This can lead to problems with drinking water and to bodies of water suffering eutrophication. Moreover nitrate leakage greatly intensifies the acidification of the soil, which in turn liberates unwanted substances such as aluminium.

OTHER POLLUTANTS

Among the other products of combustion that contribute to the pollution of the atmosphere is carbon dioxide, which may eventually affect the earth's climate if concentrations continue to increase at present rates. Hydrocarbons comprise a complex of pollutants. Every hydrocarbon compound emitted to the air is assumed to give rise to ten new substances. As a rule they break down rather quickly in the atmosphere, but they include quite a number of non-biodegradeable substances. Among the major sources are car exhausts, paint shops, and petrochemical plants. Many of the substances so discharged may be converted later into new contaminants. Under the influence of sunlight, for instance, hydrocarbons and nitrogen oxides form ozone and other photochemical oxidants. Since the 1950s the background level of ozone in northern Europe seems to have increased considerably, mainly on account of the emissions of hydrocarbons and nitrogen oxides from road traffic.

A point worth making in this context is that a combination of contaminants often reinforces their effect. It is believed that this applies to ozone, sulphur dioxide, and nitrogen oxides, which means that if all these contaminants are present at the same time, the damage will occur at lower levels than it would if each one were acting separately.

Sulphur Dioxide

In the course of the past ten years, the emissions of sulphur in Europe have gone down by about a quarter. As a result, the depositions in Scandinavia have abated by about as much. There are also signs that some acidified lakes are recovering. Although the signs are weak, they do show that the reduction of emissions is beginning to take effect.

As can be seen, several European countries have now definitely decided to reduce their emissions of sulphur, and a number of others have more or less developed plans with the same aim. Of those countries that mostly affect Scandinavia, the greatest uncertainty attaches to Great Britain, Poland, and Czechoslovakia. But in light of the pledges made by some twenty European countries, it can be assumed that the sulphur fallout over Scandinavia will probably be 30-50 per cent less by the mid-1990s than it was in the seventies.

Nitrogen Oxides

Less is known about the national emissions of nitrogen oxides. It would appear that nearly 50 per cent of the total in western Europe comes from road traffic, with petrol and diesel-driven vehicles probably accounting for about equal amounts, despite the disparity in numbers. Power production is responsible for about 30 per cent of the emissions, most of them coming from coalfired plants.

The amount of nitrogen oxides emitted by motor vehicles in any country will depend both on the composition of the fleet and the type of highway network. Older cars, driven at slower speeds on less fast roads, discharge comparatively less nitrogen oxide. In several of the countries of eastern Europe the greater part of the emissions are thus more likely to come from the combustion of coal.

Within the EEC, all member countries have decided to provide unleaded petrol, beginning in 1989. Further, any member countries who so wish will be permitted to introduce such petrol even earlier.

After considerable debate, the EEC member states agreed in 1985 to tighten the emission standards for petrol-driven cars. Denmark dissented, however, claiming that the new rules were far from adequate. In the first place, they are differentiated, with the most stringent requirements for cars with an engine volume greater than two litres, and lower for cars with 1.4 to 2 litre engines. They will be still lower for the smallest cars, with an engine volume of less than 1.4 litres. Tighter rules for the smallest cars are however to be considered in 1987. In the second place, the EEC standards are only voluntary, and Great Britain at least is already saying that it does not intend to abide by any emission rules that would necessitate the use of catalytic converters (which in any case would apply only for cars with engines larger than two litres). From this it follows that a little mini could emit a great deal more pollution than a Rolls Royce. It is also important to remember that a sizeable proportion of the cars in the EEC sphere fall into this midget category.

By contrast, in 1985 Austria, Switzerland, and Sweden decided to make the US emission standards mandatory, beginning with the 1988-89 car models. West Germany, too, recently tightened its standards for emissions of nitrogen oxides from major combustion plants, making it necessary to denitrify at oil-fired and coal-fired plants with a rated output of more than 300 MW. For smaller plants, the low NO_x- technique will be required. New plants, as well as existing plants with a certain remaining service period, are covered by the new rules.

It may be seen what effect various measures might have in reducing emissions of nitrogen oxides in Europe. This sets forth the reduction expected 15 years after adoption of the regulations. The outcome will of course vary from one country to another.

Catalytic converters on passenger cars, would reduce the emissions of nitrogen oxides from this category of vehicles by about three-fourths 15 years after the introduction of the regulations. For the whole of Europe that would mean a total reduction of such emissions by about 15 per cent.

As yet there is no effective method for reducing emissions from heavy diesel-driven vehicles. Currently the California standards are the most stringent. If those rules were introduced in Europe, the total emissions of nitrogen oxides would have fallen by about 10 per cent after fifteen years. In the field of energy generation, some result may be obtainable from inexpensive measures, such as energy saving, the use of improved combustion techniques in new power plants, and changing over to natural gas and nuclear power. Measures of this kind would not however reduce emissions to any appreciable extent—probably by no more than 5—10 per cent by the turn of the century.

Still more effective measures, which would however cost more, such as installing equipment to improve the combustion process in existing power stations as well, would reduce the total emissions in Europe by 20 per cent within fifteen years.

The EEC Commission has put forward a proposal for cutting down emissions of sulphur and nitrogen oxides from large combustion plants (with a capacity in excess of 50 MW). The proposed limits for new plants are closely in line with the West German average, and the requirements for coal-firing are comparable to the present Swedish standards. The commission proposed that the rules for new plants should come into force in 1985, and be substantially tightened thereafter.

Alongside the proposals relating to new plants, the Commission has wanted to reduce the emissions of sulphur dioxide from existing combustors by 60 per cent, and those of nitrogen oxide by 40 per cent, by 1995. If implemented, this proposal would mean a reduction of the overall emissions of nitrogen oxides in Europe by 15 per cent.

If the new West German recommendations concerning nitrogen oxide emissions from combustion plants were to be applied to the whole of Europe, the emissions would fall by about 35 per cent from present levels. Others factors determining the level of emissions are changes in modes of transportation and fuel consumption. In making the estimates it was assumed that the consumption of fuel for the generation of energy will be constant up to the year 2000, while transport work would increase by 0.6 per cent annum up to the turn of the century.

From this review it will be seen that radical measures will have to be taken in order to bring about a rapid downturn of nitrogen oxide emissions in Europe. It also shows that if decision is taken to act, say to combine tighter exhaust standards for petrol and diesel vehicles with the new West German standards, it will be feasible to cut down the emissions of nitrogen oxides by more than half over the next 15 years.

Cleaning of Car Exhausts

The levels of nitrogen oxides, hydrocarbons and carbon monoxide in automobile exhausts can be reduced in several ways. In the first place the engine can be designed so as to make the combustion process as complete as possible. Through partial recirculation of the exhaust gases the emissions of nitrogen oxides can be somewhat lessened. By means of controlled fuel-injection the fuel mixture can also be adjusted so as to reduce the emissions of unburnt residues. Measures of this kind have to some extent already been taken, and in any case do not require any major changes in standard European engines.

Emissions of pollutants can be still further reduced with so-called lean-burn engines. By using a leaner fuel mixture, such engines not only save fuel, but also emit smaller amounts of nitrogen oxides and carbon monoxide. In order to reduce the emissions of hydrocarbons as well, however, the car must also be equipped with an oxidising catalyzer. The system has nevertheless two disadvantages: lean-burn technology is not yet fully developed, and the emissions of nitrogen oxides are not so low as they can be with a three-way catalytic converter.

This last provides the most efficient means of cleaning the exhausts of petrol-driven cars. It has been in use in the United States and Japan since the mid-1970s, and at present about half of all the cars now being made are equipped with converters, which transform more than 90 per cent of the nitrogen oxides, hydrocarbons and carbon monoxide into the more innocuous nitrogen gas, carbon dioxide, and water.

The essential parts of such a cleaning system are:

1. A fuel injector, to supply fuel in well-proportioned amounts.
2. An oxygen sensor, which registers the composition of the exhaust gas near the engine, and so determines whether the amount of air in the fuel mixture is ideal.
3. An electronic control unit that receives signals from the oxygen sensor and ensures a proper mixture of new fuel, recirculated exhaust gas, and air.
4. The actual catalyzer, which converts carbon monoxide, hydrocarbons, and nitrogen oxides into carbon dioxide, water, and nitrogen gas. The casing, often of stainless steel, encloses a honeycomb core of ceramic material, coated with a thin layer of catalyzing metal over which the exhaust gases pass.

There are various estimates of the costs of a three-way catalyzer. In 1983 a government committee concluded that it would be around 3,000 kronor. Car manufacturers give varying but usually higher figures, mostly between 4,000 and 7,000 kronor.

A car equipped with a three-way catalytic converter probably consumes somewhat less fuel than one without, or at any rate not much more.

Since lead would destroy the catalyzer it is essential that unleaded petrol be freely available.

Limits Now Defined

In a report presented in 1986 by a specially appointed group of experts, the tolerance limit for depositions of sulphur on the most vulnerable soils was put at 2-7 kilograms per hectare per annum (0.2-0.7 g per square metre), which compares with a present level in southwestern Sweden of 20-30 kilograms. In central Europe the annual fallout averages 30-50 kilograms. This means that in order to save the most sensitive areas, sulphur deposition must be reduced by at least 80-90 per cent. According to the calculations of this Scandinavian group, the depositions of nitrogen per hectare should not exceed 10-20 kilograms a year (considered over a period of 25 to 50 years), if nitrogen saturation is to be avoided in forests soils. The figure for south-western Sweden is now 20—30 kilograms, while in the Netherlands it may be as much as 60 kilograms per hectare. The conclusion was that in this case depositions would have to be brought down to a half or a quarter of the present level, in order to remain within the limits of Nature's tolerance.

RESOLUTION ADOPTED BY THE GENERAL ASSEMBLY

[on the report of the second committee (A/43/905)]

Protection of global climate for present and future generations of mankind

General Assembly

Welcoming with appreciation the initiative taken by the Government of Malta in proposing for consideration by the Assembly the item entitled "Conservation of climate as part of the common herigate of mankind";

Concerned that certain human activities could change global climate patterns, threatening present and future generations with potentially severe economic and social consequences;

Noting with concern that the emerging evidence indicates that continued growth in atmospheric concentrations of "greenhouse" gases could produce global warming with an eventual rise in sea-levels, the effects of which could be disastrous for mankind if timely steps are not taken at all levels;

Recognising the need for additional research and scientific studies into all sources and causes of climate change;

Concerned also that emissions of certain substances are depleting the ozone layer and thereby exposing the earth's surface to increased ultra-violet radiation, which may pose a threat to, *inter alia,* human health, agricultural productivity and animal and marine life, and reaffirming in this context the appeal, contained in its resolution 42/182 of 11 December 1987, to all States that have not yet done so to consider becoming parties to the Vienna Convention for the Protection of the Ozone Layer, adopted on 22 March 1985, and the Montreal Protocol on Substances that Deplete the Ozone Layer, adopted on 16 September 1987, as soon as possible;

Recalling its resolutions 42/186 and 42/187 of 11 December 1987 on the Environmental Perspective on the Year 2000 and Beyond and on the report of the World Commission on Environment and Development, respectively;

Convinced that changes in climate have an impact on development;

Aware that a considerable amount of valuable work, particularly at the scientific level and in the level field, has already been initiated on climate change, in particular by the United Nations Environment Programme, the World Meteorological Organisation and the International Council of Scientific Unions and under the auspices of individual States;

Welcoming the convening in 1990 of a second World Climate Conference;

Recalling also the conclusions of the meeting held at Villach, Austria, in 1985, which, *inter alia,* recommended a programme on climate change to be promoted by Governments and the scientific community with the collaboration of the World Meteorological Organisation, the United Nations Environment Programme and the International Council of Scientific Unions;

Convinced that climate change affects humanity as a whole and should be confronted within a global framework so as to take into account the vital interests of all mankind;

1. *Recognises* that climate change is a common concern of mankind, since climate is an essential condition which sustains life on earth;

2. *Determines* that necessary and timely action should be taken to deal with climate change within a global framework;

3. *Reaffirms* its resolution 42/184 of 11 December 1987, in which, *inter alia,* it agreed with the Governing Council of the United Nations Environment Programme that the Programme should attach importance to the problem of global climate change and that the Executive Director of the United Nations Environment Programme should ensure that the Programme cooperates closely with the World Meteorological Organisation and the International Council of Scientific Unions and maintains an active, influential role in the World Climate Programme;

4. *Considers* that activities in support of the World Climate Programme, approved by the Congress and Executive Council of the World Meteorological Organisation and elaborated in the system-wide medium-term environment programme for the period 1990-1995, which was approved by the Governing Council of the United Nations Environment Programme, be accorded high priority by the relevant organs and programmes of the United Nations system;

5. *Endorses* the action of the World Meteorological Organisation and the United Nations Environment Programme in jointly establishing an Intergovernmental Panel on Climate Change of provide internationally coordinated scientific assessments of the magnitude, timing and potential environmental and socio-economic impact of climate change and realistic response strategies, and expresses appreciation for the work already initiated by the Panel;

6. *Urges* Governments, intergovernmental and non-governmental organisations and scientific institutions to treat climate change as a priority issue, to undertake and promote specific, cooperative action-oriented programmes and research so as to increase understanding on all sources and causes of climate change, including its regional aspects and specific time-frames as well as the cause and effect relationship of human activities and climate, and to contribute, as appropriate, with human and financial resources to efforts to protect the global climate;

7. *Calls upon* all relevant organisation and programmes of the United Nations system to support the work of the Intergovernmental Panel on Climate Change;

8. *Encourages* the convening of conferences on climate change, particularly on global warming, at the national, regional and global levels in order to make the international community better aware of the importance of dealing effectively and in a timely manner with all aspects of climate change resulting from certain human activities;

9. *Calls upon* Governments and intergovernmental organisation to collaborate in making every effort to prevent detrimental effects on climate and activities which affect the ecological balance, and also calls upon non-governmental organisation, industry and other productive sectors to play their due role;

10. *Requests* the Secretary-General of the World Meteorological Organisation and the Executive Director of the United Nations Environment Programme, through the Intergovernmental Pannel on Climate Change, immediately to initiate action leading, as soon as possible, to a comprehensive review and recommendations with respect to:

(a) The state of knowledge of the science of climate and climatic change;

(b) Programmes and studies on the social and economic impact of climate change, including global warming;

(c) Possible response strategies to delay, limit or mitigate the impact of adverse climate change;

(d) The identification and possible strengthening of relevant existing international legal instruments having & bearing on climate;

(e) Elements for inclusion in a possible future international convention on climate;

11. *Also requests* the Secretary-General to bring the present resolution to the attention of all Governments, as well as intergovernmental organisation non-governmental organisations in consultative status with the Economic and Social Council and well-established scientific institutions with expertise in matters concerning climates;

12. *Further requests* the Secretary-General to report to the General Assembly at its forty-fourth session on the implementation of the present resolution;

13. *Decides* to include this question in the provisional agenda of its forty-fourth session, without prejudice to the application of the principle of biennialisation.

The conference organised by the Canadian on 'The Changing Atmosphere: Implication for Global Security' in June 1988 was the first major international conference which successfully brought together scientists and policy-makers with the aim of producing positive proposals for coming to grips with the problems posed by the way we have been changing the composition of the atmosphere.

It had the advantage that those attending were not Government represen-tatives. Thus they could pursue the issues uninhibited by the limits of existing national policies. At the same time, discussions invariably combined regard for the attainable and reliance on science not always sufficiently prevalent in government and private decision making.

There is no doubt that the climate change issue has become prominent around the world. It remains at the centre of current analyses of major global environment issues. It features on the agenda of the major international political meetings. A good example of this prominence and the priority it is achieving alongside traditional economic development was the recent OECD Ministerial meeting. It gave more prominence communique dealt with the climate change issues in terms recognising the need for conventional development approaches to accommodate them:

> 'In view of increasing evidence of the risk of global warming and climate change and the necessity to respond to this issue, Ministers call for vigilant, serious and realistic assessment on a global basis of what energy policies can contribute to meeting these challenges, and of their economic and social impacts. Member governments should contribute in their energy policies to the solution of international and domestic environmental problems. As identified by IEA Ministers, they pledge to pursue in their respective energy policies greatly improved energy efficiency and conservation, new technologies and, where national decisions so contemplate, the use of nuclear power with maintained and improved safety in construction, operation and waste disposal.'

This conference has before it descriptions of much of the current action around the world on climate change. This raises the question of whether or not any real progress in terms of practical action has been made consistent with the hopes of the Toronto conference, the fate of the conclusions in the Conference Statement, and its influence on subsequent events. First, other conferences. Several major conferences since then have picked up the themes of the Toronto Conference. This is not altogether surprising, as there is generally common ground on the scientific basis of the issues, and there has been some common attendance.

No one seriously question that human activity has been changing the atmosphere, and that this in turn changes the radiation balance. The question is rather exactly what the effects are, and what they will mean for the world and its human population. Against this background, it is not surprising that major conference statements since Toronto reinforce its concerns. This will be evident from the descriptions of them available at this conference.

The high ambitions of the Conference were set out as 'A Call for Action';

> The Conference urges immediate action by governments, the United Nations and their specialised agencies, other international bodies, non-governmental organisations, industry, educational institutions and individuals to counter the ongoing degradation of the atmosphere.
>
> An Action Plan for the Protection of the Atmosphere needs to be developed, which includes an international framework convention, encourages other standard-setting agreements and national legislation to provide for the protection of the global atmosphere. This must be complemented by implementation of nation action plans that address the problems posed by atmospheric change (climate warming, ozone layer depletion, acidification and the long-range transport of toxic chemicals) at their roots.
>
> The following actions are mostly designed to slow and eventually reverse deterioration of the atmosphere. There are also a number of strategies for adapting to changes that must be considered. These are dealt with primarily in the recommendations of the working Groups.

Arguably the successful establishment of the Intergovernmental Panel on Climate Change (initiated earlier by the World Meteorological Organisation and the United Nations Environment Programme) and the achievement of the Hague Declaration reflect acceptance of this Call for Action. The issue has also achieved greater prominence since, in the United Nations General Assembly, the United Nations Environment Programme and the OECD. It is reasonable to assume that the Toronto Conference contributed to this. For example, it was evident at the conference that there would be strong moves to encourage debate on the issue at the 1988 United Nations General Assembly meeting, even though the Assembly had earlier agreed that environmental matters would wait until 1989 for the Malta sponsored resolution on 'Protection of global climate for present and future generations of mankind' which called for:

- the identification and possible strengthening of relevant existing international legal instruments having a bearing on climate;
- elements for inclusion in a possible future international convention on climate.

There is a world of difference between discussion and action. The Conference Statement explicitly recognised that it takes a long time to develop an international consensus on complex issues much as these, to negotiate, sign and ratify international environment instruments and to begin to implement them. The difficulties cannot be underestimated. At the Conference the Canadian Minister for Energy, Mines and Resources, M. Marcel Masse, pointed out that the answers being sought were not the expected outcomes from normal policy processes, and were in conflict with entrenched views and vested interests.

The Conference called for action by Governments and industry. The details and my understanding of where we are now follow.

- Ratify the Montreal Protocol on Substances that Deplete the Ozone Layer. The Protocol should be revised in 1990 to ensure nearly complete elimination of the emissions of fully halogenated CFCs by the year 2000. Additional measures to limit other ozone-destroying holocarbons should be considered.

Countries, including Australia, have ratified the Protocol. Energetic efforts to increase this number are being made. The British Government's conference in the UNEP Pacific Basin meeting in Tokyo are good examples. Some countries-Australia, Norway and Sweden spring to mind—have already enacted legislation to phase out CFCs well before the Protocol requirements. Australia has controlled exports, and it is understood that the national strategy we are developing to phase out the strongly ozone depleting halons leads the world. On the other hand, many developing countries remain sceptical. India has pointed out the special situation it believes they face. Even if the review of the Protocol now under way leads to agreement for earlier phase out, there will be considerable unevenness of implementation. The theme of assistance for developing countries to conform with these developments yet adhere to responsible development plans will recur in discussions on climate change issues.

- Set energy policies to reduce the emissions of CO_2 and other trace gases. In order to reduce the risks of future global warming. Stablising the atmospheric concentrations of CO_2 is an imperative goal. It is currently estimated to require reductions of more than 50 per cent from present emission levels. Energy research and development budgets must be massively directed to energy options which would eliminate or greatly reduce CO_2 emissions and to studies undertaken to further refine the target reductions.

The 1970's oil shocks demonstrated that with sufficient incentive, policies and practices involving greater energy efficiency and investment in alternative energy sources will be pursued. In the developed countries, the trend towards lower per capita energy use has continued, although it remains very high by global standards. There has been a steady rate of increase in developing countries, the USSR, and other planned economies. The USA, USSR and China are the greatest generators of CO_2. This illustrates the dimensions of the global problem—the USA has the highest per capital energy use, the USSR is vigorously persuasion development and has large coal and oil reserves, and China is also pursuing economic development, is the world's largest consumer of coal, and has the largest known reserves of hard coal.

Even so, some countries are reviewing energy policies. The Dutch Government fell on the failure to achieve acceptance for radical transport policies restricting energy use. The International Energy Agency and the OECD Environment Directorate have begun an examination of the issues. History suggests that national Governments will see difficulties in pursuing radical policies until there is strong support from these international examinations and international political developments. Developing countries can be expected to maintain their increasing energy use development strategies. No doubt they will be influenced by the continued, although failling, high capita use in the developed countries.

- Reduce CO_2 emissions by approximately 20 per cent of 1988 levels by the year 2005 as an initial goal. Clearly, the industrialised nations have a responsibility to lead the way, both through their national energy policies and their bilateral and multi-lateral assistance arrangements. About one-half of this reduction would be sought from energy efficiency and other conservation measures. The other half should be effected by modifications in supplies.

A preliminary analysis in Australia suggests that in our case this is possible. Successful global reduction on this scale would require much greater reductions in the developed world (currently consuming just over 50 per cent of the world's total energy use) or a strong brake on development elsewhere. I am not aware of any country which has explicitly adopted this goal, and in international debts there is still considerable wariness about accepting specific goals. For example, the European Community Environment Ministers on 8 June adopted a quite positive decision on developing a response to greenhouse issues without delay, recognising the critical place of energy for greenhouse emissions. However, they could not agree on minimum objectives for speedy stabilisation and reduction of greenhouse gas emissions.

- Set targets for energy efficiency improvements, that are directly related to reductions in CO_2 and other greenhouse gases. A challenging target would be to achieve a 10 per cent energy efficiency improvement by 2005. Improving energy efficiency is not precisely the same as reducing total carbon emissions and the detailed policies will not all be familiar ones. A detailed study of the systems implications of this target should be made. Equally, targets for energy supply should also be directly related to reductions in CO_2 and other greenhouse gases. As with efficiency, a challenging target would again be to achieve the 10 per cent energy supply improvements by 2005. A detailed study of the systems implications of this target should also be made. The contributions to achieving this goal will vary from region to region; some countries have already demonstrated a capability for increasing efficiency by more than 2 per cent a year for over a decade.

  Apart from efficiency measures, the desired reduction will require (i) switching to lower CO_2 emitting fuels; (ii) reviewing strategies for the implementations of renewable energy, especially advanced biomass conversion technologies; and (iii) revisiting the nuclear power option, which lost credibility because of problems related to nuclear safety, radioactive wastes, and nuclear weapons proliferations. If these problems can be solved, through improved engineering designs and institutional arrangements, nuclear power could have a role to play in lowering CO_2 emissions.

Again, this appears technically feasible in the developed world with the right incentives, but more problematic in the rest of the world. The nuclear option is mentioned, but even if current economic and regulatory problems are overcome, it could make only a limited contribution in this time scale in question. This proposal will influence the work of the International energy agency and the OECD Environment Directorate.

- Negotiate now on ways to achieve the above mentioned reductions.

This issue is a theme in the work of the IPCC, and arises in international conferences, and bilateral and multilateral discussions. Countries are divided about when to begin but the preponderance of views would still be to await the outcome of the work of the Intergovernmental Panel on climate Change. This will also provide a better opportunity for participation by developing countries, and for a more considered approach to the related policy issues. These include research sharing, technology transfers and funding.

- Initiate management systems, in order to encourage, review and approve major new projects for energy efficiency.
- Vigorously apply existing technologies. In addition to gains made through reduction of fossil fuel combustion, to reduce (i) emissions of accidifying substances to reach the critical load that the environment can bear; (ii) substances which are precursors of tropospheric ozone; and (iii) other non-CO_2 greenhouse gases.
- Label products. To allow consumers to judge the extent and nature of the atmospheric contamination that arises from the manufacture and use of the product.

These are restatements of known techniques for achieving efficiency and influencing consumer behaviour. Indications are that they are under notice or more active examination in several countries, but that any well developed programmes would be the exception apart from the increasing practice of labelling ozone depleting products.

The Conference Statement also calls for action by Member Governments of the United Nations, non-governmental organisations and relevant international bodies.

- Initiate the developments of a comprehensive global convention as a framework for protocols on the protection of the atmosphere. The convention should emphasise such key elements as the free international exchange of information and the support of research and monitoring, and should provide a framework for specific protocols for addressing particular issues, taking into account existing international law. This should be pursued vigorously at the International Workshop on Law ans Policy to be held in Ottawa early in 1989, the high level political conference on Climate Change in the Netherlands in the Fall, 1989, the World Energy Conference in Canada in 1989 an the Second World Climate Conference in Geneva, June 1990, with a view to having the principles and components of such a convention ready for consideration at the Intergovernmental conference on Sustainable Development in 1992. These activities should in no way impede simultaneous national, bilateral and regional actions and agreements to deal with specific problems such as acidification and greenhouse gas emissions.

Development since then are consistent with this. The UN resolution last year has already been mentioned, the Hague Declaration calls for the negotiation of the necessary legal instruments to provide an effective and coherent foundation, institutionally and financially to protect the atmosphere and to counter climate change, and the UNEP Governing Council has just resolve that work should continue towards negotiation of a climate convention as a matter of urgency.

The IPCC process includes examination of the issues that arise in consideration of such a convention. The International Workshop on Law and Policy in Ottawa hghlighted the difficulties. Its analysis of traditional approaches does not suggest any easy ways to overcome the usual problems of national sovereignty and compliance. The examples in the Toronto Conference Statement amply illustrate this. The Trail Smelter arbitration of 1935 and 1938 has not been a fruitful source for subsequent international development, Principle 21 of the 1972 Declaration of the UN conference on the Environment explicitly recognises the primacy of national rights to resource use over environmental considerations, the European agreements were hard won despite the urgency of the problems, and Part XII of the Law of the Sea Cnnvention is not regarded as a strong basis for effective environmental regulation. This is not to deny their value, but only to reinforce the long way still to go.

At the Conference Mr E H T M Nijpels, the Netherlands Minister of Housing, Physical Planning and Environment, announced that his Government would host the major Ministerial level conference on these and related issues late in 1989. The Conference is being held in November and will be considering a background paper on these issues, covering problems policy options and possible declaration elements.

- Establish a World Atmosphere Fund, financed in part by a levy on fossil fuel consumption of industrialised countries, to mobilise a substantial part of the resources needed for implementation of the Action Plan for the Protection of the Atmosphere.

As with the idea of target for CO_2 reduction there is ready acceptance of the generality but difficulty in pursuing the details. Developed countries have expressed sympathy for the need to find funding mechanisms to ensure that less developed countries can participate effectively in addressing greenhouse problems. This in turn in related to international debt problems and the need to find better ways to deal with natural disasters. The proposal does emphasise that this problem of international structural adjustment if not resolved could seriously impede development and implementation of policies to overcome greenhouse problems. The spectre of past fruitless debates on north-south issues arises again.

The ozone issue provides an example. At the meeting of the parties in Helsinki this year very few countries were prepared to consider seriously contributing to a new special fund.

- Support the work of the Intergovernmental Panel on Climate Change to conduct continuing assessments of scientific results and to initiate government-to-government discussion of responses and stragegies.

The IPCC has substantial international commitment. Unfortunately, some developing countries have been less enthusiastic, on the basis that the developed countries dominate the pace of developments will have to slow if they are to participate properly. In recognition of this, the IPCC Bureau has appointed a small group to prepare a document on ways to increase the participation of experts from the developing world, and asked UNEP and WMO to increase efforts to raise funds to support such participation.

- Devote increasing resources to research and monitoring efforts within the World Climate Programme, the International Geosphere Biosphere Programme and Human Response to Global Change Programme. It is particularly important to understand how climate changes on a regional scale are related to an overall global change on climate. Emphasis should also be placed on better determination of the role of oceans in global heat transport and the flux of greenhouse gases.

- Increase significantly the funding for research, development and transfers of information on renewable energy if necessary by the establishment of additional and bridging programmes; extend technology transfer with particular emphasis on the needs of the developing countries; and upgrade efforts to meet obligations for the development and transfer of technology embodied in existing agreements.

- Expand funding for more extensive technology transfer and technical co-operation projects in coastal zone protection and management.

These are all issues arising in IPCC and other deliberations, and within individual countries and bilaterally. For example, in Australia the Commonwealth Government has this year granted funds to the Academy of Science for planning workshops organised by its National Planning Committee for the International Geosphere Biosphere Programme, and has provided funds to UNEP to support climate impact work in the region. It has also contributed to the Commonwealth Secretariat study on sea level rises, and carried out a study on establishment of a network of sea level monitoring stations in the Pacific.

Asian countries are already showing particular interest in the issue of technology transfer and co-operation. They see this as critical to adopt measures already agreed, e.g. for the phasing out of chlorofluoro-carbons, and to the future reduction of greenhouse gas emissions. Obviously it would the problems of natural disasters and normal climate variability.

- Reduce deforestation and increase afforestation, making use of proposals such as those in the World Commission on environment and Development's (WCED) report, 'Our Common Future,' including the establishment of a trust fund to provide adequate incentives to enable developing nations to manage their tropical forest resources sustainable.

This has been a consistent response to calls for action on the greenhouse issue, along with intensification of interest on saving tropical forests in particular. For example, the meeting of European Environment Ministers already referred to endorsed the view that the European Community should play a full role in international efforts to halt the destruction of tropical rain forests, and called for action to encourage reafforestation and the increase of vegetal growth. Similarly the UNEP Governing Council meeting in Australia, the only OECD country with tropical rainforest, has successfully nominated the remaining North Queensland rainforest for World Heritage listing, and Brazil has this year announced plans aimed at rainforest conservation.

- Develop and support technical co-operation projects to allow developing nations to participate in international mitigation efforts, monitoring, research and analysis related to the changing atmosphere.

This highlights the need for development assistance programmes to be based on responsible environmental criteria, with particular regard for climate change issues. This should be simple enough, given the close relationship between climate change impacts and existing problems, and their relationship to population growth and sustainable development issues.

Undoubtedly this issue has become more prominent. For example, there is a special meeting of the OECD's Development Assistance Committee this month on aid and environment, and international financial institutions are strengthening their environmental responsiveness.

- Ensure that this Conference Statement, the Working Group reports and the full Proceedings of the World Conference, reports and the full Proceedings of the World Conference, 'The Changing Atmosphere: Implications of Global Security', are made available to all nations, to the conferences mentioned above, and to other future meetings dealing with related issues.

- Increase funding to non-governmental organisations to allow the establishment and improvement of environmental education programmes and public awareness campaigns related to the changing atmosphere. Such programmes would aim at sharpening perception of the issues, and changing public values and behaviour with respect to the environment.
- Allocate financial support for environmental education in primary and secondary schools and universities. Consideration should be given to establishing special groups in university departments for addressing the crucial issues of global climate change.

These all emphasise the need for widespread awareness of the issues and the problems involved in dealing with them. The coincidence of drought in North America with the Conference and testimony supporting the scientific basis of the greenhouse problem by Dr. James Hansen, the Director of NASA's Goddard Institute for Space Studies to the US Senate Energy and Natural Resources Committee, certainly generated great public interest in North America.

Scientific and policy making communities world wide have become absorbed by the issues. However, few countries have yet pursued detailed work of the kind undertaken in Australia, Canada, the Netherlands, New Zealand and the United States. Non-Government organi-sations and educational institutions will certainly have a critical role to play in securing an informed community and international basis for action.

My impression is that the Toronto Conference was a significant force in shaping events since. It did not begin the process of international consideration, but was instrumental in forging a link between science and policy considerations. The agenda set out in its Statement is largely what has been followed, and its more specific targets continually arise in debate. No doubt many who raise them are no longer even consciously aware of their origins. No doubt future analysis will confirm or question this conclusion.

The Conference raised issues and identified traditional approaches, but did not in its statement provide new answers. However, the detailed papers prepared do seek to come to grips with some of the problems. Many at the Conference noted that although there was representation from the developing countries, the almost overwhelming presence was from developed countries. This and the sense of urgency generated there and subsequent rapid action by the IPCC and by some European Governments must be presumed to have contributed to the less than desirable level of participation by developing countries.

Similarly the focus of much of the Conference, quite sensibly and legitimately, on science and possible future impacts, may have tended to obscure the central point that many of these impacts are with use now. Most of the Call for Action related to protection of the atmosphere-adoption strategies, which in the end may be more important for many countries, where mentioned only briefly. If we learn to deal with the impacts, mentioned only briefly. If we learn to deal with the impacts, greenhouse induced climate change will be less of a problem. The key underlying issue-population growth-was seldom mentioned.

Both these points in turn mean that the Conference proposals did not distinguish among the variety of situations faced by different countries. More widely, events have confirmed our earlier judgment that few at the Conference fully appreciated the difficulties developed countries would have in pursuing the proposals, or the extent of the northern hemisphere focus of both the science and policy considerations. That is a reminder of the need for conferences such as this one, and the Intergovernmental Meeting on Climatic Change and Sea Level rise in the South Pacific, being held in July in the Marshall Islands, and the need for strengthening regional capabilities to deal with the issues, e.g. the South Pacific Regional Environment Programme.

One of the major actors on the global environment stage is the Honourable Gro Harlem Brundtland, the Prime Minister of Norway. She chaired the United Nations sponsored World Commission on Environment and Development, and was the keynote speaker at the Conference. She proposed the following international action plan for protecting the atmosphere and, in particular, for preventing climate change:

- Firstly, we should launch immediate international discussions on the feasibility of adopting regional strategies for stabilising and reducing energy consumption and use, before the end of the century.
- Secondly, we should establish a comprehensive international research and information programme on renewable energy.
- Thirdly, we should establish an extensive technology transfer programme with particular emphasis on the needs of the developing countries.
- Fourthly, we must increase scientific research.
- Fifthly, we should consider establishing a global convention on the protection of the climate to coordinate scientific activity, technology research and transfer, information exchange and concrete measures to reduce the emissions of harmful substances.

If the judgments in this paper are near to the mark, we have a long way to go to produce concrete responses to these proposals. Without the Toronto Conference, I suspect we would have still further to go. In late 1989, Dr Norman Myers said that according to latest estimates, tropical forests have lost 142 square kilometres of their expanse during 1989. This is 1.8% of remaining forests covering roughly 8 million square kilometres. In 1979 tropical forests lost 75,000 square kilometres, so the 1989 figure represents a 90% increase in the deforesation rate.

One of the effects of such massive deforestation is the release of vast quantities of greenhouse gases, primarily carbon dioxide. Including the deforestation of closed, open and fallow forests, Myers estimates that the total carbon of 2 to 2.8 billion tons.

Myers concludes, "thus tropical forests are contributing 30% of the build up of carbon dioxide in the global atmosphere, which in turn is considered to account for nearly half of global warming".

This vicious environmental cycle continues. The faster we destroy the richest biomass on Earth the greater the threat presented by climate change. As abnormal weather conditions increase as a result, the character and biological diversity of the tropics and other regions will be disastrously imparied, and species extinction will accelerate.

WWF's major recommendations for action in the face of increasing deforestation and global warming have been clearly stated on many occasions, including; the doubling of tropical forest conservation reserves to 10% of the remaining area before the end of the decade (approximately 100 million hectares); ensuring that the world's tropical timber trade is based on "environmentally sustainable utilisation" of tropical forests which will require 45 million hectares to be under sustainable management by 1995 and a further 200 million hectares eventually needed through sustainable management to need non-fuelwood uses. WWF is also working to ensure that nations currently negotiating the climate change convention adopt the maximum relistic levels for the reduction of greenhouse gases before the end of the century.

Convention

WWF is working towards a successful conclusion to the Convention on the Conservation of Biological Diversity. This international legal instrument has great potential for the conservation of tropical forests.

The idea of a new convention to protect threatened species and their habitats and wild genetic resources first gained political momentum at the IUCN—World Conservation Union 1984 General Assembly in Madrid. A resolution sponsored by WWF and six other non-governmental organisations was passed. It asked IUCN to draft an appropriate legal document. Subsequently, the United Nations Environment Programme (UNEP) emphasised the need by calling for the "adoption of a further appropriate legal instrument, possibly in the form of a framework convention". UNEP, with support from IUCN, then began the slow process of negotiation that continues at present.

The conservation of biological diversity represented, for instance, by Brazil, Columbia, Ecuador, Mexico, Peru, Madagacar, Zaire, China, India, Indonesia, Malaysia and Australia, will benefit if the convention is eventually concluded in a from hoped for by WWF. The aim will be to maintain the broad ecological and utilitarian values to tropical forests and all other important areas for biological diversity.

Tropical forests are the world's greatest storehouse of biological diversity. At least 50%, and perhaps as much as 90% of plant and animal species occur in these forests, and yet they cover only 14% of out planet's terrestrial surface. They play a critical role in regulating hydrological, nutrient and climate cycles, and in permitting the continued existence of indigenous cultures. They provide a long list of natural products (besides tropical timber) including essential oils, fruits, nuts latexes, resins, bamboos, pesticides, dye-stuffs, and many, many more.

The forests also act as a reservoir of wild plant genes that can and do improve our cultivated plants, e.g. a naturally caffeine-free coffee has recently been found within rainforest fragments on the Comoros Islands. Pharmaceutical, products that we all use each day of our lives often owe their origins to tropical forests. The rare periwinkle plant from Madagascar has proven a winner in the battle against Hodgkin's disease and leukaemia and several other cancers. The combined annual commercial value of forest-based products (again excluding timber), food plant and pharmaceutical products is estimated to be worth billions of dollars. The Convention will commit the global community to a biological diversity conservation regime the recognises that; effective conservation of biological diversity sould mean the maintenance of the organisms (the species) and attendant ecosystem(s) such that the continuing character and natural evolutionary process of the ecosystem(s) *in situ* is sustainable and assured, at a population level (per species or per size of ecosystems) that is viable and capable of increasing on its own (with perhaps minimal needs for complementary *ex situ* preservation efforts e.g. in zoos, germ plasma banks etc.). It implies the maintenance and ensuring of the continuing evoloution of the species and requires that a population greater than the critical minimum should be maintained at all times if the species is to be conserved in its natural habitat."

The Convention will seek to raise hundreds of millions of dollars for the conservation and sustainable utilisation biological diversity (and tropical rainforests) through existing financial mechanisms and, more importantly, via new and innovative funding means, for instance by placing a levy on highly traded commodities such as coffee, cocoa and bananas. It is the developing nations that must benefit primarily from such efforts. These additional monies will need to be matched by the free transfer of technologies from developed to developing nations, particularly biotechnologies which have a direct relevance to conservation measures. The negotiation of the Convention will have to resolve the potentially divisive issues to technology transfer and financial burdens to the satisfaction of developing nations if we are to see a successful conclusion in 1992.

It is also planned to establish a Global List of areas and species important for the maintenance of biological diversity. National and international impact assessment will also feature in the Convention's growing list of conservation obligations, which will have positive implications for the sustainable management of tropical forest resources, and for mitigation of the impact of climate change.

1992 Deadline—What are the Political Chances?

Critical to the success of long-term conservation campaigns is the successful conclusion, by June 1992, of both climate change and biological diversity treaties by the United Nations Conference on Environment and Development. Similarly, the working relationships and potential overlaps of responsibility between conventions being worked on and the "Forests Convention" must be determined and satisfactorily resolved quickly.

The differing agendas of developed and developing nations has become clear in the negotiating process. If the World is to make headway in achieving crucial conservation goals before the 1992 conference, developed nations must find the means to guarantee nations of the developing world access to adequate financial assistance and appropriate technologies. Only then will developing nations be able to prosper directly from both conservation and sustainable utilisation biological resources, the global bulk to which remain within developing country national boundaries.

Michael Kennedy is a Conservation Officer with WWF-Australia.

Coral Reefs Dead Reckoning

while forest destructionist giving scientists many clues to the extent of the future impact of climate change coral reefs are already feeling the effects. Coral reefs around the world, which may serve as an early warming system for environmental degradation, are suddenly starving and abnormly cases dying because of abnormally warm seas. The phenomenon, which occurred three times in the 1980, may be more serious this year. Coral reefs are blotchy and sickly in many regions including the Florida Keys, Puerto Rico, Jamaica, the Bahamas, Bermuda, Hawaii and Okinawa.

Most scientific research indicates the current crisis is being caused by higher sea temperatures. As the water heats up, the coral polyps that build the reef with their skeletal remains eject the microscopic algae which feeds the coral and gives it golded, red and yellow hues. The phenmenon is called "bleaching" because it leaves the coral sickly and with white clothes. Without its algae partner, the coral becomes weak and stops reproducing. After several weeks it may die. Scientists know that "bleaching" is usually caused by environmental stresses such as extreme temperatures, low salinity and reduced intensity of light, but have resisted drawing a single conclusion. None the less, many marine biologists believe that coral bleaching is a sign that the delicate reef ecosystems are suffering from stress which may be one of the first biological effects of global warming, and that like canaries in a coal mine, the reefs are a danger signal.

Human activities are now thought to have the potential to alter significantly the Earth's climate on a global scale. This potential derives from increasing atmospheric concentrations of greenhouse gases which are predicted to warm the earth's lower atmosphere and surface by reducing the efficiency with which it cools to space. The amount of warming depends on the magnitude of the increase in concentration of each greenhouse gas (GHG), the radiative properties of the gases involved, and the concentration of other GHGs already present in the atmosphere. The most important GHGs are water vapour, carbon dioxide, methane, nitrous oxide, halocarbons (e.g., the chlorofluoro-carbons), and upper-tropospheric and lower-stratospheric ozone.

The atmospheric concentrations of carbon dioxide, methane, nitrous oxide, halocarbons and tropospheric ozone are increasing primarily because of energy and agricultural practices. During the last few years, however, there has been an increasing recognition that some of the predicted global warming may have been offset, especially in the northern hemisphere, because industrial activities and biomass burning have increased the concentrations of atmospheric aerosols that reflect incoming solar radiation thus tending to cool the earth's lower atmosphere and surface. In addition, the observed decrease in the concentration of ozone in the lower stratosphere at all latitudes, except the tropics, over the last two decades may have offset the greenhouse affect of the halocarbons.

Sources of Greenhouse Gases

The two primary sources of the observed increase in atmospheric carbon dioxide are combustion of fossil fuels and land-use changes; cement production is a further minor, but not insignificant source. The best estimate for global fossil fuel emissions in 1990 is 6.0 ± 0.5 GtC (1 GtC or gigatonne of carbon equals one billion or one thousand million or 10^9 tonnes of carbon). The direct net flux of carbon dioxide from land-use changes (primarily deforestation), integrated over time, depends upon the area of land deforested, the rate of reforestation and afforestation, the carbon density of the original and replacement forests, and the fate of above-ground and soil carbon. These and other factors are essential to estimate annual net emissions but significant uncertainties exist in our quantitative knowledge of them. The best estimate of annual average net flux to the atmosphere from land-use change during the decade of the 1980s is of 1.6 ± 1.0 GtC. The Food and Agriculture organisation (FAO), using information supplied by individual countries, recently estimated that the rate of global tropical deforestation in closed and open canopy forests for the period 1981-1990 was about 17 Mha/yr, approximately 50 per cent higher than in the period 1976-1980.

A total (anthropogenic plus natural) annual emission of methane of about 500 Tg (1 Tg equals 10^{12} grams equals one million tonnes) can be deduced from the magnitude of its sinks combined with its rate of accumulation in the atmosphere. While the sum of the individual sources is consistent with a total of 500 Tg methane, there are still many uncertainties in accurately quantifying the magnitude of emissions from individual sources (natural wetlands, rice paddies, enteric fermentation from cattle, landfills, coal mining, oil, natural gas). Human activities are thought to currently account for about 60-70 per cent of the total emissions. Recent methane isotopic studies suggest that approximately 100 Tg methane (20 per cent of the total methane source) is of fossil origin, largely from the coal, oil, and natural gas industries. Recent studies of methane emissions from rice agriculture, in particular

Japan, India, Australia, Thailand and China, show that the emissions depend on growing conditions, particularly soil characteristics, and vary significantly, but may be much smaller than previously estimated.

The sum of all known anthropogenic and natural sources of nitrous oxide is barely sufficient to balance the calculated atmospheric sink (stratospheric photolysis) or to explain the observed increase in the atmospheric abundance of nitrous oxide. Recently, adipic acid (nylon) production, nitric acid production and automobiles with three-way catalysts have been indentified as possibly significant anthropogenic global sources of nitrous oxide.

The worldwide consumption of chlorofluorocarbons 11, 12, and 113 is now 40 per cent below 1986 levels, substantially below the amounts permitted under the Montreal Protocol. Further scheduled emissions reductions are mandated by the 1990 London Amendments to the Montreal Protocol leading to a complete phase-out of all long-lived chlorofluoro-carbons by the year 2000. As chlorofluorocarbons are phased out, hydrochlorofluorocarbons and hydrofluorocarbons will replace them, but at lower emission rates.

About 90 per cent of atmospheric ozone resides in the stratosphere and about 10 per cent in the troposphere. Ozone is an effective greenhouse gas in the upper troposphere and lower stratosphere (i.e., 8-25 km. altitude). Significant decreases have been observed in the total column content of ozone during the last two decades at all latitudes, except the tropics, throughout the year with the downward trends being larger during the 1980s than in the 1970s.

The decreases in stratospheric ozone have occurred predominantly in the lower stratosphere (below 25 km.), where the rate of decrease has been up to 10 per cent per decade depending on altitude. The weight of scientific evidence suggests that anthropogenic chlorine- and bromine-containing halocarbons are responsible for the observed reductions in middle- and high latitude stratospheric ozone. Even if the control measures of the 1990 London amendments to the Montreal Protocol were to be implemented by all nations, the abundance of stratospheric chlorine and bromine will increase over the next several years. Consequently, ozone depletion at these latitudes is predicted to continue unabated through the 1990s.

There is evidence to indicate that ozone levels in the troposphere up to 10 km. altitude above the few existing ozonesonde stations at northern middle latitudes have increased by about 10 per cent per decade over the past two decades. The abundance of carbon monoxide, an ozone precursor in the troposphere, appears to be increasing in the northern hemisphere at about 1 per cent per year; however, there is little information on the global trends of other tropospheric ozone precursors (non-methane hydrocarbons and oxides of nitrogen). Each of these ozone precursors have significant natural and anthropogenic sources, but their detailed budgets remain uncertain.

A key issue is to relate emissions of GHGs and their precursors to *future* concentrations of GHGs in order to assess their impact on the radiative balance and thereby the Earth's climate. A number of different types of carbon cycle and tropospheric chemistry models have been developed for this purpose. However, it should be noted that *all* carbon cycle models are subject to considerable uncertainty because of an inadequate understanding of the processes controlling the uptake and release of carbon dioxide from the oceans and terrestrial ecosystems. Similarly, tropospheric chemistry models exhibit substantial differences in their predictions of changes in ozone, and in other chemically important active gases due to emissions of methane, non-methane hydrocarbons, carbon monoxide and, in particular, oxides of nitrogen, because of uncertainties in the knowledge of background chemical composition and our inability to represent small-scale proceses occurring within the atmosphere.

Gases can exert a radiative forcing both directly and indirectly—direct forcing occurs when the gas itself is a greenhouse gas, and indirect forcing, when chemical transformation of the original gas produces a gas or gases with greenhouse impacts. The concept of the Global Warming Potential (GWP) has been developed for policy-makers as a measure of the possible warming effect on the surface-troposphere system arising from the emission of each gas relative to carbon dioxide.

The indices are calculated for the contemporary atmosphere and do not take into account possible changes in chemical composition of the atmosphere. Changes in radiative forcing due to carbon dioxide, on a mass (e.g., kilogram) basis, are non-linear with changes in the atmospheric carbon dioxide concentrations. Hence, as carbon dioxide levels increase from present values, the GWPs of the non-carbon dioxide gases would be higher than those evaluated here. For the concept to be most useful, both the direct and indirect components of the GWP need to be quantified.

Direct Global Warming Potentials

The direct components of the Global Warming Potentials (GWPs) have been calculated, taking into account revised estimated lifetimes, for a set of time horizons ranging from 20 to 500 years, with carbon dioxide as the reference gas. The table below shows values for a selected set of key gases for the 100-year time horizon. The carbon cycle model used in these calculations probably somewhat underestimates both the direct and indirect GWP values for all non-carbon dioxide gases. The magnitude of the bias depends on the atmospheric lifetime of the gas, and the GWP time horizon.

Indirect Global Warming Potentials

Because of our incomplete understanding of chemical processes, it would be prudent to assume that most of the indirect GWPs reported in IPCC (1990) are likely to be in substantial error. However, it is clear that the indirect GWP for methane is positive and could be comparable in magnitude to it direct value. In contrast, the indirect GWPs for chlorine and bromine halocarbons are likely to be negative because they are likely to be the cause of the observed global ozone depletion in the lower stratosphere.

The concept of a GWP for short-lived, inhomogeneously distributed constituents, such as carbon monoxide, non-methane hydrocarbons, and oxides of nitrogen may prove inapplicable, although, as noted above, it is known that these constituents will affect the radiative balance of the atmosphere through changes in tropospheric ozone and the hydroxyl radical which controls the atmospheric lifetime of many tropospheric species such as methane and hydro-chloro-fluoro-carbons. Similarly, a GWP for sulfur dioxide is viewed to be inapplicable because of the non-uniform distribution of sulphate aerosols.

- The GWP values are dependent upon the time horizon chosen because of the different atmospheric lifetimes of the gases, e.g., the direct GWPs for methane for 20 and 500 year time horizons are 35 and 4, respectively.

Relative Importance of Greenhouse Gases

Combining the 1990 emissions of GHGs with their global warming potentials suggests that the most important greenhouse gas directly influenced by human activities is carbon dioxide, with methane being the second most important. More than half of the enhanced greenhouse affect can be attributed to carbon dioxide, with methane being responsible for up to another quarter; hence, the emphasis in this analytical framework on these two gases.

Future Emissions of Greenhouse Gases

Future emissions of GHGs will depend upon a wide range of economic, demographic and policy conditions and are inherently controversial to predict because they reflect different views of the future. Considerable uncertainties arise from the evolution of the types and levels of human activities (including economic growth and structure), technological advances, and human responses to possible environmental, economic and institutional constraints. Consequently, predicting future emissions of GHGs from fossil fuel and biotic sources is inherently difficult because it requires embodying a wide array of assumptions on factors such as population growth, economic growth, structural changes in economies, role of nuclear power, fossil fuel availability, energy prices, technological advances, the rate of diffusion of renewable energy technologies, changes in land-use patterns (particularly rates of deforestation), food demand, agricultural productivity, and land tenure policies.

It is, clear, however, that if present trends persist, it is very likely that the resulting impacts on the global atmosphere will lead eventually to changes of the global climate that would seriously perturb human societies and perhaps even endanger human life. Further, the response time of the climate system is such that, by the time significant changes are detected, it may be too late to rectify the situation.

This is why even though there is still much scientific disagreement on the extent and likely consequences of global warming, many countries are committed to putting precautionary policies in place, of which the GEF is a part. The idea is that, as further evidence on global warming and its consequences is gathered, the investments of the GEF will leave the international community better placed to reduce carbon accumulations to safe levels, over the long-term, should the need arise. The approach is equivalent to an insurance policy, but it can also be described as "preventive maintenance" in engineering parlance, i.e., taking steps to avoid breakdowns that necessitate major repair and even disaster. The various measures (technologies, practices and policies) that the GEF should support are those that would need to be turned to on a large scale in a scenario in which carbon accumulations have to be restricted appreciably.

It is important at this stage to make a distinction between the objective of *reduction of net emissions* and the objective of *stabilisation of accumulations*. If what is needed is, say, for the OECD and CIS countries to reduce gradually their net CO_2 emissions from energy production and use by 20 per cent relative to today's levels, and for developing countries to reduce the rate of growth of energy consumption from 6 per cent per year to 4 per cent per year (a major reduction), then this objective could be accomplished by reforming energy pricing policies and other measures to improve energy efficiency.

However, even with such improvments in energy efficiency, global CO_2 emissions each year would still be twice their present levels in 40 years time and carbon accumulations would likewise be twice their present levels by the middle of the century. In short, by focussing only on the *reduction* of CO_2 emissions, the global warming problem would only have been delayed by a decade or two, but would have been substantially unaddressed. Delaying global warming cannot prevent it.

The reduction of net emissions is a necessary step, but not sufficient; in addition, the magnitude of reduction should be sufficient to achieve the stabilisation, if not reduction, of GHG accumulations in the atmosphere at some safe level over the long term. Thus, the effectiveness of interventions cannot be divorced from objectives and benefits cannot be separated from targets. However, virtually all these interventions involve investments and therefore there has to be strategy for reducing net GHG emissions into the atmosphere.

It will take many months of intensive governmental negotiations to provide a clear picture of what a politically feasible and environmentally effective climate change treaty will look like. Although there is a considerable, if recent, history in drafing international environmental agreements, the breadth and complexity of climate change will take the world community well beyond the scope of prior experience.

Nonetheless, it is possible to identify at the outset a number of features which will be essential to a meaningful agreement. A convention will need to set long-term global goals and shorter-term binding national obligations. It will need to be comprehensive-addressing sources and sinks of the various greenhouse gases—without delaying action until all issues are resolved. It will need to catalyse a "bottom up" process that allows individual nations to devise their own strategies for curbing emissions. And it must provide a mechanism to transfer the resources-human, technological and financial-needed by much of the developing world to respond to the threat of changing global climate.

An Overall Objective

A "framework" convention on climate, by its very nature, will not immediately resolve all issues leading to atmospheric change, but it must establish a clear basis for addressing them over time. To provide such a basis, the convention should state an overall long-term goal of stabilising atmospheric concentrations of greenhouse gases at levels which will minimise ecological risk. While it is difficult to state such levels with precision, there is wide agreement that total greenhouse gas concentrations (expressed as CO_2 equivalents) must be held significantly below 500 parts per million. This should limit long-term temperature rise to between 10^0 C and 20^0 C above pre-industrial levels and sea-level rise to about 20-50 centimetres above current levels. Only by keeping change to within these limits can we hope to save ecosystems from devastating change.

The term "framework" convention ordinarily implies just that—a broad agreement on principles that leaves specific details and concrete obligations until later, usually to be taken up in negotiations for subsequent protocols to the convention. At least this has been the past experience with lutant-specific protocols (So_2 and No_2) ler the Geneva transboundary air tution convention, and with the Protocol (on CFC) to the Vienna layer convention.

WWF, therefore, has called on industrialised nations to agree to curb their CO_2 emissions by 20% by the year 2005, based on 1988 levels. This commitment should be made at the 1992. Brazil conference, as part of the Convention on Climate change.

The management of biotic resources plays a significant role in the global build-up of greenhouse gases, especially carbon dioxide and methane, and it is crucial that these emissions be addressed in the convention or its protocols. Stemming rates of deforestation must be a key objective, although ecologically sound afforestation programmes can serve as "sinks" for carbon, and will likely play a role in an overall strategy and emissions accounting.

A purely "climate driven" forest policy, if, however, carried to its theoretical extreme, would result in massive eucalyptus plantations across and tropics to soak up fossil-fuel emissions. It is then, essential to ensure the convention leads to balanced forestry policies that take into account other concerns such as biological diversity and the rights of indigenous peoples.

It remains to be seen whether the climate negotiations will provide an adequate forum to forge broad new forest policies, or whether such undertakings are better suited to a proposed freestanding convention devoted solely to forests. Because forests are on the climate agenda, however, WWF is urging policies which emphasise saving existing forests whose disappearance in the tropics is feeling the greenhouse effect and contributing to species loss. Specific WWF targets include halting net deforestation by the year 2000, doubling the area of tropical moist forest in protected reserves (to about 100 million hectares or 10% of total remaining), and basing the entire international timber trade on sustainable utilisation by 1995.

National Plans

Although an international convention is in many respects a "top down" exercise, a key to its success will be the ability to generate "bottom up" policy reforms and initiatives at the national, state and local levels. To play a productive role, the convention should call for each country to prepare a national strategy for addressing global warming, within one year of becoming party to the convention. The national plans would be presented to the other parties and would state how the country intends to need its obligations under the convention, and could set additional goals.

The process of preparing national plans should serve to identify new policy options and scenarios, enhancing willingness to set more ambitious emissions targets and articulating specific measures, projects and reforms which can be supported by a special climate fund established for that purpose.

Banking on Change

The provision of new and additional aid will be a necessity to enable developing countries to meet their obligations under a climate convention, as well as a prerequisite to their undertaking climate obligations in the first instance. Although funding for climate-related measures will likely be made available through a number of multilateral and bilateral arrangements, establishing a central fund within the convention itself should offer the incentive vital for broad participation by developing countries.

Money for the fund could possibly be generated by a carbon tax in the industrialised countries and should be provided primarily in the form of out right assistance, as opposed to loans. Priority areas for funding include technology transfer and development, forest protection, and promotion of renewable energy resources and energy efficiency programmes—possibly as projects identified in the national plans discussed above.

The first round of formal negotiations on a climate convention were held near Washington DC in early February. No draft text was produced, and little progress was made in establishing working groups to tackle various aspects of the problem. In particular, the information of the single working group on "sources and sinks", made it appear unlikely that a separate protocol on CO_2 would be forthcoming or that forests would get the kind of close attention needed for genuine policy reforms.

Nevertheless, the approach of the June 1992 UN conference is sure to exert increasing pressure on governments to complete and agreement with the coming year. Unwavering NGO support may be decisive in whether an agreement is reached in time and whether the strength of the convention is equal to the task of protecting the global community.

The measures needed to control the greenhouse effect are affordable and within reach, according to a team of international energy experts. Their study Carbon Emissions Control Strategies was published by WWF late in 1990.

"Control measures to hold carbon dioxide emissions constant would cost most industrialised nations little or nothing", according to William U. Chandler, a senior scientist with Battelle Memorial Institute. He concluded that efforts to cut emissions by 20% by the year 2005 would cost only a small fraction of economic output. Twenty-five energy experts and economists from eight developed nations participated in the two-year study. These nations—Canada, France, Hungary, Japan, Poland, the Soviet Union, the UK and the Us—release half the world's energy-related carbon dioxide emissions. Each country study was prepared by a team of leading experts from that country.

According to Chandler, two thirds of the emissions reduction would be produced by energy efficiency measures. The remainder could come from fuel switching, primarily to natural gas but also other sources, which would include renewable energy.

Chandler recommended three measures to produce best results: market reform in former planned economies, energy taxes to motivate energy savings and fuel substitution, and mandatory efficiency standards on appliances and automobiles.

The study included the first work of its kind from the Soviet Union and Eastern Europe. A Soviet team leader, Igor Bashmakov of the Energy Research Institute of the Soviet Academy of Sciences, wrote that energy efficiency is the most important means of curbing emissions growth in his country. But he said that economic reform must be introduced to make its work. In addition, demilitarisation of the Soviet economy will be necessary to free funds for improving the Soviet economy. He says that the world's governments must balance market force and government regulations because the carbon dioxide problem is not soluble without both.

The Polish case study leader Stanislaw Sitnicki suggested strong medicine. For economic, as well as environmental reasons, he would halt new energy supply investments until that nation captures its massive energy efficiency potential. He added that further work is needed to determine how to share the burden of carbon emissions reduction and to evaluate the benefits and costs of emissions control.

William Eichbaum, Vice President of WWF-US said that "With these results, leaders can take action to protect the planet from climate change without risking the health of the world's economies. Leaders need not choose between economic growth and environmental protection," he said. He provided a blueprint for possible future action. "These results mean that climatic change can be slowed with measures that are, for reasons unrelated to climate, in many nations' self-interest. Structural change and energy efficiency scenarios both significantly reduce carbon emisions and increase per capita incomes.

Implementation Procedures

All institutions based on international co-operation face free riding by signatiories who obtain the benefits of an international agreements while avoiding the costs by non-compliance. The likelihood that signatories to a Climate Change Convention might try to avoid meeting their commitments poses the question of monitoring, verification and enforcement of compliance.

Analysts concur that a multi-gas agreement will be much harder to monitor and likely impossible to verify. Taxes, tradeable permits, and abatement services all require monitoring to ensure that the terms and conditions of the scheme to fulfil commitments under a Climate Change

Convention are being met. Monitoring, however, must be scientifically credible. Monitoring of non-CO_2 greenhouse gases such as methane from paddy fields or from diverse, mobile point sources such as cattle is not feasible due to the uncertainty as to emission rates. While rice production is relatively well known, methane emissions vary greatly with soil type, nutrients, light, and temperature. Estimates for rice paddy in Spain and Italy vary by more than 100 per cent.

Also, sinks for and terrestrial reservoirs of greenhouse gases must also be monitored, both to ensure that sinks endowed as property rights are maintained, and to verify and claims made as to additional carbon fixation. Yet rates of re-and de-forestation are highly contentious, and satellite-based remote sensing cannot yet provide adequate monitoring and verification of biotic carbon sinks. The deforestation rate in Brazil, for example, is highly controversial (estimates range by a factor of five). Similar arguments apply to other greenhouse gases such as nitrous oxide. For all these reasons, therefore, I conclude that it is only meaningful to cost monitoring and verification of carbon dioxide released from fossil fuels at the outset of the implementation of the Convention.

In Chapter 5, however, I assumed that property rights are created in proportion to national carbon sinks that are the basis of determining permissible emissions in future years. It is possible to monitor and verify the status of the forest stocks and thus carbon reservoirs in those forests although it is not feasible to track the carbon *flows* to and from them. A monitoring system is feasible that would use remote sensing and *in situ*, ground-based validation to determine the fulfilment of commitments made to maintain or to expand these reservoirs. I assume, therefore, that biotic stock of carbon will be included eventually under a protocol for monitoring and verification, although not the carbon emissions from these sources.

Verification is the international control of compliance with aggreed measures and behaviour by means of tools and procedures agreed upon in an instrument of international law—for example, a protocol on compliance to a Climate Change Convention. Verification can be defined as having different densities depending upon the level of distrust between parties to the agreement and the technical of obtaining information with an adequate level of confidence on the other.

I assume that the verification procedures that are adopted in a verification protocol will be multilateral rather than bilateral in implementation (although they may rely heavily on national/unilateral monitoring and verification capabilities such as satellite systems). Assuming that all parties will be accorded equal treatment in the protocol, it is reasonable that all parties also will be subject to monitoring and verification to suppose implementing organisation established under the convention.

Anthropogenic sources of CO_2 vary greatly with respect to characteristics that affect greatly their suitability for monitoring and verification. Some are stationary, unit copiously and continuously, and are suitable for direct, quantified monitoring. Power stations and large factories exemplify this type of emitter. Other sources are stationary and numerous but only emit intermittently very small quantities of gas. Fireplaces and open fires are typical. There are also very many mobile point sources that are sproadic emitters such as vehicles and livestock. Finally, there very diffuse sources such as non-commercial fuels based on animal wastes.

Only the first category is suitable to direct monitoring and verification. There are, for example, well-developed techniques for determining gaseous emissions such as ultrasonic instruments in the off gas stack which measures the effluent density and velocity and there by volume to within 3-5 per cent accuracy. The other sources all exhibit characteristics that would make information collection enormously onerours due to their number or the lack of observational methods.

Good statistics are available for energy production and consumption balances for most countries. However, to convert these energy data to emissions on an international basis, energy balances must be made more complete and accurate, carbon content and conversion into emissions must be made more precise, and statistics must be collected according to consistent and compatible ground rules. The parties to the Convention must agree on the types of date, required disaggregation and detail, and common reporting rules for national reports. Fortunately, the Intergovernmental Panel on climate Change has already produced a set of guidelinges of this nature, which are being refined and updated. In particular, rules are needed to determine whether emissions are based on energy production or consumption. The latter is particularly problematic because of the difficulty of ascertaining conversion losses.

The implementing organisation must be able to verify the accuracy of date such as ash content of coal or oxidation rates of conversion in power plants, etc. It cannot hope to collect the requisite date independently but only to analyse the date supplied by parties to the convention by cross-checking the reported fuel cycles and conversion to emissions with 'spot' checks including on-site visits to 'check the books' of very large, stationary emitters. In most nations, however, the latter checks would only cover 1-5 per cent of a given country's emissions. It is crucial to an effective greenhouse regime that verification be conducted routinely by subjecting national reports to independent, critical scrutiny and assessment, treated as an expert technical rather than political process.

A verification system that combines data analysis with spot checks would likely enable the implementing organisation to detect an emission infringement of the Convention by a party that deviates 10 per cent or more from the party's commitments. Similarly, a verification system that uses remote satellite and air-based sensing with local inspections should be able to detect departures from declarations to maintain or to expand biotic carbon pools such as forest reserves, to within a five per cent deviation from commitment, (The verification protocol would have to define the ground resolution at which it requires monitoring, the calibration and interpretational rules to be followed, and the density of selective observation needed for confidence to exist that parties are complying with their commitments.)

The cost of the verification system will consist of the direct costs of the implementing organisation engaged in checking the annual national reports of complicance and field inspections, plus the indirect costs of obtaining independent sources of information needed to cross-check national claims about emissiosn or the status of carbon stocks or sinks. It is reasonable to assume that the implementing organisation will not have to meet the capital or direct operating costs of remote sensing satellites. Rather, these costs will be covered in the budgets of the space agencies in Europe, Japan, and the United States which already pay for the huge cost of earth-observing satellites.

By way of comparison, the International Atomic Energy Agency's (IAEA) safeguards department currently consists of 450 persons including 190 field inspectors. The implementing organisation for verifying a Climate Change Convention would probably require about twice as many staff given the much larger number of facilities to be visited and much broader international scope of the verification system compared with that applied in the nuclear field.

In 1987 the IAEA safeguards applied to about 230 tonnes of plutonium, 30,000 tonnes of enriched uranium, and 50,000 tonnes of depleted uranium, thorium or uranium. In 1983, the IAEA safeguards agreements applied to a total of 881 installations such as power reactors and other fuel cycle facilities. A carbon monitoring system will apply to billions of tonnes of fuels, and millions of heactares of forest at hundreds of thousands of sites—a much bigger task.

The safeguards surveillancoe and meterials balance inspectorate system for the sensitive nuclear materials run by the IAEA cost about US$30 million per year. The cost of a system that verifies compliance with a Climate Change Convention might thererfore approach $100 million per year. It is doubtful that more than three times the IAEA's budget would be provided to the implementing organisation at a time when the IAEA already finds it difficult to obtain funds for such a politically sensitive field to concern to great powers.

The source of the funds for the implementing organisation and its activities would either be charges that follow the UN scale of payments of a special formula similar to that developed by the IAEA in 1971 (and later revised) that levies sates on a per capita income basis (with a ceiling) and a cap on contributions by poor states.

The previous sections have argued that the greenhouse arena is characterised by complexity due to multiple gases (unless limited to carbon dioxide); an effectively infinite number of point and mobiie pollution sources; mostly national information on energy use which is subject to distortion, withholding, and differing reliability, varying analytical methods, and underlying assumptions; reliance on extrapolation from existing energy statistics rather than new monitoring of greenhouse gas emissions; and a long lead time before an effective monitoring system and verification could be created.

I analysed three mechanisms to achieve agreed reductions and to fund the South's excess' incremental abatement costs: carbon taxes, tradeble permits, and trade in abatement services. Each of these mechanisms poses different demands on a verification system. A carbon tax system, for example, requires that a baseline emission be set and updated each year to confirm that states are reducing emissions to agreed targets. A tradeable permit system within an overall global emissions target demands that trading be monitored continuously in addition to establishing national emissions relative to an agreed baseline. Trade in abatement services requires that claimed reductions by one country actually have been achieved in another country. Monitoring complicance of such claims could be politically difficult for an international monitoring system.

In all three cases, achieving a high degree of certainty seems to require an extensive monitoring system and bureaucracy. Yet most states do not (yet) perceive the stakes in the greenhouse regime to demand monitoring and verification like that imposed on flows of special nuclear materials. Only a small international bureaucracy based on national reports and data cross-checking seems politically feasible at this time.

The history of international arms control and environmental agreements offers six important lessons for a greenhouse verification system. First, environmental costs and benefits do not accure as fast as the costs and benefits of abandoning arms control agreements and the stakes are not perceived as central to the immediate security of the state nor (usually) to regime survival in that state. States may therefore be less demanding of a verification system for environmental agreements than in other domains. Morevoer, when states co-ordinate because of self-interest, there is little reason to defect or cheat and little or no verification or enforcement is needed. If the costs of carbon abatement are as low as suggested in the studies reported in this book—at least for the first 20 per cent reduction and therefore the first decade or two of an agreement—then verification measures are needed mostly to build confidence in the regime rather than to raise the question of non-compliance and enforcement.

Second, it is inevitabel and proper that enforcement responsibility will be lodged primarily at the same level as implementation responsibility, that is, within nation states. The bulk of the monitoring and verification should be conducted at this level, rather than internationally. Third, most military control regimes were created in confrontational contexts under conditions of secrecy and with little or no participation. These characteristics led to many problems of implementation for arms control agreements. This experience implies that a greenhouse regime should strive for maximal transparency and openness, including a strong role for non-governmental organisations in monitoring compliance.

Fourth, some states have skillfully used verification issues in the past to block international agreements (most notoriously, the United States with regard to the Complete Test Ban Treaty). If the analysis in previous sections is correct, then this problem should not arise in the greenhouse gas arena." Fifth, there are important precedents for monitoring and verifying international atmospheric agreements, at the regional level in Europe, and globally in ozone depletion convention. This experience should provide some good signposts for the greenhouse gas agreements (in North-East Asia) that could be developed to supplement a global greenhouse regime.

Sixth, the history of international monitoring and inspection of nuclear power provides some useful lessons. The IAEA's history suggests that an international secretariat should be created to audit national reports and ensure that they are bona fide, consistent and follow internationally recognised procedures. An independent technical committee could be appointed to define the reporting requirements of states to the parties to a Climate Change Convention. The same technical committee could also explore with states qualitative anomalies (such as refusal to allow an on-site inspection of emission rates or a claimed efficiency improvement) and quantitative discrepancies (such as inconsistencies between national reports and international statistics) that might arise from time to time. Article 14 of the Convention states that disputes between parties should be settled by negotiation or by any peaceful means that they care to select, including arbitration by the International Court of Justice, and/or in accordance with procedures yet to be adopted by the parties to the treaty. Thus, the Convention provides little guideness as to what methods of dispute resolution should be incorporated into a protocol. It casts no light at all on the appropriate means of enforcing compliance with treaty commitments.

At a meeting in The Hague in 1989, twenty-four national leaders called for a 'new institutional authority' to set and implement environmental standards authoritatively. Moreover, there is no compulsory dispute settlement jurisdiction. The major stumbling block had been the socialist bloc rejection, and more recently, US rejection of compulsory third party arbitration.

In relation to disputes involving the failure of developing countries to comply with emission reduction targets, it would likely be highly counter-productive to try to enforce agreed targets by economic sanctions. These would worsen the very technological and economic dificulties that cripple many developing countries' ability to comply. It would also impose substantial costs on states that meet their commitments and there by reduce the benefits to new signatories considering joining the Convention. Moreover, using the trade system to enforce climate change policies would likely prove to be unmanageable because every product and service that is traded internationally results in greenhouse gas emissions. Consequently, alternatives to supranational regulation have emerged. States recognise each other's licensing rather than ceding licensing powers to an international authority. Such reciprocal recognition schemes operate in many areas including phytosanitary certificates for exports, shipping oil pollution prevention certificates, marine waste disposal permits, hazardous meterials trade, and trade in endangered species. States also harmonise standards and standard-setting laws and procedures, often by adopting models from overseas. (For example, the environmental impact assessment, green labelling and pollution taxes).

Many states have also committed themselves, outside treaties, to providing early warning and notification, for example, with respect to banned chemicals, exchange of standards, or a adjustments to trade-controlled items (such as endangerd species). Regimes also creat transnational networks of lower level 'operational' national officials who short-circuit hierarchical communication across boundaries, or who communicate directly with international organisations that may then reintroduce environmental information at much higher political levels in the same nation state. Such structures can defuse and even prevent disputes arising in the first place.

Non-adversarial techniques have also developed to resolve international disputes from escalating to interstate conflicts. Local legal challenges have been mounted across borders thereby achieving settlement without involving the states themselves. (In Europe and North America, this technique requires that legal systems grant status to foreign parties in local judicial or administrative procedures). The filing of complaints and the launching of infringement hearing are two other techniques that have been used (in the Monetreal Protocol and in the European Economic Community, respectively).

Reporting requirements are also in important means of imposing national 'discipline' on treaty parties, especially when combined with international expert auditing and public debate in committees or annual conferences. Such procedures are already well developed in the occupational health and safety agreements administered by the International Labour Organisation. Similar procedures are used by the international Monetary Fund, and multilateral funding agencies also conduct national and sectoral audits as preconditions for or requirements of development loan agreements. Environmental auditing, however, has a weak tradition although the Montreal Protocol requires substantial reporting to permit monitoring of compliance and administration of various aspects of the agreement.

As noted earlier, it seems inevitable that the national reporting requirements of a Climate Change Convention will be the heart of a monitoring and verification system, and the key to effective enforcement through self-regulation. National reporting will the core confidence-building measure that will build widespread commitment to the regime. It can be supplemented by international auditing of the kind referred to above, but such reports (as occurs with the International Energy Agency reports) buttress only the normative power of domestic proponents of fulfilling treaty commitments and have no direct legal connotations.

These measures may be supplemented by other measures that increase the incentives of signatories to comply. Large emitters, for example, can offer to match the abatement of new signatories. They can also threaten to punish offenders by reducing their own abatement by as much or more than that of the defector, thereby reducing the benefits of free riding by increasing the costs of climate change. States can also lock in their own commitments so as to reduce the uncertainty facing a potential signatories as to shether they will reap the benefits of reduced climate change.